ANNUAL REPORTS ON

NMR SPECTROSCOPY

ANNUAL REPORTS ON
NMR SPECTROSCOPY

Edited by

G. A. WEBB

Department of Chemistry, University of Surrey, Guildford, Surrey, England

VOLUME 15

1983

ACADEMIC PRESS

A Subsidiary of Harcourt Brace Jovanovich, Publishers

London • New York
Paris • San Diego • San Francisco • São Paulo
Sydney • Tokyo • Toronto

ACADEMIC PRESS INC. (LONDON) LTD.
24–28 Oval Road,
London, NW1 7DX

U.S. Edition Published by

ACADEMIC PRESS INC.
111 Fifth Avenue
New York, New York 10003

ISBN 0-12-505315-0
ISSN 0066-4103

Printed in Great Britain by J. W. Arrowsmith Ltd.
Bristol BS3 2NT

LIST OF CONTRIBUTORS

DAVID J. CRAIK, *Victorian College of Pharmacy, 381 Royal Parade, Parkville, Victoria 3052, Australia.*

DETLEF DEININGER, *Sektion Physik, Karl-Marx-Universität, DDR-7010 Leipzig, Linnéstrasse 5, German Democratic Republic.*

POUL ERIK HANSEN, *Institut I, Life Sciences and Chemistry, Roskilde University Centre, PO Box 260, 4000 Roskilde, Denmark.*

WOLFGANG MEILER, *Sektion Physik, Karl-Marx-Universität, DDR-7010 Leipzig, Linnéstrasse 5, German Democratic Republic.*

HARRY PFEIFER, *Sektion Physik, Karl-Marx-Universität, DDR-7010 Leipzig, Linnéstrasse 5, German Democratic Republic.*

E. A. WILLIAMS, *General Electric Company, Corporate Research and Development, Schenectady, New York 12301, USA.*

PREFACE

The ever growing extent of the application of NMR to chemical problems is again reflected in the amount of data presented in the four chapters comprising this volume.

The first two chapters deal with substituent and isotope effects, respectively, on nuclear shielding. These accounts contain a number of chemically important concepts which are developed and their influence on observed chemical shifts evaluated. The third chapter deals with recent developments in ^{29}Si NMR and brings up to date a previous account on this topic in Volume 9 of this series.

The final chapter deals with the relatively new, and exciting, field of NMR studies of adsorbed species. It is a pleasure for me to express my gratitude to all the authors for their assistance with the preparation of this volume and for the promptness with which they have delivered their manuscripts.

University of Surrey,
Guildford, Surrey,
England

G. A. WEBB
July 1983

CONTENTS

Substituent Effects on Nuclear Shielding

DAVID J. CRAIK

Isotope Effects on Nuclear Shielding

POUL ERIK HANSEN

Recent Advances in Silicon-29 NMR Spectroscopy

E. A. WILLIAMS

NMR of Organic Compounds Adsorbed on Porous Solids

HARRY PFEIFER, WOLFGANG MEILER AND DETLEF DEININGER

Substituent Effects on Nuclear Shielding

DAVID J. CRAIK

Victorian College of Pharmacy,
381 Royal Parade, Parkville,
Victoria 3052, Australia

ANNUAL REPORTS ON NMR SPECTROSCOPY
VOLUME 15   ISBN 0-12-505315-0

I. INTRODUCTION

The important role of nuclear magnetic resonance (NMR) spectroscopy in chemistry arises largely from the consequences of nuclear shielding. The fact that nuclei in different electronic environments have different nuclear shieldings, and hence different chemical shifts, makes NMR a powerful probe of electronic structure. Empirical rules relating chemical shifts to substituent ("σ") constants, electron densities, electronegativities, and a variety of other empirical parameters have proven of great benefit to problems of organic structural elucidation, and to fundamental studies of molecular electron distributions. This review focuses on one specific application in the latter category – the study of substituent *electronic* effects on chemical shifts. The aim is not to provide a compendium of substituent effects on chemical shifts to aid in structural assignments, but to show how fine detail relating to the distribution and polarization of electrons in molecules may be determined from chemical shift studies.

The study of substituent electronic effects on NMR chemical shifts dates back to the 1950s, following soon after the discovery of the NMR phenomenon itself. Early studies of relationships between ^{1}H chemical shifts and substituent properties demonstrated the potential of chemical shifts as structural and electronic probes.[1,2] However, first practical applications of NMR as a probe of substituent electronic effects arose from ^{19}F chemical shift measurements.[3] Investigations of substituent effects on ^{19}F shifts in *meta-* and *para*-substituted fluorobenzenes[4] in the early 1960s clearly demonstrated the differential transmission of polar and resonance effects at *meta* and *para* ring positions. In studies of this type, the fluorine atom served to some extent as a probe of electronic effects in the *adjacent* molecular skeleton. The ability to measure ^{13}C chemical shifts at natural abundance[5,6] represented a major advance, as it allowed chemists to examine substituent effects directly at molecular *backbone* sites, and hence to trace the transmission of substituent effects in a carbon skeleton.[7] Studies of substituent effects on chemical shifts have now been extended to a variety of other nuclei, including ^{15}N, ^{17}O, ^{33}S, ^{77}Se and ^{199}Hg, to mention a few.

In direct proportion to the number of studies reported, this review concentrates mainly on substituent effects on ^{13}C chemical shifts; however, representative studies of some of the more esoteric NMR nuclei are also discussed. In general, the types of molecular systems to be examined here are those in which the probe site is well separated (i.e. by more than three bonds) from the substituent, so that steric effects are minimized. The substituent is to be regarded as a perturbing influence on a relatively rigid molecular framework and should not induce structural (geometrical) distortions of the probe site; i.e. its influence should be purely electronic. Substituted benzenes or naphthalenes fit the above requirements and indeed much of the discussion will centre on these types of systems. Since substituent effects on side-chain chemical shifts in aromatic systems have been the subject of another recent review,[8] the emphasis here is on other probe sites.

II. NUCLEAR SHIELDING THEORY

The term "nuclear shielding" succinctly describes the effect of surrounding electrons on a magnetic nucleus. Equation (1) shows the well known relationship between the effective magnetic field seen by a nucleus, $\mathbf{B}_{eff}$, (which determines its chemical shift) and the external magnetic field ($\mathbf{B}_0$). Here, $\boldsymbol{\sigma}$ is the shielding tensor, so named because of the way that the electrons "shield" the nucleus from the external magnetic field.

$$\mathbf{B}_{eff} = (1 - \boldsymbol{\sigma})\mathbf{B}_0 \tag{1}$$

In principle, the magnetic shielding of a given nucleus depends on the molecular orientation with respect to the external magnetic field, but for the systems examined here, i.e. liquids or solutions in which the individual molecules undergo fast random motion, the effective or average value of the shielding tensor need only be considered. This scalar quantity may be represented as a sum of several terms:

$$\sigma^A = \sigma^{AA}_{dia} + \sigma^{AA}_{para} + \sum \sigma^{BA}_{inter} + \sigma' \tag{2}$$

The electronic origin of the shielding arises because the circulation of electrons produces electrical currents and associated secondary local magnetic fields, which may either enhance or reduce the external field. The origins of these induced local magnetic fields are evident from the equations[9-13] describing the individual components of the shielding. For example, the diamagnetic term, σ^{AA}_{dia} arises from the circulation of electrons in spherically symmetrical (e.g. s) orbitals on a given atom A:

$$\sigma^{AA}_{dia} = \frac{\mu_0 e^2}{12 m \pi} \sum_i \left\langle \frac{1}{r_i} \right\rangle \tag{3}$$

Here $\langle 1/r_i \rangle$ is the mean inverse distance of the electrons from the nucleus and the summation is made only over electrons on atom A, i.e. this is an intra-atomic term.

The paramagnetic term, σ^{AA}_{para}, takes into account deviations of the electron distribution from spherical symmetry and hence is important for nuclei surrounded by p electrons, e.g. ^{13}C, ^{15}N, ^{19}F, etc.:

$$\sigma^{AA}_{para} = -\frac{\mu_0 e^2 \hbar^2}{6 m^2 \pi} \langle \Delta E \rangle^{-1} \langle 1/r^3 \rangle_{2p} \sum q \tag{4}$$

In equation (4), $\sum q$ represents a summation of terms from the charge-density bond-order matrix, and accounts, in part, for the relationship between chemical shifts and electron densities. However, the term related to the mean distance of the 2p electrons from the nucleus, $\langle r^{-3} \rangle_{2p}$, is also closely related to electron densities: an increase in electron density leads to a decrease in the effective nuclear charge and hence an expansion of the 2p orbitals. This results in a decrease in $\langle r^{-3} \rangle_{2p}$, and reduces the paramagnetic contribution to the screening. Because of its negative sign, a decreased magnitude of σ_{para}^{AA} increases the total shielding. In equation (4), ΔE is the average excitation energy[12] and represents an approximation to a summation of energy differences between ground and excited states. A consideration of equations (2) and (4) shows that a decrease in ΔE leads to a decrease in nuclear shielding. Since ΔE is not readily calculated, its role in determining chemical shifts is still not well defined. This is particularly so in ^{13}C NMR, and in studies of substituent effects on ^{13}C chemical shifts it is often assumed that ΔE remains constant for all members of a structurally related series. On the other hand, in a series where the electronic environment of the observed nucleus changes dramatically from one member to another, gross chemical shift changes have been related to ΔE. For example, ^{59}Co chemical shifts in a variety of cobalt complexes are linearly proportional to $1/\Delta E$,[14] where ΔE is estimated from absorption spectral data.

The third term in equation (2), $\sum \sigma_{inter}^{BA}$, represents the interatomic contribution to chemical shifts. Local intramolecular anisotropic effects enter via this term, as do delocalization effects in cyclic systems.

Finally, the term σ' in equation (2) is used to consider collectively all of the additional factors which can contribute to chemical shifts. Intermolecular effects due to solvent are one such type of interaction. Webb and co-workers[15–17] have made a number of experimental and theoretical studies of the effects of solvent on chemical shifts. In particular, they have demonstrated the significant contribution of solvent effects to nitrogen screenings. Solvent effects on ^{13}C chemical shifts have also attracted much recent attention,[18–20] and solvent effect studies of less common NMR nuclei, such as ^{17}O, are also beginning to appear.[21]

In studies of substituent effects on chemical shifts it is possible to keep many of the terms in equation (2) constant. For example, if all spectra in a series are measured under uniform conditions, the final term in this equation remains constant. To maintain this condition, it is important that spectra are obtained for dilute solutions. This minimizes complications from intermolecular association or other interactions which might produce additional contributions to the magnetic shielding of the probe nuclei. The interatomic or anisotropic term of equation (2) remains constant if the fixed probe group is sufficiently distant from the changing substituent. Finally, substituent-induced changes in the diamagnetic term are often relatively small, and therefore the paramagnetic term is usually most important in determining substituent effects on chemical shifts for nuclei other than protons. Hence, in much of the following discussion, trends in chemical shifts are related to changes in terms which make up the paramagnetic component of the screening.

In closing this section, it should be noted that the aim has not been to provide a rigorous discussion of chemical shift theory, but instead to summarize briefly the types of terms which can contribute to substituent effects on chemical shifts. More detailed definitions of all of the terms discussed above are available elsewhere.[9-13]

III. MODES OF TRANSMISSION OF SUBSTITUENT EFFECTS

The currently accepted mechanisms describing the transmission of substituent effects have been reviewed recently by a number of authors,[22-25] and hence only brief comments are given here. A perennial problem in this area is that of defining precisely the various mechanisms by which substituents may exert their effects. This difficulty is particularly acute in NMR studies of substituent effects, since chemical shifts are such sensitive probes of electronic effects that they often reflect contributions from several competing transmission mechanisms. On the other hand, reactivity studies, which often depend on gross differences between a charged transition state and a neutral ground state, are more usually dominated by either simple electrostatic or resonance effects.

Traditionally, substituent effects are separated into *polar* and *resonance* components.[26] The latter effects are related to the ability of a substituent to delocalize or transfer charge, whilst the former effects refer to a substituent's ability to polarize the electron distribution of a probe site (i.e. polar effects involve no formal charge transfer between the substituent and the probe site). Resonance effects are simply represented in terms of canonical structures such as those shown in [1]. Here the π electron-donating fluoro group increases the π electron density at the *ortho* and *para* ring positions, and shifts the corresponding ^{13}C resonances to low frequency (high field).

[1]

In contrast to resonance effects, polar effects may be transmitted partially by one or more different mechanisms. Structures [2a], [2b] and [2c] show three types of polar effects of the fluoro substituent.

In [2a] the electronegative nature of the fluorine causes a decrease in σ electron density at the adjacent carbon, with a smaller effect at successive carbon sites. This is the classical through-bond inductive effect,[27] and is now thought to be important only at sites very close to the substituent[22,23] (i.e. less than two or three bonds distant). The deshielding of the ^{13}C resonance of fluoromethane compared with methane can be attributed in large part to this effect. The question of whether the effect at successive carbons is simply reduced in magnitude (as shown in [2a]) or whether it alternates in sign, as suggested by MO calculations,[28] is as yet unresolved. In any case, since we are concerned mainly with substituent effects at sites relatively distant from the substituent, this through-bond inductive effect does not play an important role in the following discussion.

One mechanism which is important in determining longer range substituent effects on chemical shifts is the field effect, arising from the substituent's electrical dipole.[22-25,29,30] The through-space transmission of this effect is shown in [2b]. Many authors who have examined field effects on chemical shifts have followed the approach of Buckingham,[31] who showed that the effect of electric fields may be expressed as a power series:

$$\delta = aE + bE^2 + \cdots \tag{5}$$

Usually only the first term of this series is considered when the substituent is more than a few bonds distant from the probe site. Effects on chemical shifts arise from substituent-induced polarization of electron density along the bonds to the atom being examined. Because of this constraint, electric fields perpendicular to the bond direction have no effect, and only the electric field component resolved along the bond direction is important. In a rigid molecule this feature can impart a geometric dependence to chemical shifts. Examples of this are discussed in later sections.

Recently, the π-polarization mechanism [2c] has also been shown to be extremely important in determining chemical shifts in a variety of unsaturated systems. This effect is transmitted through space like classical field effects; however, in the π-polarization mechanism the probe is an unsaturated π system. In [2c] the electron-withdrawing dipolar fluoro substituent polarizes the π electron cloud of the benzene ring.

The importance of π-polarization can be seen from studies of systems containing a charged substituent on a side-chain attached to a phenyl ring.[32] Structures [3a]–[3c] show the π-polarization effect for monopolar and dipolar substituents, as deduced from ^{13}C NMR substituent chemical shifts, in $PhCH_2NH_3^+$, $PhCH_2CH_2Cl$, $PhCH_2CH(NH_3^+)$ CO_2^- and related systems.[32]

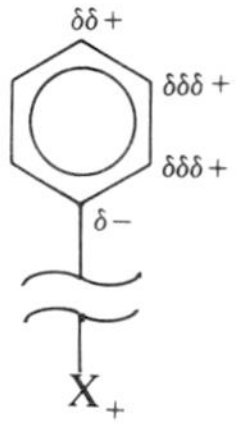

[3a] 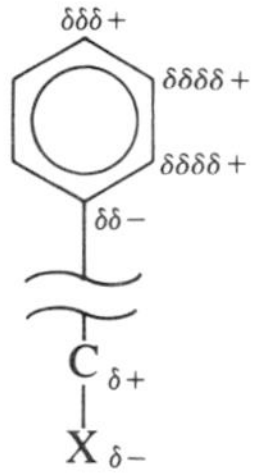   [3b] 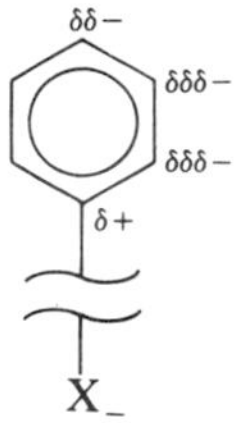   [3c]

Other evidence for π-polarization comes from studies of side-chain chemical shifts, where this mechanism can be used to explain "reverse" chemical shifts observed at C-α for many series containing vinyl,[33,34] carbonyl[8,35] or other unsaturated side-chains.[36,37] For example, in *meta*- and *para*-substituted acetophenones [4][35] *electron-withdrawing* polar substituents such as CN, CF_3 or NO_2 cause *low frequency* shifts of the carbonyl resonance because of the induced polarization pattern shown. The dipoles of these substituents induce a transfer of π electron density in the carbonyl group from the oxygen to the carbon, where the increased electron density leads to a low frequency shift.

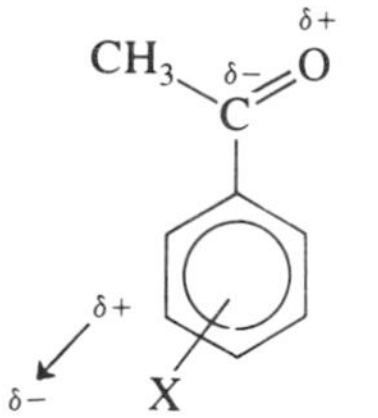

[4]

The substituent effects discussed above represent the most important electronic effects on chemical shifts. The role of other factors such as orbital repulsion effects[22] or inductomesomeric effects[22] is less clear.

IV. CORRELATION ANALYSIS

A number of correlation schemes for the analysis of substituent effects have been proposed over the last half century, following the pioneering work of Hammett. In the last 10–15 years NMR spectroscopy has provided a broad, accurate data base that has in many ways shaped the development of improved correlation procedures. The general aim has been to use relationships between chemical shifts and substituent parameters (σ constants) to obtain one or more transmission coefficients (commonly denoted by the symbol ρ), which are used to quantify the ways in which substituents transmit their effects.

Since this review is primarily concerned with NMR applications, rather than studies of substituent effects in themselves, it is not appropriate to discuss and compare the various correlation methods that have been applied.[38,39] Instead it will simply be noted that, in general, correlations will be discussed in terms of the *dual substituent parameter* (DSP) method[40,41] (or extended versions of this procedure).[42] In this method, the property (P) being examined is related to a linear combination of polar and resonance substituent parameters (σ_I and σ_R respectively) as shown by equation (6):

$$P - P_0 = \rho_I \sigma_I + \rho_R \bar{\sigma}_R \qquad (6)$$

The bar above the σ_R symbol denotes the fact that any one of the four resonance scales (σ_R^-, σ_R^0, σ_R^{BA}, or σ_R^+) can be used for a given correlation, depending on the electron demand placed on the molecular site being examined. The "goodness of fit" of a DSP correlation is judged from the parameter $f = SD/RMS$, where SD is the standard deviation of the fit, and RMS is the root-mean-square size of the experimental data. The smaller the f value the better the fit, with f values of 0.0–0.1 representing excellent correlations, $f \sim 0.1$–0.2 moderately good ones, and f values greater than 0.3 representing only crude trends. In order to achieve a meaningful separation of polar and resonance effects it is necessary to obtain experimental data for a wide range of substituents, or "basis set", which represents an independent set of σ_I and σ_R values.[41] The currently recommended[43] basis set is as follows: two strong donors (NMe_2, NH_2 or OMe); two halogens (not both Cl and Br); Me and H, one acceptor (NO_2, CF_3 or CN); and one carbonyl acceptor (COMe or COOR).

Two other multiparameter correlation schemes that have been applied to the analysis of a variety of reactivity and chemical shift data are those of Swain and Lupton[44] and Yukawa and Tsuno.[45] Although these methods have some limitations relative to the DSP method, all three multiparameter correlation methods represent an improvement over the single parameter Hammett treatment, especially for the analysis of NMR chemical shift data. Single parameter treatments cannot, in general, cope with differing relative transmission of polar and resonance effects at different probe sites.[46]

Finally, although there is now a vast body of literature dealing with empirical correlations between NMR chemical shifts and a variety of substituent parameters, it is perhaps easiest to demonstrate the general acceptance of substituent chemical shifts as valid probes for the physical organic chemist by citing the recent proposal[47] of an undergraduate teaching laboratory experiment in which students prepare substituted styrene derivatives, measure their NMR spectra and correlate the side-chain ^{13}C chemical shifts with Hammett-type substituent constants. The transition of this type of approach from the research laboratory to the teaching laboratory proves the utility of chemical shifts as a measure of substituent electronic effects.

V. CONVENTIONS AND DEFINITIONS

Chemical shifts for all nuclei are expressed using the accepted sign convention that positive values imply deshielding, associated with the shift of a resonance to lower magnetic field (higher frequency). Usually, we will be concerned not with absolute chemical shifts, but with *substituent chemical shifts* (SCS). An SCS is defined as the change in chemical shift (in parts per million) of an observed resonance that occurs when a hydrogen atom in one part of the molecule is replaced with a substituent group. For example, the *para* carbon SCS for the NH_2 substituent in benzene is shown to be −9.8 ppm in [5].

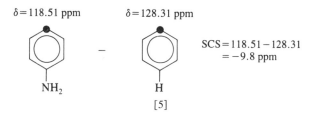

$\delta = 118.51$ ppm　　　$\delta = 128.31$ ppm

NH₂　　　　　H

SCS = 118.51 − 128.31
= −9.8 ppm

[5]

Where electron densities are quoted in the text, they will also be generally expressed in terms of substituent effects. The terms Δq_π, Δq_σ and Δq_t will be used to refer to substituent-induced changes in π, σ and total ($= \pi + \sigma$) electron density respectively. Since these changes are usually small (hundredths of an electron or less), numerical values are often scaled up by a factor of 10^3 or 10^4. Positive Δq values refer to a substituent-induced increase in electron density. Care should be taken to note that *electron* densities are used in this review, instead of *charge* densities, which have the opposite sign.

In the following sections, studies of various nuclei are arranged in order of atomic number. No attempt has been made to cover exhaustively all nuclei which have been investigated by NMR, as the emphasis is biased heavily towards those nuclei important in substituent effect studies. In the period 1975–8, a number of reviews[39,48,49,51,63] related to the use of chemical shifts as probes of electronic structure and substituent effects appeared, and so the present article deals mainly with literature reports after 1978, although some earlier background material is given.

VI. ^{1}H CHEMICAL SHIFTS

While there have been many reported studies of substituent effects on ^{1}H chemical shifts,[2,49–83] systematic correlations of these shifts with substituent parameters or electron densities have been of limited *general* utility. This arises because of the small chemical shift range of this nucleus, coupled with significant contributions to ^{1}H shielding from anisotropic and ring current effects. Nevertheless, in carefully designed systems these complications are minimized and useful correlations have been obtained.

A. Aromatic compounds

An abundance of data are available relating ^{1}H chemical shifts in monosubstituted benzenes to those in more highly substituted benzenes.[50] In one such study, Beistel and coworkers[52] summarized many of the early studies of substituent effects in these series, and derived empirical relationships for the prediction of proton chemical shifts in disubstituted benzenes based on the "internal chemical shifts", Δ_{XY} defined as follows:

$$\Delta_{XY} = \delta_Y - \delta_X \qquad (7)$$

Here δ_X is the chemical shift of protons *ortho* to X, and δ_Y is the shift of protons *ortho* to Y in a disubstituted benzene XC_6H_4Y. Relationships of the form

$$\Delta_{XY} = m\Delta_{XZ} + b \qquad (8)$$

were derived[52] to relate internal shifts in one series (fixed substituent = Y) to those in another (fixed substituent = Z). Initially these correlations were applied only to halogen series (Y, Z = Cl, Br, etc.), but were later extended to other families of disubstituted benzenes.

A number of other empirical schemes[2,53–62] have also been developed for making assignments in polysubstituted aromatics. Many of these are based on additivity of substituent effects.[58–61] Other authors have taken a more theoretical approach, and sought to determine the origin of substituent effects on ^{1}H chemical shifts. For example, Hehre, Taft and Topsom[63] have examined relationships between ^{1}H shifts and calculated electron densities in monosubstituted benzenes. In general, *meta* ^{1}H shifts can be moderately well correlated with an equal combination of σ electron densities at the hydrogen atom and its attached carbon (i.e. $\Delta q_\sigma^C + \Delta q_\sigma^H$).[63] On the other hand, the *para* ^{1}H shifts are more closely related to Δq_t^C at the attached carbon, and the inclusion of contributions from Δq_σ^H on the hydrogen atom leads to a poorer correlation.[63] This is indicated by the DSP analyses below:[63]

$$^1\text{H SCS} = 0.27\sigma_I + 1.25\sigma_{R^0} \qquad f = 0.13 \qquad (9)$$

$$10^2 \Delta q_\sigma^H = -0.15\sigma_I - 0.17\sigma_{R^0} \qquad f = 0.18 \qquad (10)$$

The sign difference in the two equations reflects the convention of quoting electron densities here. The equations show that donor substituents increase electron density on the hydrogen atom and bring about low frequency shifts of the *para* ^{1}H signal. The critical feature of the two equations is the large difference in the ratio of polar and resonance effects for the ^{1}H SCS values and electron densities on the hydrogen atom, showing that these terms are not linearly related.

Although most studies of substituent effects in aromatic series are concerned with substituted benzenes, some work on naphthalenes,[64] phenanthrenes,[65] dehydroannulenes,[66] ethylenes[67] and a number of heterocyclic derivatives[68] have appeared. The general principle that ^{1}H shifts should be determined at infinite dilution (if purely electronic effects are to be examined) can be confirmed from the data of Lucchini and Wells,[64] who carried out a concentration study of several 1- and 2-substituted naphthalenes, and found considerably different magnitudes of solute–solute induced shielding for different substituted naphthalenes and for different protons within each substrate.

B. Side-chain series

Correlations of ^{1}H SCS values in the side-chains of aromatic compounds with single parameter Hammett treatments have been concluded to be of little value.[69,70] However, Ewing[39] has pointed out that this reflects the limitation of single parameter approaches (only one blend of polar and resonance effects is inherent in σ constants), rather than the nature of the data itself. This had earlier been stressed[41] for a variety of reactivity and SCS data. Dual parameter treatments of side-chain ^{1}H shifts have been more successful.[33,37,71–77]

One of the main reasons for studying side-chain systems is to investigate long-range polar substituent effects. In rigid systems it is sometimes possible to monitor orientational effects arising from differential field transmission.

Hamer, Peat and Reynolds[33] have used this approach to separate contributions of π-polarization and classical field effects to ^1H SCS values in substituted styrenes [6].

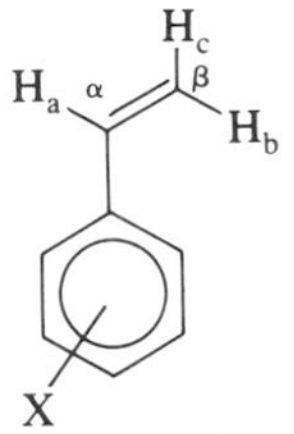

[6]

It was found[33] that observed substituent effects on ^1H chemical shifts of the side-chain protons could be interpreted in terms of three transmission mechanisms: (a) resonance delocalization by the substituent; (b) π-polarization of the vinyl side-chain which brings about changes in electron density at C-α and C-β. This induces related changes in hydrogen atom electron densities and hence ^1H chemical shifts; and (c) the through-space field effect. Since contributions to ^1H chemical shifts arising from π-polarization of the adjacent π system are equal for protons H_b and H_c, the term $\delta(H_b) - \delta(H_c)$ reflects differences in electric field polarization of the two C—H bonds. These suggestions are borne out by observed larger ^1H SCS values[33] for H_c compared with H_b. In terms of the Buckingham[31] approach, the resolved component of the electric field along the C—(H_c) bond is larger.

Analysis of the H_b and H_c SCS values using the Swain–Lupton treatment was carried out[33] and the resultant equations are shown below:

$$H_b \text{ SCS} = 0.10F + 0.41R \tag{11}$$

$$H_c \text{ SCS} = 0.17F + 0.42R \tag{12}$$

These equations indicate the resonance effects account for about 70% of the total substituent effect at H_c. Resonance effects at H_b are of similar magnitude, but make a greater percentage contribution ($\sim80\%$) to the total shifts because polar effects are smaller. Of the 30% polar contribution at H_c, field effects and π-polarization contribute approximately equally. Electron densities calculated using the CNDO/2 method were also correlated with Swain–Lupton substituent parameters, and confirmed the same overall trends observed in the ^1H chemical shifts. A correlation of the shift difference, $\delta(H_b) - \delta(H_c)$, with the difference in calculated electron densities for the two β hydrogen atoms has a slope of 18 ppm per electron.[33] Again, the same general trends were later confirmed with more sophisticated *ab initio* MO calculations.[73]

The phenylacetylene series [7] is also ideally suited to studies of electronic substituent effects in ^{1}H chemical shifts.[37,76]

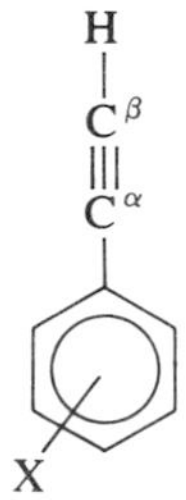

[7]

In this series, the β side-chain proton undergoes a shift of 0·46 ppm as the *para* substituent is changed from NO_2 to NH_2. As expected, donor substituents induce low frequency shifts and acceptors high frequency ones, consistent with observed substituent effects on hydrogen atom electron densities calculated by the semi-empirical CNDO/2 MO method.[37] The ^{1}H shifts are also closely related to ^{13}C shifts at the adjacent C-β atom:[37]

$$\text{SCS (C-}\beta) = 15.3 \text{ SCS (H)} \qquad r = 0.999 \qquad (13)$$

Ab initio calculations have recently been performed[77] on this series and the correlation between ^{1}H chemical shifts and these electron densities is shown in Fig. 1. The slope is $\sim$40 ppm per electron.

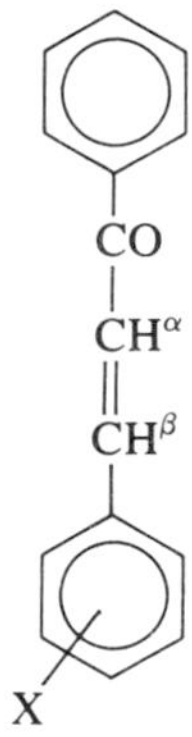

[8]

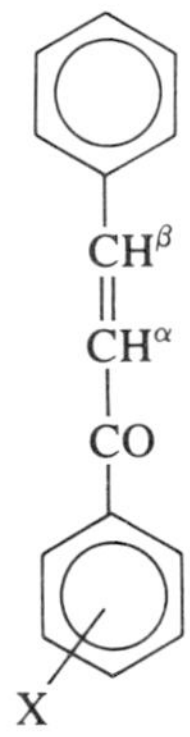

[9]

Many larger side-chain series have been examined;[78,79] however, a full discussion of observed trends will not be attempted here. Instead, an illustrative example from the chalcone series[75] is discussed briefly. Of the two chalcone series [8] and [9], substituent effects on the H-α and H-β side-chain protons are larger in series [8].[75] Correlations of the ^{1}H chemical

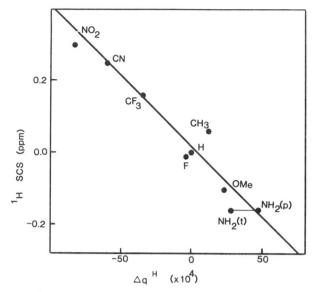

FIG. 1. Plot of ^{1}H SCS in *para*-substituted phenylacetylenes[37] versus calculated *ab initio* electron densities[77] on the hydrogen atom. Slope = 40 ppm per electron. NH$_2$(t) and NH$_2$(p) refer to tetrahedral and planar geometries for the NH$_2$ group respectively.

shifts with Hammett σ parameters, and with Swain–Lupton F and R values have been reported,[75] with the latter method giving better results. DSP analyses of the side-chain ^{1}H shifts are given in Table 1. The reversal in the sign of ρ_I for H-α and H-β shows that the shifts are strongly affected by π-polarization of the adjacent vinyl group.

Of the correlations in Table 1, only that for H-α in series [8] is statistically significant. Although the crude correlations for the other sites show the

TABLE 1

DSP correlations of side-chain ^{1}H SCS values in series [8] and [9].[a]

Series		ρ_I	ρ_R	f
[8]	H-α	0.11	0.38	0.12
	H-β	−0.06	0.07	0.66
[9]	H-α[b]	−0.13	−0.05	0.39
	H-β	0.03	0.06	0.45

[a] SCS data from reference 75. Correlations were carried out using σ_I and $\sigma_R{}^o$. The relatively poor fits in these correlations are due to the extremely small range of the SCS data.

[b] NO$_2$ substituent omitted.

expected sign alternation in ρ_I, perhaps the main purpose of the table is to illustrate the extremely small range in ^{1}H shifts in most side-chain derivatives. It is only when there is a significant resonance interaction with a side-chain proton (such as for H-β in styrenes) that reliable correlations with substituent parameters are expected.

One other recent study which is of interest concerns ^{3}H NMR studies of substituent effects in benzamides and N-alkylbenzamides.[80] This is perhaps the only published systematic study of substituent effects on tritium chemical shifts, and although it is of much technical interest, the conclusions derived are basically the same as those expected from a related ^{1}H NMR study.[81] Nevertheless, in some instances, the use of the tritium isotope makes it easy to obtain assignments in overlapping spectral regions.[80]

C. Aliphatic compounds

Many early correlations of ^{1}H SCS values in aliphatic compounds have been summarized elsewhere.[2,70] Relatively few reports have appeared recently;[82,83] however, in one study Gasteiga and Marsili[82] examined relationships between ^{1}H chemical shifts and charges on hydrogen determined by the method of iterative partial equalization of orbital electronegativity.[84] Their discussion is limited to σ-bonded systems and non-conjugated π systems to avoid ring current effects in aromatic compounds. Proton shifts in the former systems are determined largely by the diamagnetic term, described by σ_{dia}^{AA} (equation (3)). The authors summarize previous attempts to correlate proton shifts with empirical,[85] semi-empirical[86] and non-empirical methods.[87] In general it is found that proton chemical shifts correlate linearly with charges on the hydrogen atom, although this charge is *not* related to charge at the adjacent carbon atom. This is in contrast to findings in aromatic series (monosubstituted benzenes,[63] styrenes[33] and phenylacetylenes[37]).

VII. ^{11}B CHEMICAL SHIFTS

An extensive survey of NMR applications of many of the nuclei of the periodic table has recently been provided.[88] It is stressed again here that this review concentrates on systematic studies of *substituent* effects. Only a limited number of such studies on ^{11}B chemical shifts have appeared,[89-92] despite the high sensitivity of this nucleus to NMR detection. In general, ^{11}B shifts are an order of magnitude more sensitive to substituents than ^{1}H chemical shifts. In ring-substituted phenylboronic acids [10], for example, an ^{11}B shift change of nearly 7 ppm is observed[89] when the *para* substituent is changed from NO_2 to NH_2.

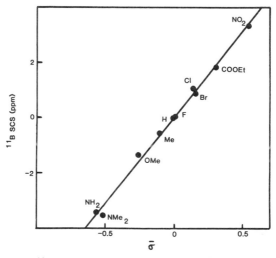

[10]

A DSP analysis of the ^{11}B SCS values shows that resonance effects and polar effects contribute approximately equally:

$$^{11}\text{B SCS} = 4.2\sigma_\text{I} + 4.5\sigma_\text{R}^{\text{BA}} \qquad f = 0.08 \qquad (14)$$

Figure 2 shows a graphical representation of the DSP correlation given in equation (14). The linearity of this plot, which may be regarded as a

FIG. 2. DSP plot of ^{11}B SCS values in *para*-substituted phenylboronic acids[89] versus an effective substituent constant $\bar{\sigma} = \sigma_\text{I} \cos\theta + \sigma_\text{R} \sin\theta$, where $\tan\theta = \lambda = \rho_\text{R}/\rho_\text{I}$.

comparison of observed versus calculated SCS data, demonstrates the systematic sensitivity of ^{11}B chemical shifts to electronic substituent effects. In general, the trends observed in substituted phenylboronic acids are readily extended to other similar boron-containing compounds.[90,91] For example, resonance and polar effects have been shown to be important in determining ^{11}B chemical shifts in potassium tetra-arylborates.[91] Relationships between π and total electron densities and ^{11}B chemical shifts have also been established in a variety of boron compounds.[92] General parallels have also been found between ^{11}B and ^{13}C shifts in isoelectronic boranes and carbo-cations.[93]

VIII. ^{13}C CHEMICAL SHIFTS

A. Monosubstituted benzenes

The ^{13}C resonance of a solution of 3% benzene in $CDCl_3$ has a chemical shift 128.31 ppm to high frequency of TMS.[42] Introduction of a substituent into the ring removes the degeneracy of the NMR signal by perturbing the *ipso*, *ortho*, *meta* and *para* carbons to different extents. For example, a resonance-withdrawing substituent such as NO_2 decreases the shielding at the *ortho* and *para* carbons by removing π electron density as shown in [11].

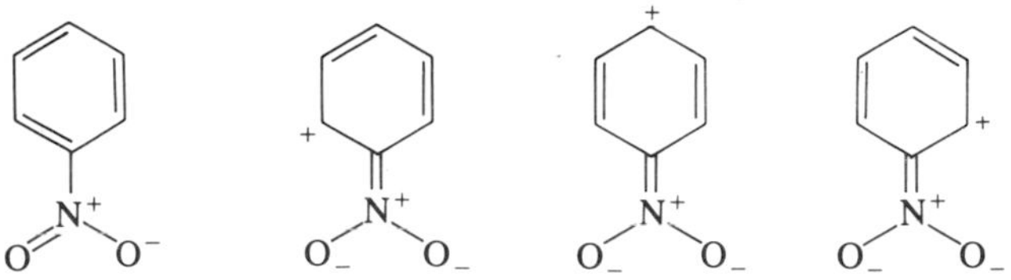

[11]

The *ipso* site is influenced by the electronegative nature of the adjacent nitrogen and is also deshielded. These simple observations reflect the fact that different sites in a molecule are influenced by different blends of substituent effects, i.e. resonance effects contribute mainly at the *ortho* and *para* positions, and through-bond inductive effects contribute at the *ipso* position. π-Polarization plays an important role at all ring positions.

Table 2 summarizes SCS values for 13 common substituents in monosubstituted benzenes. This represents the most uniform set of SCS data currently available for this series and was derived for dilute solutions of the monosubstituted benzenes in an inert solvent, measured under uniform conditions.[42] Similar tables, which are available in standard texts[94–96] on ^{13}C NMR, have been used extensively to aid assignments (assuming additivity of SCS values) in polysubstituted aromatics. In fact the assumption of additivity is a useful approximation, although major deviations have been observed, as discussed later.

Note that for the simple substituents in Table 2, the *ipso* site has the largest range of SCS values. This trend also holds true for a large variety of other substituted benzenes. Ewing[97] has compiled probably the most complete survey of substituent effects on ^{13}C shifts in monosubstituted benzenes. Data from this compilation are represented in graphical form in Fig. 3, which illustrates the successive decrease in SCS values at the *ipso*, *para*, *ortho* and *meta* positions.

Many of the early studies[98–106] of correlations between ^{13}C chemical shifts in monosubstituted benzenes and substituent parameters have been summarized recently.[39] DSP analyses[42] of the uniform SCS data given in Table

TABLE 2

^{13}C substituent chemical shifts (ppm) for monosubstituted benzenes in deuterochloroform.^{a}

X	Ipso	Ortho	Meta	Para
NMe$_2$	22.28	−15.66	0.74	−11.69
NH$_2$	17.91	−13.25	0.88	−9.80
OMe	31.21	−14.45	0.88	−7.68
F	34.51	−13.00	1.13	−4.39
Cl	5.93	0.31	1.40	−1.90
Br	−5.87	3.16	1.71	−1.50
Me	9.52	0.68	−0.09	−3.05
H^b	0.00	0.00	0.00	0.00
CF$_3$	2.31	−3.13	0.40	3.41
CN	−15.96	3.72	0.70	4.35
COOEt	2.14	1.17	−0.09	4.42
COMe	8.69	0.15	−0.09	4.67
NO$_2$	19.95	−4.85	0.93	6.22
CHO	8.06	1.36	0.63	6.07

^{a} Reference 42. Positive shifts are to high frequency. Digital resolution 0.05 ppm.
^{b} The parent derivative has a shift 51.30 ppm from the central peak of CDCl$_3$ (128.31 from TMS).

2 show that resonance effects dominate the shifts at the *para* position, confirming the importance of π electron density changes brought about by interactions such as shown in [11]:

$$C_p \text{ SCS} = 4.6\sigma_I + 21.5\sigma_{R^o} \qquad f = 0.03 \qquad (15)$$

$$C_m \text{ SCS} = 1.6\sigma_I - 1.3\sigma_{R^o} \qquad f = 0.25 \qquad (16)$$

Polar effects also have a significant influence on *meta* and *para* SCS values. Good correlations with σ_I and σ_{R^o} are not expected at the *ortho* and *ipso* positions due to steric[107] and through-bond effects.

Note that the decrease in ρ_I at the *meta* position relative to the *para* position is consistent with the polar component of the shifts being determined largely by π-polarization (see, for example, [2c]). The negative sign of ρ_R at the *meta* position is of much interest, but has not yet been fully explained. The observation of reverse resonance effects at non-conjugating sites in aromatic compounds is, however, a common observation, and for some series it has been suggested[108] that this may be a reflection of contributions from π_{orb} effects.[22]

A number of correlations of ^{13}C chemical shifts with either π or total electron densities in monosubstituted benzenes have been reported. Early studies[98] suggest that the shifts at all ring positions are better correlated

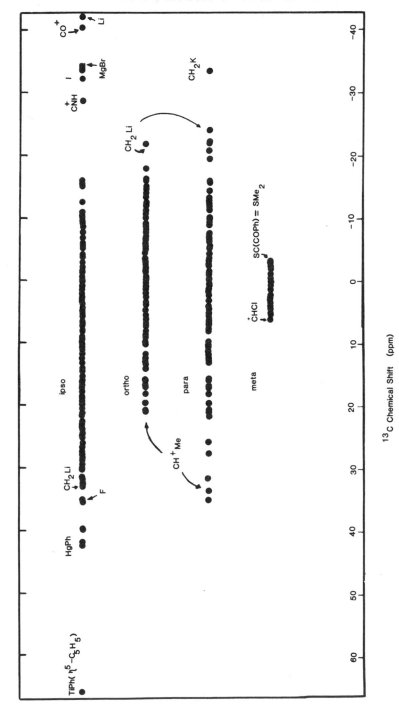

FIG. 3. Representation of the different ranges of *ipso, ortho, meta* and *para* substituent chemical shifts (SCS) in monosubstituted benzenes. SCS data from reference 97.

with total rather than π electron densities (calculated by the CNDO/2 method). However, if only the *para* site is considered, π and total electron densities give fits of equally good quality. More recent studies suggest that, at sites where steric and anisotropic effects are not present, π electron densities are the controlling influence on ^{13}C shifts. Hehre, Taft and Topsom[63] have summarized many of the available data and have presented plots of *meta* and *para* SCS values versus total electron densities calculated by *ab initio* MO theory. The slope is 150 ppm per electron for the *meta* position and 450 ppm per electron for the *para* position. The *meta* plot is slightly more scattered. A plot of *para* SCS values versus Δq_π is linear with a slope of 200 ppm per electron.[63] The *ab initio* results thus confirm the trend in the earlier semi-empirical calculations that π and total electron densities give equally good correlations (in terms of quality of fit) with C_p SCS values. The two electron density components do, however, yield considerably different slopes (200 and 450 ppm per electron for Δq_π and Δq_t respectively: estimated error in these slopes is $\pm 10\%$). The correlations are of equally good quality because of a very precise relationship between substituent-induced changes in the π and σ components of electron density, which also implies a precise relationship between π and total electron densities (since $\Delta q_t = \Delta q_\sigma + \Delta q_\pi$):

$$\Delta q_\sigma = -0.57 \Delta q_\pi \qquad r = 0.999 \qquad (17)$$

Similar relationships have previously been observed in CNDO/2 studies of monosubstituted benzenes.[109] Since the π and σ effects oppose each other, changes in total electron densities are smaller than changes in π electron densities. This explains the larger slope for C_p SCS/Δq_t correlations compared with C_pSCS/Δq_π correlations. Since it is not possible to accurately predict *a priori* the correct magnitude of the slope for SCS/electron density correlations, the *para* site cannot be used to differentiate between total and π electron densities as probes of SCS values. The greater importance of π electron densities can, however, be implied from correlations of π and total electron densities versus SCS values in unsaturated side-chains. In a number of cases,[8,110,111] π electron densities predict the correct direction of shifts, while total electron densities predict the wrong direction (in these cases, σ/π interactions significantly different from those given in equation (17) are observed). Specifically, the C-α site in ring-substituted benzonitriles[110] and benzoyl fluorides[111] can be used to investigate differences between π and total electron densities as probes of chemical shifts. These series are discussed in later sections.

A practical benefit of the large number of correlations between SCS values and either substituent constants or electron densities is that the derived equations may be used to estimate substituent constants (or Δq_π values) for substituents not included in the original analyses. To do this,

SCS values for the new substituted derivative are measured, and substituted into the previously derived correlation equation. For Hammett substituent constants, an equation involving C_p SCS values is normally used; however, two equations are required to derive σ_I and σ_{R^o} values for a given substituent. In this case, C_p SCS and C_m SCS correlation equations may be used. It should be stressed that substituent constants obtained in this way[112-117] are limited by the accuracy of the correlation equations used to derive them, and care should be taken to measure SCS values for the new substituent under similar conditions to those used in the original correlation procedure. Recent examples of the use of correlation equations to derive substituent constants for groups not studied extensively by other methods include the determination of σ_p values for a variety of phosphorus-containing substituents,[113] σ_p^+ and σ_I and σ_{R^o} values for $-BX_2$ substituents (X = Cl, F, 9-bicyclononane, OH, and Me)[114] and σ_I and σ_{R^o} values for substituted methyl groups (CH_nX_{3-n}, X = F, Cl, Br, I).[115]

In the latter study, correlation equations similar to equations (15) and (16) were used to derive σ_I and σ_{R^o} values for halomethyl groups based on ^{13}C chemical shifts for a wide range of halotoluenes.[115] These ^{13}C chemical shifts are reported in Table 3. As was pointed out,[115] the statistical fit of the *meta* SCS data for monosubstituted benzenes (equation 16) is significantly poorer than for the *para* SCS data, and so the derived values

TABLE 3

^{13}C chemical shifts (ppm) in α-substituted toluenes.[a]

X^b	C-1	C-2	C-3	C-4	C-α	C-β
CH_3	137.88	129.08	128.27	125.36	21.53	
CH_2CH_3	144.25	127.88	128.33	125.61	20.12	28.85
$CH(CH_3)_2$	148.89	126.44	128.33	125.80	26.60	34.21
$C(CH_3)_3$	151.10	125.28	128.11	125.47	34.75	31.48
CH_2F	136.27	127.52	128.64	128.77	84.60	
CHF_2	134.37	125.49	128.62	130.67	114.76	
CF_3	130.77	125.24	128.76	131.78	124.38	
CH_2Cl	137.48	128.55	128.61	128.36	46.24	
$CHCl_2$	140.36	126.09	128.76	129.89	71.81	
CCl_3	144.27	125.52	128.38	130.35	97.76	
CH_2Br	137.77	129.00	128.76	128.38	33.59	
$CHBr_2$	141.90	126.49	128.63	129.81	41.19	
CBr_3	146.97	126.49	128.06	130.14	36.5	
CH_2I	139.25	128.73	128.73	127.82	5.77	
CHI_2	145.25	126.31	128.20	129.05	-0.59	

[a] Reference 115. Shifts relative to TMS.

[b] The data represent ^{13}C shifts in monosubstituted benzenes, where X is the substituent at C-1.

of σ_I for the CH_nX_{3-n} substituents are necessarily less precise than the derived values of σ_{R^0}. Nevertheless, reasonable agreement with substituent constants derived from other sources was obtained.[115] Brownlee and Sadek[116] have suggested that more accurate values of σ_{R^0} can be derived in this way if σ_I values can be obtained from independent methods.

In the study of borylbenzenes noted above,[114] as well as deriving substituent constants for BX_2 groups, ring electron densities for a number of these substituents are correlated with *ortho* and *para* ^{13}C chemical shifts. Using C_p SCS data for the four BX_2 groups, where $X = F$, Cl, OH and 9-bicyclononane, better correlations are obtained with Δq_t than with Δq_π; however, the range of substituents is too limited to conclude any general superiority of Δq_t as a measure of ^{13}C SCS values. Odom *et al.*[118] have also examined ^{13}C chemical shifts in a variety of phenylboranes.

Several other authors have utilized ^{13}C chemical shifts in substituted benzenes as a probe of electronic properties of less common substituent groups.[119,120] In an illustrative example, Ricci *et al.*[119] found that the difference between *para* and *meta* ^{13}C chemical shifts ($\delta_{para} - \delta_{meta}$) for the positively charged substituents NMe_3^+, $CH_2NMe_3^+$ and PMe_3^+ cannot be related solely to σ_{R^0} values. The term $\delta_{para} - \delta_{meta}$ can, however, be related to a combination[102] of σ_I and σ_{R^0} values. This illustrates the fact that although values of $\delta_{para} - \delta_{meta}$ provide a rough measure of resonance effects,[101] complete cancellation of polar effects does not occur (see, for example, equations (15) and (16)).

Increasing attention is being directed towards studies of solvent effects on chemical shifts, and some understanding of their role is beginning to emerge. In an early study of solvent effects on ^{13}C shifts in substituted benzenes,[98] the measurement of C_p SCS values allowed σ^+ for a number of substituents to be estimated in different media (e.g. CCl_4 and trifluoracetic acid). This study showed that ^{13}C shifts can be used to monitor changes in a substituent's electronic effect (and hence its σ value) as a function of solvent; however, changes in the transmission of substituent effects in different media were not taken into account. An attempt to examine these effects was made in a recent study[121] in which only solvents and substituents not likely to undergo specific mutual interactions were included. Table 4 shows DSP analyses of C_p SCS values for monosubstituted benzenes in a number of solvents.[121] The trends in the ρ_I and ρ_R values may be taken as a measure of differential transmission of substituent effects in the different solvents. These values increase with increasing solvent polarity, and correlate well with π^* parameters.[122]. The increase in ρ_I can be explained[121] in terms of larger π-polarization effects which occur when substituent–solvent dipole–dipole interactions increase.[123] On the other hand, increases in ρ_R can be attributed[121] to a greater relative importance of resonance forms [12] and [13] in more polar solvents.

TABLE 4

DSP analyses of C_p SCS values in monosubstituted benzenes in various solvents.[a]

Solvent	π^{b}	ρ_I	ρ_R	λ	n	SD	f
c-C_6H_{12}	0.00	3.35	20.55	6.13	18	0.07	0.01
CCl_4	(0.19)	3.73	20.72	5.55	20	0.06	0.01
$CDCl_3$	(0.51)	4.54	21.54	4.74	15	0.11	0.02
Acetone	0.68	4.95	21.81	4.41	19	0.22	0.04
DMF	0.88	5.32	21.93	4.12	17	0.27	0.05
Me_2SO	1.00	5.48	22.21	4.05	17	0.28	0.05

[a] Reference 121.
[b] Reference 122. Values in parentheses are those applicable to NMR shifts.

In general, solvent effects on the polar component of C_p SCS values are found[121] to be larger than effects on the resonance component, in agreement with other studies.[124,125]

Finally, a comparison of two studies discussed above[98,121] shows that there is considerable difficulty involved in assigning changes in SCS values with solvent to either changes in inherent substituent effects or to changes in the transmission of these substituent effects. Further work is necessary in this area.

B. Disubstituted benzenes

These systems have been studied using a variety of spectroscopic techniques, with one emphasis being on the investigation of mutual substituent–substituent interactions. ^{13}C chemical shifts have played an important role in examining these effects. To avoid complications from steric interactions, the discussion here is limited to *meta* and *para* disubstituted benzenes. Because of the interest in examining mutual conjugation effects, most studies to date have concentrated on *para* derivatives. Substituent chemical shifts in a wide range of *para* disubstituted benzenes [14] are given in Tables 5–8.[42]

In a given series, Y is regarded as the fixed substituent, and X varies over a wide range of electron donating and withdrawing groups. There are thus four sites (relative to X – *ipso* (C-1), *ortho* (C-2), *meta* (C-3) and *para* (C-4)) at which the influence of the common substituent Y on SCS values of X can be measured. A scan across the rows of Tables 5–8 shows

Y
|4
3
2
1
X

[14]

that the largest variation (i.e. from one Y series to another) in SCS values of a given X substituent occurs at the sites *para* to X (adjacent to Y) and *ipso* to X (i.e. *para* to Y). First we examine the effect of the fixed substituent Y in modifying *para* SCS values of X. Although it is clear from Table 5 that Y considerably changes the response of C-4 chemical shifts to substituent effects of X (variations of up to 8 ppm are observed), there is not yet complete agreement on the origin of this effect. In one study of non-additivity,[123] it was suggested that group Y modifies the extent of π-polarization of the *para* substituent, by affecting σ electron density at C-4. For the Y groups NH_2, OMe, F, CN, NO_2, a reasonable correlation is noted between σ electron density at C-4 and ρ_I obtained from a DSP correlation of C-4 SCS values. A later investigation[110] with a larger range of Y groups showed that this is not a general relationship, however,

In another treatment of non-additivity, Lynch[126] proposed that non-additivity of ^{13}C shifts in *para* disubstituted benzenes can be best treated using a simple proportionality approach in which the chemical shifts obey the relation

$$S_X(Y) = a + b \times SCS_X(H) \tag{18}$$

where $S_X(Y)$ is the chemical shift of the carbon *para* to X in series Y, and $SCS_X(H)$ is the corresponding substituent chemical shift of X in monosubstituted benzenes (i.e. Y = H). The parameter a represents the chemical shift of the measured site in series Y when X = H. If $b = 1$ then additivity prevails, otherwise a "proportionality relationship"[126] exists. The correlation of a large number of *para* carbon (C-4) chemical shifts with equation (18) gave proportionality relationships ($b \neq 1$) of reasonable precision, as indicated in Table 9.[126] Trends in the slopes, b, for different Y series such as (a) $F < OMe < NH_2$, $NO_2 < H < Me$; (b) $F < H < Cl < Br < I$; and (c) $CMe_3 < SiMe_3 < GeMe_3 < SnMe_3$ are used to suggest that physical properties of the Y group, such as electronegativity or ionization potential might be responsible for modifying the SCS behaviour of the *para* carbon site. Such effects are presumed to modify the average excitation term, ΔE, of the paramagnetic screening expression (equation (4)). Within series (b) and (c) above, reasonable fits are obtained for correlations between the slopes, b, and inverse of the ionization potentials of the key atom of Y.[126]

TABLE 5

Para ^{13}C SCS values (ppm) of X in 1,4-disubstituted benzenes[a] in $CDCl_3$.

1-Substituent (X)	4-Substituent (Y)													
	NMe_2	NH_2	OMe	F	Cl	Br	Me	H	CF_3	CN	COOEt	COMe	NO_2	CHO
NMe_2	-6.56	-8.35	-7.48	-7.20	-12.43	-13.93	-11.76	-11.69	-12.82	-15.11	-13.15	-11.79	-11.10	-11.36
NH_2	-5.78	-7.67	-6.83	-6.54	-11.12	-12.27	-10.10	-9.80	-10.61	-12.54	-10.44	-9.32	-9.13	-8.94
OMe	-4.88	-6.35	-5.68	-5.63	-8.74	-9.67	-7.96	-7.68	-7.77	-8.48	-7.62	-6.80	-6.76	-6.41
F	-3.15	-3.86	-3.84	-4.07	-5.15	-5.96	-4.56	-4.39	-4.17	-3.81	-3.74	-3.54	-3.84	-3.41
Cl	-1.61	-1.36	-1.36	-1.60	-1.71	-2.17	-1.65	-1.90	-1.54	-1.59	-1.60	-1.70	-1.82	-1.70
Br	-1.12	-0.83	-0.89	-1.05	-1.07	-1.41	-1.12	-1.50	-0.90	-1.13	-1.16	-1.26	-1.21	-1.31
Me	-1.79	-2.52	-2.28	-1.82	-3.21	-3.40	-3.15	-3.05	-2.79	-3.14	-2.72	-2.43	-2.14	-2.23
H	0.00	0.00	0.00	0.00	0.00	0.00	0.00	0.00	0.00	0.00	0.00	0.00	0.00	0.00
CF_3	1.55	3.13	2.48	1.79	3.88	3.96	4.18	3.41	3.35	3.69	3.20	2.62	1.70	2.28
CN	1.79	4.22	3.24	2.14	5.24	5.51	5.78	4.35	3.91	4.31	3.79	2.77	1.75	2.33
COOEt	2.57	4.47	3.64	2.79	4.89	5.34	5.53	4.42	3.62	3.88	3.64	3.06	2.23	2.82
COMe	2.72	4.86	3.84	2.82	5.14	5.83	5.92	4.67	3.71	3.93	3.74	3.11	2.09	2.57
NO_2	3.42	6.21	5.05	3.38	7.07	7.52	8.06	6.22	5.29	5.94	5.39	4.37	2.77	3.69
CHO	3.64	6.27	5.05	3.66	6.04	7.28	7.62	6.07	5.00	5.29	4.66	4.13	2.87	3.64
δ-H[b]	73.58	69.21	82.51	85.81	57.23	45.43	60.82	51.30	53.61	35.34	53.44	59.99	71.25	59.36
δ-H[c]	150.59	146.22	159.52	162.82	134.24	122.44	137.83	128.31	130.62	112.35	130.45	137.00	148.26	136.37

[a] Reference 42. Positive shifts are to high frequency; digital resolution 0.05 ppm.
[b] Shifts of the parent derivative (X = H) relative to the centre peak of $CDCl_3$.
[c] Shifts of the parent derivative (X = H) relative to TMS.

TABLE 6

Ortho ^{13}C SCS values (ppm) of X in 1,4-disubstituted benzenes[a] in $CDCl_3$.

1-Substituent (X)	4-Substituent (Y)													
	NMe_2	NH_2	OMe	F	Cl	Br	Me	H	CF_3	CN	COOEt	COMe	NO_2	CHO
NMe_2	−13.66	−13.65	−14.52	−15.47	−15.88	−15.93	−15.03	−15.66	−17.36	−17.68	−17.58	−17.68	−18.74	−18.05
NH_2	−12.49	−12.54	−12.89	−13.49	−13.53	−13.37	−13.02	−13.25	−14.62	−14.69	−14.47	−14.62	−15.93	−14.92
OMe	−14.08	−14.44	−14.47	−14.69	−14.53	−14.34	−14.47	−14.45	−14.77	−14.31	−14.76	−14.61	−15.26	−14.65
F	−13.68	−13.61	−13.47	−12.76	−13.03	−12.85	−13.38	−13.00	−12.78	−12.22	−12.87	−12.68	−12.86	−12.64
Cl	−0.22	−0.14	0.09	0.46	0.11	0.17	0.00	0.31	0.38	0.62	0.39	0.49	0.27	0.44
Br	2.61	2.77	3.00	3.45	3.04	3.10	3.00	3.16	3.34	3.60	3.30	3.59	3.35	3.45
Me	0.49	0.48	0.67	0.79	0.59	0.77	0.67	0.68	0.57	0.76	0.72	0.91	0.48	0.83
H	0.00	0.00	0.00	0.00	0.00	0.00	0.00	0.00	0.00	0.00	0.00	0.00	0.00	0.00
CF_3	−2.73	−2.52	−2.29	−1.82	−2.92	−3.14	−3.09	−3.13	−2.80	−2.88	−2.90	−2.60	−2.49	−2.88
CN	4.22	4.47	4.72	5.17	3.60	3.33	3.69	3.72	3.92	3.72	3.92	4.22	4.19	3.89
COOEt	2.08	2.33	2.23	2.56	1.18	0.97	1.26	1.17	1.21	1.00	1.16	1.50	1.40	1.22
COMe	1.45	1.50	1.31	1.40	−0.13	−0.25	0.09	0.15	−0.15	−0.41	−0.10	0.24	0.05	−0.29
NO_2	−2.96	−2.87	−3.33	−3.15	−4.82	−5.01	−4.71	−4.85	−4.64	−4.74	−4.76	−4.37	−4.37	−4.70
CHO	2.81	3.09	2.72	2.72	1.13	0.87	1.40	1.36	1.16	0.81	1.26	1.45	1.21	1.12
δ−H[b]	52.04	52.18	52.18	52.43	52.70	53.01	51.21	51.30	51.70	52.00	51.21	51.21	52.23	51.93
δ−H[c]	129.05	129.19	129.19	129.44	129.71	130.02	128.22	128.31	128.71	129.01	128.22	128.22	129.24	128.94

[a] Reference 42. Positive shifts are to high frequency; digital resolution 0.05 ppm.
[b] Shifts of the parent derivative (X=H) relative to the centre peak of $CDCl_3$.
[c] Shifts of the parent derivative (X=H) relative to TMS.

TABLE 7

Meta ^{13}C SCS values (ppm) of X in 1,4-disubstituted benzenes[a] in CDCl$_3$.

1-Substituent (X)	4-Substituent (Y)													
	NMe$_2$	NH$_2$	OMe	F	Cl	Br	Me	H	CF$_3$	CN	COOEt	COMe	NO$_2$	CHO
NMe$_2$	2.74	1.50	1.11	0.06	0.21	0.19	0.55	0.74	1.14	1.24	1.65	2.04	2.63	2.19
NH$_2$	2.89	1.59	0.89	0.27	0.43	0.49	0.68	0.88	1.49	1.63	2.04	2.23	2.86	2.61
OMe	2.02	1.24	0.86	0.41	0.66	0.72	0.87	0.88	1.72	1.88	1.94	2.04	2.40	2.24
F	1.32	0.89	0.89	1.37	1.28	1.42	1.24	1.13	2.44	2.58	2.52	2.38	2.83	2.49
Cl	1.18	1.12	1.32	1.37	1.20	1.28	1.31	1.40	1.61	1.28	1.41	1.12	1.43	1.17
Br	1.44	1.59	1.82	1.86	1.57	1.65	1.80	1.71	1.70	1.32	1.51	1.31	1.55	1.22
Me	0.54	0.14	-0.11	-0.47	-0.40	-0.25	-0.10	-0.09	-0.05	-0.12	0.00	-0.15	0.05	-0.05
H	0.00	0.00	0.00	0.00	0.00	0.00	0.00	0.00	0.00	0.00	0.00	0.00	0.00	0.00
CF$_3$	-1.30	-0.97	0.08	0.62	0.47	0.58	0.29	0.40	0.73	0.60	0.44	0.10	0.61	0.20
CN	-1.32	-0.74	0.84	1.48	1.01	1.14	0.78	0.70	0.95	0.70	0.53	0.14	0.81	0.15
COOEt	-2.01	-1.31	-0.40	0.04	-0.01	0.05	-0.05	-0.09	0.14	0.11	-0.10	-0.34	0.00	-0.19
COMe	-2.11	-1.46	-0.25	0.23	0.09	0.34	0.14	-0.09	0.44	0.41	0.24	0.00	0.39	0.00
NO$_2$	-2.15	-1.75	0.12	1.07	0.89	1.12	0.73	0.93	1.57	1.40	1.16	0.83	1.41	0.78
CHO	-1.76	-1.04	0.43	0.99	0.76	0.92	0.78	0.63	0.88	0.80	0.68	0.19	0.78	0.39
δ-H^b	35.64	38.05	36.85	38.30	51.61	54.46	51.98	51.50	48.17	55.02	52.47	51.45	46.45	52.66
δ-H^c	112.65	115.06	113.86	115.31	128.62	131.47	128.99	128.31	125.18	132.03	129.48	128.46	123.46	129.67

[a] Reference 42. Positive shifts are to high frequency; digital resolution 0.05 ppm.
[b] Shifts of the parent derivative (X = H) relative to the centre peak of CDCl$_3$.
[c] Shifts of the parent derivative (X = H) relative to TMS.

TABLE 8

Ipso ^{13}C SCS values (ppm) of X in 1,4-disubstituted benzenes[a] in CDCl$_3$.

1-Substituent (X)	4-Substituent (Y)													
	NMe$_2$	NH$_2$	OMe	F	Cl	Br	Me	H	CF$_3$	CN	COOEt	COMe	NO$_2$	CHO
NMe$_2$	27.41	26.31	25.09	23.53	22.58	22.67	23.56	22.28	20.44	19.73	20.44	20.34	19.48	19.85
NH$_2$	21.25	20.05	19.26	18.44	18.46	18.58	18.45	17.91	17.64	17.79	17.96	18.11	17.91	18.11
OMe	35.43	34.18	33.23	31.76	31.76	31.82	31.99	31.21	30.30	30.12	30.43	30.39	30.04	30.19
F	39.00	37.77	36.57	34.84	34.82	34.97	35.75	34.51	32.90	32.31	32.88	32.67	31.67	32.10
Cl	5.20	4.61	4.89	5.17	6.13	6.37	5.78	5.93	6.42	6.84	6.41	6.41	6.78	6.50
Br	−8.11	−8.43	−7.85	−7.44	−6.13	−5.78	−6.21	−5.87	−5.31	−4.70	−4.95	−4.70	−4.57	−4.66
Me	9.45	9.22	9.25	9.34	9.78	9.90	9.42	9.52	10.29	10.95	10.63	10.78	11.36	11.07
H	0.00	0.00	0.00	0.00	0.00	0.00	0.00	0.00	0.00	0.00	0.00	0.00	0.00	0.00
CF$_3$	1.18	1.50	2.23	2.53	2.68	2.91	2.58	2.31	2.26	1.88	1.51	1.36	1.39	1.24
CN	−19.83	−18.71	−16.76	−15.38	−15.65	−15.59	−16.05	−15.96	−15.68	−16.00	−16.50	−16.69	−16.24	−16.74
COOEt	0.67	1.50	2.21	2.79	2.45	2.74	2.48	2.14	1.94	1.59	1.36	1.22	1.31	0.73
COMe	8.59	9.11	9.59	9.54	8.90	8.93	9.33	8.69	7.91	7.12	7.33	7.14	6.84	5.97
NO$_2$	20.54	20.63	20.89	20.50	20.04	20.24	20.88	19.95	18.25	17.36	17.76	17.38	16.50	16.75
CHO	8.39	8.92	9.34	9.04	8.27	8.25	8.89	8.06	6.93	6.05	6.46	6.75	5.53	5.63
δ−H[b]	39.61	41.50	43.62	46.91	49.40	49.80	48.25	51.30	54.71	55.65	55.72	55.97	57.52	57.37
δ−H[c]	116.62	118.51	120.63	123.92	126.41	126.81	125.26	128.31	131.72	132.66	132.73	132.98	134.53	134.38

[a] Reference 42. Positive shifts are to high frequency; digital resolution 0.05 ppm.
[b] Shifts of the parent derivative (X = H) relative to the centre peak of CDCl$_3$.
[c] Shifts of the parent derivative (X = H) relative to TMS.

TABLE 9

Proportionality relationships of the 4-carbon shifts in *para*-disubstituted benzenes to the shifts in monosubstituted benzenes[a]

Y	a	b	r	f	N
F	163.03	0.619	0.9930	0.12	10
OMe	160.15	0.714	0.9945	0.11	10
NH_2	146.97	0.825	0.9925	0.13	10
NO_2	147.78	0.820	0.9893	0.13	10
CF_3	131.00	1.049	0.9972	0.07	9
Me	138.13	1.103	0.9976	0.08	10
Cl	134.77	1.119	0.9992	0.04	10
CN	113.03	1.161	0.9921	0.13	10
$CH=CH_2$	137.42	0.918	0.9981	0.06	11
$CMe=CH_2$	141.02	0.963	0.9992	0.04	8
$C(CMe_3)=CH_2$	143.08	0.990	0.9985	0.05	9
$C\equiv CH$	122.39	1.034	0.9979	0.06	11
Et	141.61	1.087	0.9991	0.05	9
C_6H_5	141.61	0.978	0.9982	0.07	8
CMe_3	150.97	1.073	0.9960	0.09	8
$SiMe_3$	140.00	1.251	0.9974	0.06	7
$GeMe_3$	142.24	1.305	0.9998	0.04	6
$SnMe_3$	141.99	1.345	0.9966	0.07	7
$OCH=CH_2$	156.40	0.765	0.9963	0.08	8
$SCH=CH_2$	134.76	1.536	0.9982	0.06	9
$SeCH=CH_2$	129.77	1.537	0.9988	0.07	9
$CONH_2$	134.25	0.998	0.9924	0.14	7
Br	122.58	1.241	0.9981	0.06	10
I	94.32	1.474	0.9965	0.09	8

[a] Reference 126. a and b are the intercept and slope respectively, defined by equation (18). r is the correlation coefficient, $f = SD/RMS$ and N is the number of data points in the correlation.

The localized nature of the effect of Y is confirmed by noting that correlations of C-1 and C-2 SCS data to equation (18) give slopes $b = 1$, indicating additivity of the shifts at these positions.[126] This conclusion is, however, not consistent with the results of a more recent study in which considerable variations in *ipso* (C-1) SCS values with different *para* groups are observed.[127] The range of shifts at the *meta* position, C-3, is too small to derive meaningful correlations.[126]

This treatment provides a simple means of correlating shifts in disubstituted benzenes; however, the method does not take adequate account of enhancement (conjugation) and saturation interactions between substituents, as evidenced by the fact that data points for compounds containing two strong acceptor substituents had to be excluded from the correlations.[126]

Systematic deviations from the above linear equation can be illustrated by plotting chemical shifts for the *para* position in disubstituted benzenes against those for the corresponding position in monosubstituted benzenes. Although in many cases it is possible to represent this relationship with a linear plot, deviations larger than experimental error do occur. The systematic nature of the deviations becomes clearer with the observation that for some plots, most notably C_p $SCS_X(NO_2)$ versus C_p $SCS_X(H)$, the relationship is better represented as bilinear (although this is still an approximation) rather than linear. Figure 4 shows a plot of C_p SCS values of X in the series

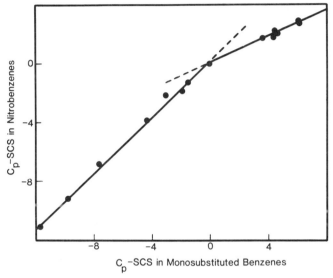

FIG. 4. Plot of C_p SCS values in *para*-substituted nitrobenzenes versus those in monosubstituted benzenes.[42]

$Y = NO_2$ versus C_p SCS values of X in the series $Y = H$.[42] For π acceptor substituents X, the slope is 0.46, whereas for π donors it is 0.92.[42] In general terms, the smaller slope for π acceptor substituents arises because of π-repulsive saturation interactions[128] of these substituents with the electron-withdrawing NO_2 group. On the other hand, π donor substituents undergo conjugative interactions with this group. However, if these were the only effects of the common NO_2 group, the slope for donor substituents X would be greater than unity, and for acceptor substituents it would be less than unity.[42] The fact that both slopes are less than unity arises because *the common nitro group reduces the sensitivity of the adjacent carbon to substituent effects of X*.[42]

The reverse situation occurs for the case when the fixed substituent is a π donor such as NH_2, as illustrated in Fig. 5. Here the slope for *para* donor

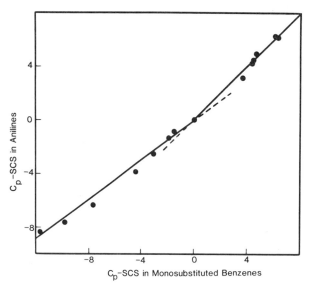

FIG. 5. Plot of C_p SCS values in *para*-substituted anilines versus those in monosubstituted benzenes.[42]

substituents is less than the slope for *para* acceptor substituents, reflecting the reversal in the nature of repulsive and conjugative interactions compared with the nitrobenzene series.

These general observations are not meant to imply that SCS values in disubstituted benzenes can be treated quantitatively using the above bilinear approach. Instead the discussion is used to show the general importance of the two major effects:[42] (a) mutual substituent interactions which modify the electron distribution at C-4 relative to monosubstituted benzenes, and (b) modification of the shift–charge ratio[42] (SCR) at C-4 brought about by the Y group.

The effects of mutual conjugation or repulsion of *para* substituents are best treated with an extended DSP equation in which the effective σ_R^0 value for X ($\bar{\sigma}_R = \sigma_R^0/(1 - \varepsilon\sigma_R^0)$) varies depending on the electron demand exerted by the Y group. This equation, designated DSP-NLR (DSP non-linear resonance), is shown below,[42] and correlations of C_p SCS values using this equation are given in Table 10:

$$SCS = \rho_I\sigma_I + \rho_R\sigma_R^0/(1 - \varepsilon\sigma_R^0) \qquad (19)$$

The fits are considerably improved compared with the usual DSP equation,[42] showing that conjugative and repulsive interactions of X with the Y group have an important effect on C_p SCS values of X. The extent of these interactions can be gauged from the sign and magnitude of the demand parameter ε. Table 10 shows that ε is positive for π donor Y

TABLE 10

DSP–NLR (equation (19)) analyses of C_p SCS data for 1,4-disubstituted benzenes in dilute $CDCl_3$ solutions.[a]

Y	ρ_I	ρ_R	ε	SD	f
NMe_2	1.6	13.7	0.25	0.16	0.05
NH_2	4.1	21.0	0.56	0.28	0.06
OMe	3.1	17.8	0.45	0.26	0.06
F	1.7	14.7	0.14	0.26	0.07
Cl	5.3	24.2	0.05	0.26	0.04
Br	5.4	26.7	0.05	0.32	0.04
Me	6.0	25.9	0.34	0.26	0.04
H	4.5	21.9	0.06	0.15	0.03
CF_3	4.6	18.0	−0.42	0.16	0.03
CN	5.5	18.5	−0.60	0.29	0.04
COOEt	4.6	17.3	−0.48	0.27	0.05
COMe	3.4	15.3	−0.49	0.29	0.06
NO_2	1.8	12.2	−0.72	0.16	0.03
CHO	2.9	13.5	−0.60	0.29	0.06

[a] Reference 42.

groups, essentially zero for Y = H, and negative for π acceptor Y groups. These trends mean that donor substituents X become better π donors (i.e. their effective σ_R value becomes larger in magnitude) in the presence of Y acceptor groups, and poorer donors when Y is an electron-donating group. Opposite trends apply for X acceptor groups.

The additional effect of the common substituent Y in modifying the *sensitivity* of the adjacent ^{13}C shifts to the nature of the *para* substituent can be treated empirically, and *shift–charge ratios* (SCRs) of the *para* site are found to vary significantly, depending on the Y substituent.[42] SCR values for a number of Y groups are given in Table 11, and although some rationalization of trends in these data has been given,[42] all of the factors which contribute to SCRs are not yet completely clear. As expected, there is a parallel between the SCR values in Table 11 and the *b* values in Table 9. Inamoto *et al.*[129] have also noted a general relationship between these latter values and their inductive substituent parameter (ι).[130] However, it has been pointed out[42] that although crude relationships between SCR or *b* values and electronegativities or ionization potentials may exist for *limited* series, no *general* relationship with a single substituent property has yet been found.

Afanas'ev and Trojanker[131] have also examined C_p SCS values in disubstituted benzenes to determine the influence of Y on substituent effects of X. These authors use a dual parameter approach, but criticize σ_I and σ_R^o

TABLE 11

Shift charge ratios (SCR) for C−4 in 1-X,4-Y-disubstituted benzenes.

Y	SCR[a]	r[b]
NMe$_2$	113	
NH$_2$	167 (179)	(0.997)
OMe	142 (157)	(0.996)
F	121 (152)	(0.995)
Cl	199	
Br	220	
Me	211 (205)	(0.996)
H	189 (189)	(0.991)
CF$_3$	161 (163)	(0.996)
CN	170 (183)	(0.993)
CO$_2$Et	156	
COMe	140	
NO$_2$	113 (132)	(0.982)
CHO	124	

[a] Derived empirically by the method described in reference 42. Values in parentheses correspond to correlations of SCS values versus calculated (*ab initio*) π electron densities.

[b] Correlation coefficients of SCS–Δq_π plot.[42]

values, claiming that they do not properly separate polar and resonance effects. Instead they use σ^* and σ^r values derived previously for aliphatic compounds.[132] The original σ^* values used are those derived by Taft[133] to describe inductive effects in aliphatic derivatives, whereas the σ^r values appear to have a more arbitrary origin:[132]

$$SCS = \rho^*\sigma^* + r^*\sigma^r \qquad (20)$$

Fits to equation (20) of C_p SCS data for monosubstituted benzenes in various solvents are claimed[131] to be equal to or better than those from a DSP treatment with σ_I and σ_{R^0}; however, this claim appears to be inappropriate since the f (= SD/RMS) values using the approach of Afanas'ev and Trojanker are consistently between three and five times poorer than those[121] using the Taft DSP approach for a larger number of substituents. It is disturbing to note that fits to equation (20) yield negative ρ^* values and positive r^* values for all cases of C_p SCS data examined.[131] Negative ρ^* values imply that polar withdrawing substituents contribute to a low frequency ^{13}C SCS at C-*para*. This is clearly not expected to be the case, and the reason for the anomalous correlation appears to be that the σ^* and σ^r substituent constants are not appropriate for analysis of SCS data in substituted benzenes.

Non-additivity effects at the *ipso* site (relative to the changing substituent X) have also been extensively investigated. In contrast to the observations of Lynch,[126] who suggested that additivity prevails at the *ipso* site (C-1), it is apparent from Table 8 that the *ipso* SCS of a given X substituent changes with the Y group. Changes in *ipso* SCS values become even more apparent when the *para* Y group contains a cationic centre. Olah and Forsyth[134] have shown that the *ipso* SCS for the methyl group in substituted toluenes varies markedly with the electron demand of the *para* substituent. For example, the methyl *ipso* SCS of 8.8 ppm in 4-methylaniline changes to 20.2 ppm in the 4-methylstyryl cation.[134] More recently, the variations in *ipso* SCS values of OCH_3, CH_3, F, Cl, Br and CF_3 substituents have been examined in 25 series of disubstituted benzenes.[127] *Ipso* SCS values for these substituents vary markedly with changes in the electron demand of the *para* group. The Br substituent, for example, is found to induce shielding of the *ipso* site in neutral series, but induces deshielding in carbocations where there is a high electron demand for resonance stabilization from the bromine atom.[135] On the other hand, the fluoro substituent becomes significantly less deshielding with an increase in electron demand from the *para* group.[127]

In examining *ipso* SCS variations here, it is convenient to make a slight change in nomenclature with respect to structure [14], and now regard Y (= OCH_3, CH_3, F, Cl, Br, CF_3) as the substituent whose properties (e.g. *ipso* SCS) are being investigated, and X as a perturbing influence. This is useful for demonstrating a duality in the analysis of *ipso*/*para* SCS values discussed later in this section.

Since the *para* carbon shift of X in monosubstituted benzenes may be taken as a measure of the electron demand of this group, plots of this parameter versus C_{ipso} SCS values of Y (Fig. 6)[127] demonstrate the important influence of electron demand on *ipso* SCS values. For F, OMe and CF_3 substituents, an increase in electron demand leads to a decreased *ipso* SCS (decreased deshielding or increased shielding) while the opposite trend is noted for Cl, Br, and Me substituents.[127] Membrey, Ancian and Doucet[136] had previously noted similar trends; however, these authors did not observe the discontinuity in the direction of the plots for CF_3 and Me Y groups. This trend was noted for Me and other alkyl groups in a later communication.[137] Membrey *et al.*[136] treated variations in *ipso* shielding in terms of the proportionality approach of Lynch.[126] They examined *ipso* shielding of Y (series 14) in terms of equation (21).

$$SCS_{i,Y}^X = SCS_{i,Y}^H + (b-1)SCS_{p,X}^H \qquad (21)$$

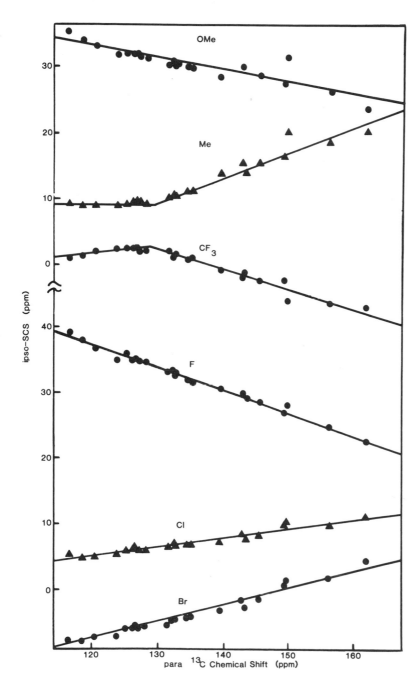

FIG. 6. Plots of C-*ipso* SCS values for Y = OCH$_3$, CH$_3$, F, Cl, Br and CF$_3$ substituents in 1-X,4-Y-substituted benzenes versus *para* ^{13}C shifts of X-monosubstituted benzenes.[127]

Here, $SCS_{i,Y}^{X}$ represents the *ipso* substituent chemical shift of Y in series X, $SCS_{i,Y}^{H}$ is the *ipso* SCS of Y in monosubstituted benzenes, and $SCS_{p,X}^{H}$ is the *para* SCS of X in monosubstituted benzenes. In this equation, deviations from additivity of the *ipso* shifts of Y (described by $SCS_{i,Y}^{X} - SCS_{i,Y}^{H}$) are factored into two terms, $(SCS_{p,X}^{H})$, characterizing the electronic influence of X on the benzene ring, and $(b-1)$, characterizing the susceptibility of Y to the nature of X.[136] Least squares analyses of a range of SCS data in terms of equation (21) yield only moderately good correlations ($r \sim 0.825 - 0.988$);[136] however, the derived slopes, $(b-1)$, obtained for some Y series are consistent with values of b derived by Lynch[126] from a study of the effects of Y on C_p SCS values of X. Membrey *et al.*[136] point out the equivalence of their approach to that of Lynch, expressed in the form

$$SCS_{p,X}^{Y} = b \times SCS_{p,X}^{H} \qquad (22)$$

The equivalence arises because the *ipso* SCS of a given Y group in an X, Y-disubstituted benzene represents the same ^{13}C chemical shift (expressed relative to a different parent derivative) as the *para* SCS of a given X group in the same series. Other authors[138] have also discussed related non-additivity effects in *para* disubstituted benzenes.

Kelly and coworkers[127] attribute the trends in the sign and magnitude of the slopes in Fig. 6 to the shift–charge ratio (SCR) of the Y group. This conclusion is based on the observation of a good linear correlation between the slopes of Fig. 6, and SCR values derived from an independent study[42] of C-4 chemical shifts in 1-X,4-Y disubstituted benzenes. As a general principle, it can be concluded that greatest non-additivity in 1,4-disubstituted benzenes occurs when the *ipso* substituent has an SCR much different from H (i.e. b different from unity) and the *para* substituent has a large electron demand.

Further evidence of the non-additivity of ^{13}C SCS values in disubstituted benzenes has come from studies of chlorobenzenes and chlorophenols,[139] where it has been confirmed that at least part of non-additivity arises from the effects of changes in ring electron distribution from substituent interactions (e.g. repulsion or saturation). The effects of σ and π backbonding of Cl and OH substituents were investigated in these studies.[139]

Compared to studies of substituent effects at *ipso* and *para* sites in *para* disubstituted benzenes, there have been relatively few systematic analyses of these effects at *ortho* and *meta* sites. Substituent effects at *meta* sites are generally much smaller than at the other ring positions. Since these shifts display only a small resonance dependence[42] (e.g. equation (16)), the DSP–NLR equation does not provide a significant improvement over the fits obtained with the usual DSP equation. The small range of shifts makes it difficult to determine the effect of Y on C_m SCS values of X; however, to a first order approximation the shifts may be regarded as additive (i.e. Y has a negligible effect on C_m SCS values of X).[42]

SCS values in *meta* disubstituted benzenes have not yet been analysed in as much detail as have shifts in the *para* series. Also currently unavailable are extensive and systematic studies of solvent effects on ^{13}C shifts in disubstituted benzenes. These studies should provide additional insight into medium effects on the transmission of substituent effects. Tsuno and co-workers[140] have shown, for example, that ^{13}C shifts of 4-substituted toluenes vary with concentration and solvent. (CCl_4, MeOH and DMF solutions at a large range of concentrations were studied.) Derivatives containing more polar substituents are most affected, with C-1 and C-4 varying significantly. Changes have been attributed to specific substituent–solvent interactions.[140] Opposite dilution shifts for C-1 and C-4 have been interpreted in terms of a π-polarization mechanism.[140]

C. Trisubstituted benzenes

Non-additivity of SCS values is readily apparent in many of these derivatives. However, a quantitative treatment of this effect is more difficult than in disubstituted benzenes because (except in 1,3,5 derivatives) *ortho* steric interactions perturb electronic substituent effects. From a consideration of a limited number of derivatives,[141] it is apparent that the greatest non-additivity occurs at positions C-2 and C-6 in 1,2,3-substituted species. Large non-additivity also is present in 1,2,4-substituted species, whereas additivity is most closely approached in 1,3,5-substituted derivatives. These trends are consistent with at least part of the non-additivity being caused by the modification of substituent donor/acceptor properties by steric interactions with nearby groups. Relles has discussed related effects.[142]

D. Polycyclic aromatics

^{13}C NMR in polycyclic aromatic compounds has been the subject of a recent review[143] and so only a relatively brief update of substituent effects in these compounds is given here. Because of its structural rigidity, and the large number of sites that can be probed via ^{13}C NMR, the naphthalene framework is extremely valuable for the study of substituent effects. Until recently, *reliable* SCS data for all carbons were not available because of uncertainties in assignments. However, the use of several complementary assignment techniques,[144–155] including specific deuteration[144,148,153] has allowed SCS values for a wide range of 1-X and 2-X naphthalenes (series [15a] and [15b] respectively) to be derived and these are reproduced in Table 12.[144]

A comparison of the SCS data in Table 12 for carbons 1–4 with the SCS data in Table 2 for monosubstituted benzenes shows that the SCS values

^{13}C substituent chemical shifts in substituted naphthalenes[a]

Substituent	Carbon number									
	1	2	3	4	5	6	7	8	9	10
1-Substituted naphthalenes										
NH$_2$	+14.06	−16.27	+0.39	−9.10	+0.52	−0.13	−1.13	−7.21	−9.94	+0.78
OCH$_3$	+27.54	−22.07	+0.04	−7.70	−0.46	−0.57	−0.68	−5.92	−7.86	+1.02
F	+30.95	−16.42	−0.24	−4.26	−0.38	−0.98	+0.32	−7.37	−8.01	+1.44
Br	−5.08	+4.04	+0.32	−0.01	+0.38	+0.85	+1.28	−0.61	−1.49	+1.14
CH$_3$	+6.34	+0.55	−0.30	−1.37	+0.61	0.00	−0.13	−3.81	−0.88	+0.06
H[b]	0.00	0.00	0.00	0.00	0.00	0.00	0.00	0.00	0.00	0.00
COCH$_3$	+7.39	+2.66	−0.64	+4.92	+0.35	+0.43	+2.04	−2.05	−3.49	+0.35
CF$_3$		−1.08	−1.48	+5.04	+1.04	−1.02	+2.02	−3.56	−3.85	+1.05
CN	−17.83	+6.66	−0.84	+5.23	+0.58	+1.59	+2.66	−3.12	−1.27	−0.42
NO$_2$	+18.58	−1.95	−1.95	+6.62	+0.58	+1.40	+3.51	−4.90	−8.48	+0.74
2-Substituted naphthalenes										
NH$_2$	−19.35	+18.30	−7.61	+1.26	−0.20	−3.38	+0.52	−2.12	+1.43	−5.53
OCH$_3$	−22.14	+31.82	−7.09	+1.49	−0.23	−2.21	+0.55	−1.14	+1.10	−4.52
F	−17.04	+34.84	−9.58	+2.39	−0.05	−0.75	+1.01	−0.57	+0.70	−3.93
Br	+1.66	−6.18	+3.41	+2.01	−0.07	+0.42	+1.04	−1.04	+1.01	−1.66
CH$_3$	−1.08	+9.62	+2.30	−0.66	−0.30	−0.88	+0.03	−0.30	+0.19	−1.79
H	0.00	0.00	0.00	0.00	0.00	0.00	0.00	0.00	0.00	0.00
COCH$_3$	+2.21	+8.67	−1.99	+0.44	−0.16	+2.56	+0.87	+1.64	−1.01	+2.01
CF$_3$	−1.96		−4.18	+1.14	+0.14	+2.42	+1.52	+1.24	−1.05	+1.35
CN	+6.10	−16.56	+0.35	+1.17	+0.03	+3.08	+1.72	+0.39	−1.37	+1.00
NO$_2$	−3.42	+19.55	−6.70	+1.49	−0.04	+3.80	+1.98	+1.95	−1.66	−2.20

[a] Reference 144; measured for 0.5–1.0 M solution in CDCl$_3$. Positive SCS values indicate decreased shielding.
[b] The absolute shifts for naphthalene in DCCl$_3$ are 127.96 ppm (C−1); 125.88 ppm (C−2); 133.55 ppm (C−9).

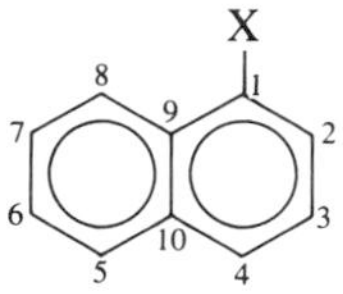

[15a] 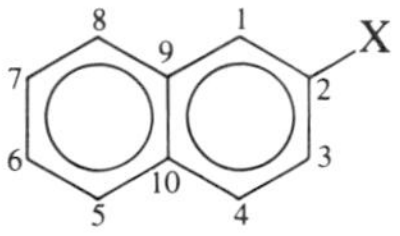   [15b]

at corresponding positions are quite similar, although a number of systematic differences do occur. For example, *ipso* SCS values in 1-X-naphthalenes are generally smaller in magnitude than in monosubstituted benzenes, except for CN. On the other hand, naphthalene C-4 SCS values for π donor substituents are smaller or similar to those in monosubstituted benzenes, while SCS values for electron withdrawing substituents are larger. DSP analyses[144] (Table 13) yield reasonable fits of the SCS values with ρ_I and

TABLE 13

DSP analyses of SCS values in substituted naphthalenes.[a]

Carbon no.	ρ_I	ρ_R	f	n
		1-Substituted naphthalenes		
3	−1.80	−1.63	0.49	9
4	5.92	19.98	0.12	9
5	0.82	0.59	0.80	9
6	2.23	0.41	0.18	9
7	4.10	3.89	0.15	9
10	1.36	−1.66	0.46	9
		2-Substituted naphthalenes		
4	2.95	−2.00	0.21	9
5	−0.04	0.36	0.78	9
6	4.01	7.74	0.06	9
7	2.85	0.37	0.16	9
8	1.28	4.32	0.40	9
9	−1.30	−3.80	0.30	9
10	0.41	11.23	0.12	9

[a] Reference 144; correlations using the σ_R^0 resonance scale.

ρ_R^0 for all carbons, except at the proximate sites (C-1, C-2, and C-9 in series [15a] and C-1, C-2, and C-3 in series [15b]). Almost negligible substituent effects are observed at C-5 in series [15a], in contrast to ^{19}F measurements in the corresponding fluoronaphthalenes.[40,156] This has been taken[144] as evidence that ^{13}C and ^{19}F shifts in the respective series are controlled by different mechanisms. π-Polarization is thought to dominate the polar component of the ^{13}C shifts, whereas significant through-space effects have been used to account in part for modified ^{19}F SCS values.[144]

The pattern of ρ_R values in Table 13 confirms that resonance effects are better transmitted to the unsubstituted ring in 2-X-naphthalenes compared with 1-X-naphthalenes. In series [15b], large ρ_R values are observed for the conjugating sites C-6, C-8 and C-10 whereas almost negligible values are seen for C-5 and C-7. Relatively small, negative values are detected at the two positions (C-4 and C-9) formally *meta* to the substituent.

An increase in solvent polarity from CDCl$_3$ to acetone brings about a slight increase in derived ρ_I values[144] for most ring positions, in accord with parallel observations in earlier studies of solvent effects on ^{13}C shifts in substituted benzenes.[123,124] In general this can be attributed to increased π-polarization resulting from enhanced substituent polarity in more polar solvents.[123]

Total electron densities calculated by the INDO MO method provide a fair description of the observed SCS values; however, as in other series, the slopes of SCS–Δq relationships change significantly with the molecular site (slopes vary from 187 to 324 ppm per electron).[152]

Substituent effects on ^{13}C shifts in 1-(2-naphthyl)-1-R-ethyl cations [16] have also been recently examined.[157]

[16]

This series was used to investigate the relative abilities of R = CH$_3$, C$_6$H$_5$ and c-C$_3$H$_4$ to delocalize positive charge. Earlier, Olah and coworkers[158–160] used ^{13}C shifts of the carbenium centre in ions of the form (C$_6$H$_5$)-(CH$_3$)(R)C$^+$ to monitor the electron delocalizing of phenyl, methyl and cyclopropyl R groups. However, local influences[157,158] resulting from a change of substituent at this site may yield misleading results. Kitching, Adcock and Aldous[157] therefore suggested that the use of remote sites in the naphthalene framework [16] would provide a better probe for measuring the relative charge delocalizing abilities of the R groups. Using remote ^{13}C, ^{19}F or ^{1}H nuclei as probes of these effects, it is concluded that least demand from the charged site for resonance with the 2-naphthyl group occurs when R = cyclopropyl.[157] The utility of the naphthalene framework for investigating substituent properties (e.g. σ_I or σ_R^0 constants) of a variety of other substituent groups has also been demonstrated.[161–164]

Adcock and Abeywickrema[165] have recently reported ^{13}C SCS data for a series of 1-X-substituted 4-methyl-1, 2, 3, 4-tetrahydro-1, 4-ethanonaphthalenes [17] as well as their corresponding 6- and 7-fluoroderivatives.

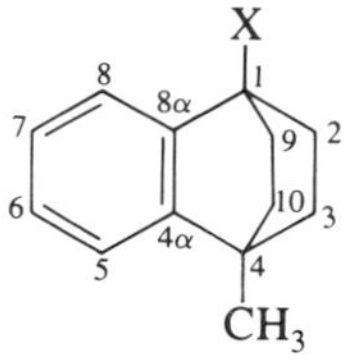

[17]

In this series, the geometrical relationship between the C—X bond and the unsubstituted ring is similar to that in 1-X-naphthalenes and hence a comparison of SCS data is of interest. It might be expected that polar effects would be similar, but resonance effects would be quite different for these two related series. This is because resonance effects rely on transmission through the π electron framework whereas polar effects are largely transmitted through space. SCS values for the arene carbons in a range of substituted derivatives of series [17] are given in Table 14. These values show distinct differences from the corresponding shifts in naphthalenes. A comparison of ^{13}C chemical shifts of the NH_3^+ substituent relative to those of CH_3 is shown in structures [18] and [19].

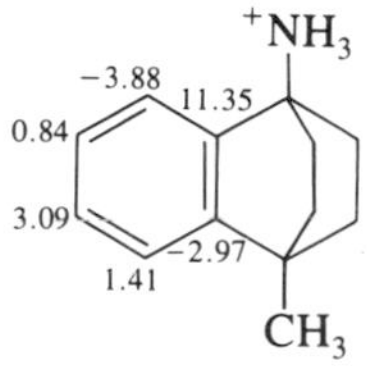

[18] 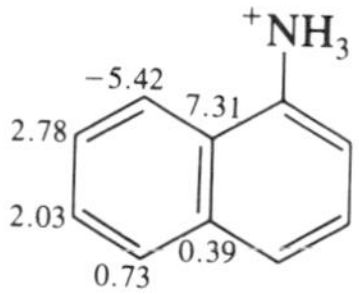   [19]

Previous studies[163,166] have shown that ^{13}C shift differences of these isoelectronic substituents allow a cancellation of steric and resonance effects and hence yield a qualitative indication of the π-polarization effect induced by the electric field of the positive charge. Similar overall patterns are seen in [18] and [19] although the magnitudes of effects at individual carbons are quite different. This suggests that polarization of both rings in naphthalene brings about additional changes in π-electron populations not present when only a single ring is polarized, as in [18].

Shapiro[167] has examined ^{13}C chemical shifts in structurally related 9-substituted fluorenes (series [20], X = OH, Cl, Br, I).

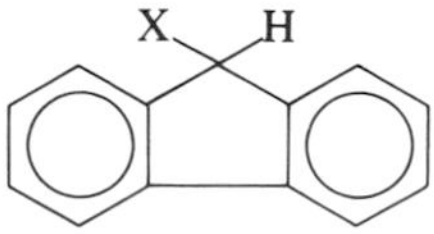

[20]

TABLE 14

^{13}C substituent chemical shifts (SCS) of arene ring carbons in series [17].[a]

X	C-5	C-6	C-7	C-8	C-8a	C-4a
			Substituent chemical shift			
NH$_2$	−0.34	−0.11	0.05	−3.72	2.27	−1.32
OCH$_3$	−0.08	−0.06	−0.06	−3.03	−0.24	−1.90
OH	−0.40	0.10	0.14	−3.80	1.15	−2.16
F	−0.31	0.61	0.35	−4.49	−1.64	−2.96
Br	−0.44	1.24	0.70	1.69	−1.67	−1.66
I	−0.58	1.62	0.97	6.47	−0.97	−2.00
CH$_3$	−0.16	−0.28	−0.03	−2.83	1.89	−0.12
H	0.00	0.00	0.00	0.00	0.00	0.00
COOH	0.33	0.67	0.23	−0.94	−3.20	−0.92
COOCH$_3$	0.18	0.41	0.05	−0.98	−2.35	−0.79
CN	0.47	1.69	0.90	−0.90	−5.91	−2.14
NO$_2$	0.38	1.62	0.83	−3.02	−5.45	−2.25

[a] Reference 165.

Substantial substituent effects are observed at most ring positions and these are interpreted in terms of dominant contributions from resonance (hyperconjugative) and π-polarization effects.

The difference in the electronic structures of alternate and non-alternate molecular π systems makes a comparison of SCS effects in the $C_{10}H_8$ isomers of naphthalene and azulene of interest. As in naphthalenes, a variety of techniques are required to make reliable assignments in the latter series. The structures of 1- and 2-substituted azulenes (series [21a] and [21b]) are shown below.

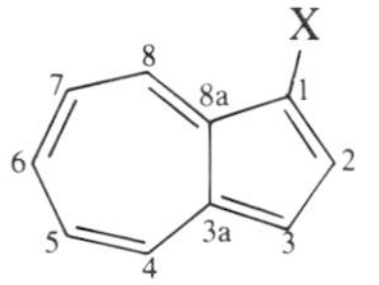

[21a] 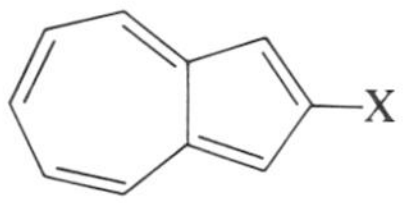   [21b]

Two recent independent studies of thse systems[108,168] reached the same general conclusion, namely that substituent effects at positions close to the substituent (C-1, C-2, C-8a) are fairly similar to those at the corresponding positions in 1-X-naphthalenes, whereas substituent effects at the more distant sites are generally larger in azulenes than in naphthalenes. Apart from systematic absolute shift differences, the assignments in these two studies are in agreement, except for a reversal in the assignments for C-4 and C-8 in the 1-COMe derivative. The assignment of Wells, Penman and

Rae[168] is presumably the correct one, based on proton–carbon coupling patterns and comparisons with deuterated derivatives, although it conflicts with other reports.[108,169]

Ironically, Wells *et al.*,[168] who measured SCS data for a large and varied range of 1-X-substituents meeting the DSP basis set requirement,[41] did not perform a DSP analysis, but Holak *et al.*,[108] who did not have sufficient SCS data in this series adequately to separate polar and resonance effects, did carry out a DSP analysis. DSP analyses, using the experimental data of Wells *et al.*[168] for series [21a] have been performed here and are given in Table 15 along with the correlations[108] for series [21b]. In both series,

TABLE 15

DSP analyses of SCS values in substituted azulene.[a]

Carbon	ρ_I	ρ_R	f
	1-X-Azulene[b] (series [21a])		
4	4.9	1.2	0.33
5	7.4	9.3	0.23
6	5.3	−0.1	0.34
7	8.0	13.0	0.30
8	1.2	9.7	0.40
	2-X-Azulenes[c] (series [21b])		
2	−24.3	−72.7	0.17
4	6.3	13.1	0.19
5	6.1	−1.0	0.08
6	7.1	13.0	0.16

[a] For comparison, all correlations are given using the σ_R^0 resonance scale. Slightly better fits are obtained for C-5 and C-7 in 1-substituted derivatives using σ_R^-. Similarly, SCS data for C-5 and C-6 in 2-substituted derivatives are better correlated with σ_R^+.

[b] SCS data from reference 168.

[c] SCS data from reference 108. 1 substituent excluded from correlations.

fair correlations are obtained for most carbons in the seven-membered ring. In the 1-substituted derivatives, series [21a], polar effects at C-5 and C-7 are similar, but resonance effects are somewhat larger at C-7. These trends agree roughly with those derived[108] for a more limited data set. However, an important difference is observed between the DSP correlation in Table 15, derived from the extensive SCS data of Wells *et al.*,[168] and the DSP correlations of Holak *et al.*[108] Negligible resonance effects at C-4 and C-6 are predicted by the correlations in Table 15, while Holak *et al.* predict significant resonance effects at these sites. This difference can be rationalized by noting that Holak *et al.* did not have SCS data for any strongly electron-donating substituents and hence unreasonable DSP correlations were derived. This illustrates the importance of including a basis set[41] of substituents in any DSP correlation. Although the fits in Table 15 (which include a basis set of substituents) are only fair, the predicted negligible resonance component for C-4 and C-6 can be confirmed by noting that NH_2 and OMe substituents, which are resonance donors but polar withdrawing groups, produce high frequency shifts at these sites.[168] This illustrates the lack of significant transmission of resonance effects via the pathway

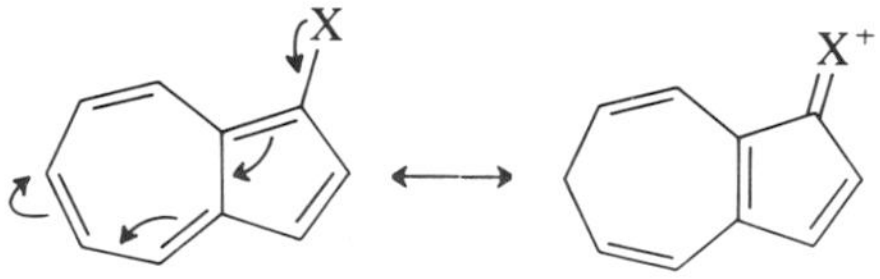

[22]

shown in [22]. The observed resonance contribution at C-8 may be due to *peri*-type interactions with the substituent. On the other hand, the significant ρ_R values seen at C-5 and C-7 demonstrate the importance of the transmission pathway shown in [23].

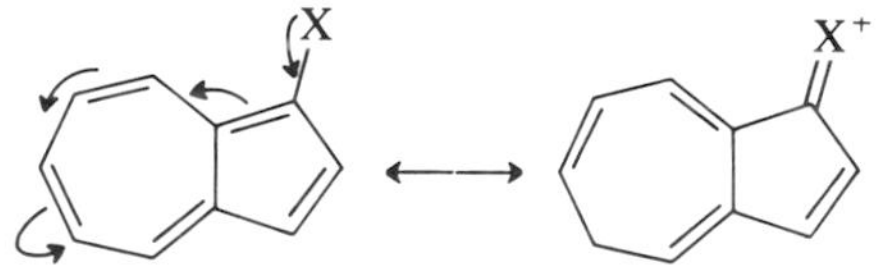

[23]

In the 2-substituted derivatives, series [21b], polar and resonance effects contribute significantly to the shifts at all sites in the seven-membered ring, except for C-5 and equivalent C-7 which have negligible resonance effects, reflecting their non-conjugating nature. Interestingly, substituent effects at C-4 (and C-8) and C-6 in this series are almost identical.

Contributions to the polar component of the SCS values from π-polarization are confirmed from CNDO/2 calculations of π electron densities in 1- and 2-azulylmethyl cations.[108] π electron density differences between azulenes containing the CH_2^+ substituent in an orthogonal conformation and azulene itself are taken as a measure of π-polarization effects and correspond closely with ρ_I (or f in the Swain–Lupton treatment) values derived from the SCS data. The combined experimental and theoretical data indicate that electron-withdrawing dipolar substituents cause a drift in electron density from the seven-membered ring to the five-membered one with a node near the ring junction.[108]

9-Substituted acridines [24] have been investigated[170] in a study initiated because of the pharmacological activity of these compounds. This series is discussed here because of its similarity to the naphthalenes examined above.

[24]   [25]

As expected, absolute ^{13}C chemical shifts in the parent derivative (X = H) are similar to those in quinoline.[170] In addition, a comparison of substituent effects in 9-X-acridines and 1-X-naphthalenes reveals a strong similarity in the ^{13}C SCS values at corresponding positions.[170] However, correlations with the triparameter equation of Smith and Proulx,[171]

$$SCS = aF + bR + cQ + d \qquad (23)$$

are generally poor, except for C-2. Here F and R are the usual Swain–Lupton substituent parameters,[44] and Q is a semi-empirical term.[172] Since good correlations might be expected at the distant C-4 and C-3 sites, but are not obtained, it is possible that the SCS data were not measured at sufficiently low concentrations to be suitable as monitors of electronic effects only. ^{13}C chemical shift effects in 9-substituted acridinium ions [25] were also studied, but because of differences in the solvents used to measure ^{13}C spectra for series [24] and [25], it is difficult to compare SCS values in the two series to investigate the effect of protonation. However, variations in *ipso* SCS values do appear to confirm trends noted in earlier studies. For example, the *ipso* SCS of the CH$_3$ substituent increases markedly (from 6.3 to 11.6 ppm) on protonation, in agreement with trends noted in disubstituted benzenes[127] when the electron demand at the *para* site increases. In the acridines, the change in electron demand is reflected by a change in the absolute ^{13}C shift at C-9, which increases from 136.1 to 150.6 ppm on protonation. Similarly, a slight decrease in the *ipso* SCS of OMe upon protonation of the *para* nitrogen is noted to be consistent with trends for this substituent in substituted benzenes,[127] when electron demand is increased.

Other polycyclic aromatics which have been examined recently include 1- and 2-substituted anthraquinones [26a] and [26b].[173] Multi-parameter correlations using equation (23) demonstrate significant polar and resonance effects at most distant carbon sites.[172] Polar susceptibility coefficients in series [26a] are consistent with the charge redistribution shown below, [27], resulting from π-polarization of the distant phenyl and carbonyl groups.

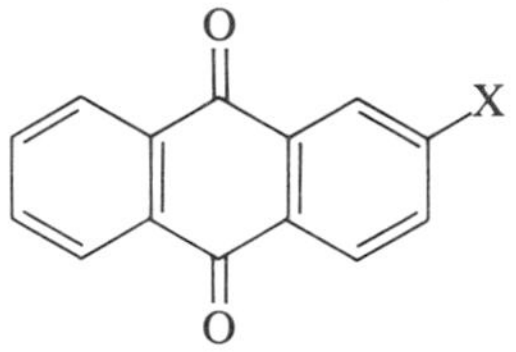

[26a] 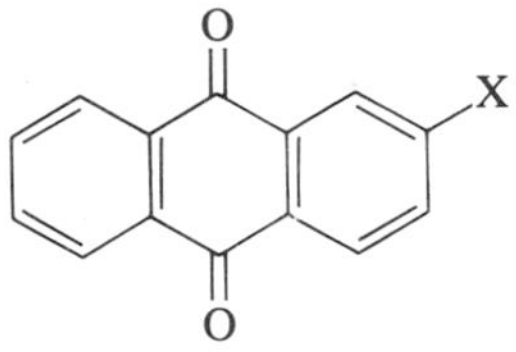   [26b]

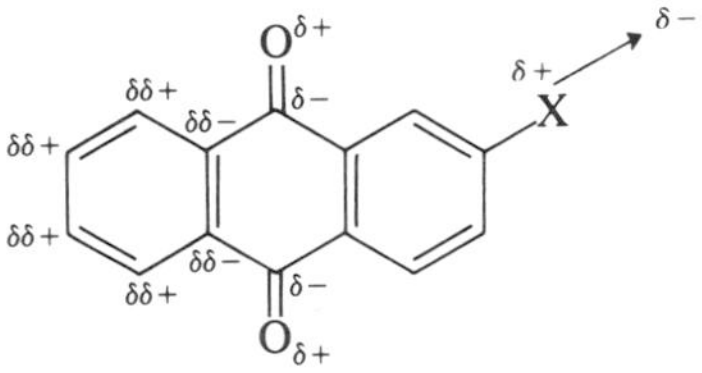

[27]

There have been few reported studies of substituent effects in bridged aromatic macrocycles. These studies will be of great interest because of the possibility of examining π-polarization in distorted phenyl rings. Some preliminary SCS data can be derived from investigations on monosubstituted [2.2]paracyclophanes [28].[174] Substituent effects of the OMe and CO_2Me substituents are similar to those in monosubstituted benzenes, indicating that the π systems of the two benzene rings in these series are sufficiently well separated to make mutual interactions small.

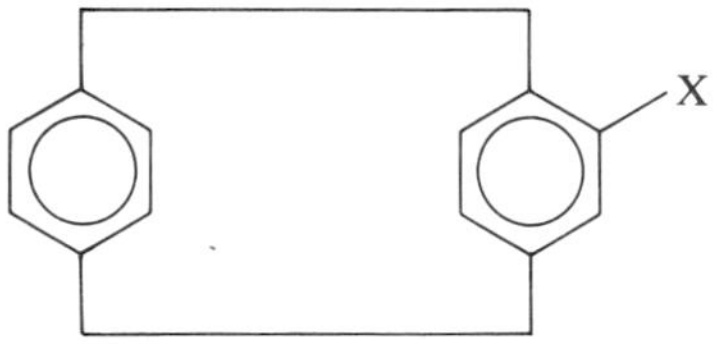

[28]

E. Non-benzenoid aromatics

2-Substituted tropones [29] will be examined here as a representative of this class of compounds.

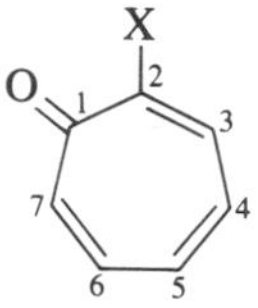

[29]

Compounds in this series containing π electron-donating substituents display generally larger substituent effects on ^{13}C shifts (Table 16)[175] at a given site relative to the corresponding position in monosubstituted benzenes. A reasonably linear correlation is obtained between the shifts at C-5 in 2-X-tropones and C-4 (*para*) in monosubstituted benzenes. For positions closer to the point of attachment, the substituents Cl, CH_3 and NH_2 fall on one correlation line and $NHCOCH_3$, $NHCH_3$, $OCOCH_3$ and OCH_3 fall on a second correlation line. This has been attributed to differential steric–electronic interactions of these two classes of substituents with the adjacent carbonyl group.[175]

TABLE 16

^{13}C SCS values[a] for carbons 1–5 in 2-substituted tropones.

Substituent	C-1	C-2 (C-*i*)	C-3 (C-*o*)	C-4 (C-*m*)	C-5 (C-*p*)
$NHCH_3$	−11.3	+14.5 (+21.7)	−27.9 (−16.2)	−6.4 (+0.7)	−12.8 (−11.8)
NH_2	−11.5	+15.7 (+18.0)	−23.1 (−13.3)	−4.5 (+0.9)	−10.9 (−9.8)
$NHCOCH_3$	−8.8	+4.9 (+11.0)	−15.2 (−9.9)	+1.2 (+0.2)	−4.0 (−6.5)
OCH_3	−7.6	+23.3 (+31.4)	−23.6 (−14.4)	−2.0 (+1.0)	−6.8 (−7.7)
$OCOCH_3$	−9.4	+15.8 (+23.0)	−7.7 (−6.4)	−2.3 (+1.3)	−0.3 (−2.3)
Cl	−7.8	+7.0 (+6.2)	+2.5 (+0.4)	−3.2 (+1.3)	−0.5 (−1.9)
CH_3	−0.8	+10.3 (+8.9)	+3.6 (+0.7)	−0.9 (−0.1)	−2.2 (−2.9)

[a] Reference 162. SCS values for carbons C-6 and C-7 are not given because assignments have not been firmly established. Values in parentheses are corresponding SCS values in monosubstituted benzenes.

F. Substituted alkenes and alkynes

In directly substituted ethylenes, the carbon to which the substituent is attached may undergo considerable changes in bond angle with different substituents, and hence SCS values at C-α and C-β do not solely reflect electronic effects. Nevertheless, SCS data in many such derivatives[176] show general patterns consistent with polarization of the π electrons, as well as short-range inductive effects, and resonance effects. Theoretical investigations have confirmed the importance of π-polarization in these systems.[73,177,178] For example, Seidman and Maciel[178] examined the effect of a distant charge on ethylene and acetylene molecules, and found polarization of π electrons to be more important than polarization of σ electrons, although both effects are significant. "Isolated molecule" (IM) calculations[73,177] have also been used to demonstrate the importance of through-space polarization of π electrons. In contrast to the approach of Seidman and Maciel,[178] who use a point charge as the perturbing influence, IM calculations utilize a real substituent group. For example, the dimer shown below [30] (where X is varied over a range of common substituents) has been used to simulate through-space π-polarization of the vinyl group in styrenes.[73]

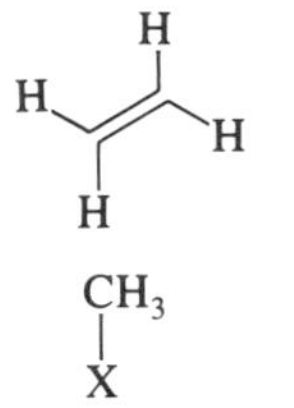

[30]

Since we are concerned here with electronic substituent effects, the remaining discussion in this section centres on alkenes where the substituent group is located some distance from the unsaturated bond. In an extensive study meeting this requirement Hill and Guenther[179] recently reported ^{13}C chemical shifts for a number of substituted 1-butenes [31] and 1-pentenes [32] and for the aromatic carbons of related series [33] and [34]. Some substituted derivatives of these series were previously examined by other authors.[32,180,181]

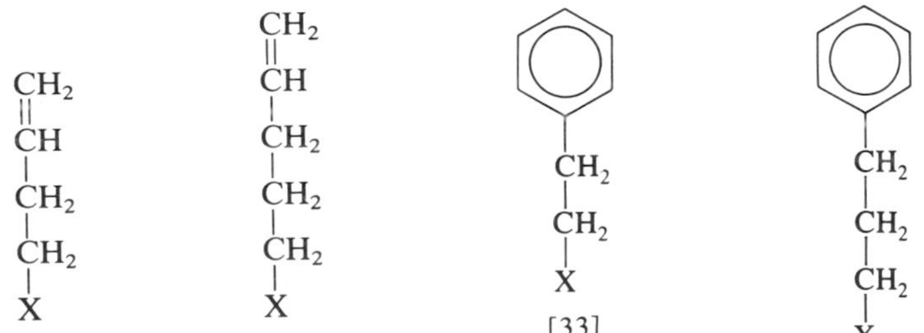

[31]   [32]   [33]   [34]

^{13}C shifts for a wide range of alkenes were measured in diethyl ether at concentrations of 10–15% by volume.[179] Electron-withdrawing substituents cause low frequency shifts of the C-2 carbon resonance (this is the unsaturated site closer to the substituent), and high frequency shifts for C-1. This is consistent with a dominant π-polarization mechanism. The internal consistency of the data can be seen from the fact that the C-1 and C-2 SCS values are linearly related, with a slope close to unity.[179] SCS values for the *ipso* and *para* carbons in series [33] and [34] are of smaller magnitude, but follow similar trends. For series [31] and [32], a wide variety of correlations between the chemical shift difference $\Delta\delta = \delta(\text{C-1}) - \delta(\text{C-2})$, and chemical shifts in other related series, substituent constants, and calculated electric field components have been attempted;[179] although correlations of high statistical precision are not generally obtained, moderately linear trends do emerge. The conformational flexibility of many of these systems may in part account for the lack of very precise correlations. Electric field effects are found to be particularly sensitive to the chosen geometry.[179] One complication associated with this conformational flexibility is the possibility of direct association between the substituent, and the probe site, although the authors express the opinion that this would be minimized in ether solvent.[179]

A preliminary study of solvent effects was carried out for series [31]; $\Delta\delta$ values for the CH_3 substituent are found to increase with the π^* solvent polarity parameter[122] which reflects principally the solvent molecular dipole moment. This trend is attributed to specific solvent–double bond interactions,[179] whereas previously observed increases in polar components of substituent effects with solvent polarity have been interpreted largely in terms of increased π-polarization arising from increased substituent polarity.[121,124] Hill and Guenther suggested that values of $\Delta\delta$ for other substituted butenes [31] *relative* to pentene discussed above (i.e. $X = CH_3$) provide a means of isolating the solvent influence on the *transmission* of substituent effects from its *direct effect* on the double bond carbon shifts. The differences $\Delta\delta = \delta(C\text{-}2) - \delta(C\text{-}1)$ for the substituents CO_2H, Br and CN, relative to the same value for $X = CH_3$, decrease in the solvent order cyclohexane > pentane > diethyl ether > DMF > DMSO. It is argued that this order is consistent with an expected attenuation of effects transmitted by a field effect (e.g. π-polarization) obeying the equation

$$E = q/\varepsilon_r R^2 \qquad (24)$$

The electric field at the probe site decreases with an increase in the effective dielectric, ε_r. Here E is the electric field acting at the probe site, q is the charge of the substituent and R is the intervening distance. In their electric field calculations, Hill and Guenther[179] derive q from a summation of point charges on an assumed substituent geometry. Even though the order of attenuation effects appears to be correct, the magnitude is markedly less than that expected on the basis of the different solvent dielectric constants. This may reflect differences between the *effective* dielectric of the solute and cavity, and the bulk solvent.

In an attempt to reduce contributions to the observed ^{13}C shifts in series [31] from through-bond or other effects present in the underlying σ framework, corrected shifts are derived by subtracting shifts in the saturated analogues, and these are shown in Table 17. These values give slightly better correlations with σ_I and electric field calculations; however, correlation coefficients are still on the order of 0.95–0.97. Theoretical studies have also supported the need to use corrected charges to rationalize ^{13}C shifts.[182]

One of the most important conclusions to be derived from this study concerns the ^{13}C chemical shifts in series [35] and [36].

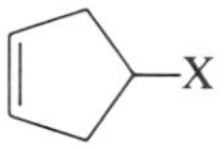

[35]

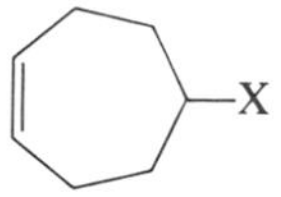

[36]

TABLE 17

Corrected ^{13}C substituent chemical shifts (ppm) in 4-X-butenes.[a]

	(C-1)	(C-2)	$\Delta\delta$
MgBr	−5.60	0.54	6.14
CH$_2$MgBr	−1.91	2.06	3.97
Si(CH$_3$)$_3$	−1.02	−1.01	0.01
H	0.00	0.00	0.00
CH$_2$NH$_2$	0.15	0.63	0.48
CH$_2$OH	0.26	0.64	0.38
CH$_3$	0.77	0.74	−0.03
CO$_2$C$_2$H$_5$	1.45	−0.71	−2.16
CH$_2$Br	1.54	−0.76	−2.30
CO$_2$H	1.58	−0.66	−2.24
CH$_2$Cl	1.84	−1.20	−3.04
CH$_2$CN	3.43	−0.41	−3.84
OH	2.14	1.00	−1.14
NH$_2$	2.18	0.80	−1.38
Br	3.62	−1.57	−5.19
CN	3.72	−2.56	−6.28
Cl	3.74	−1.40	−5.14
NO$_2$	5.07	−2.32	−7.39

[a] Corrected by subtracting the SCS of the corresponding saturated derivative. Reference 179.

In these series the C—X bond is perpendicular to the double bond, and a purely through-space mode of transmission of π-polarization effects (which directly perturb π electron density) predicts zero net change in ^{13}C shifts. In sharp contrast, the experimental data[179] indicate significant substituent effects in series [35]; a range of about 5 ppm is observed, with electron-withdrawing substituents bringing about a low frequency shift of the equivalent C-1 and C-2 carbons.[179] (The effect of the methyl group appears to be anomalous.) Similar trends, of smaller magnitude, are seen for series [36]. Interestingly, the observed shifts generally parallel the algebraic sum of C-1 and C-2 shifts of the open chain series [31], suggesting that some contributions from through-bond transmission may be important. This provides confirmatory evidence for the concept of transmission via "molecular lines of force".[46] These results may also indicate the importance of σ bond polarization of C—C and C—H bonds, most of which are not perpendicular to the substituent dipole.[178]

A number of other series containing C=C double bonds held rigidly in a saturated framework, with respect to the substituent, have been investigated. When the substituent dipole is not orthogonal to the C=C bond, opposite signs of SCS effects are observed at the two ends of the π system, as expected on the basis of π-polarization. This can be seen, for example, in 1-camphenes [37][183] and in 9-thiabicyclo[3.3.1]non-2-enes [38].[184]

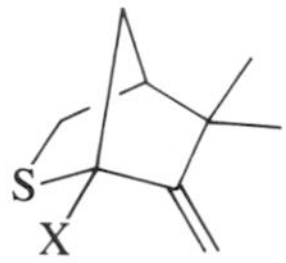

[37]

[38]

Most studies concerned with substituent effects in triple-bonded systems have related to phenylacetylenes, the data for which will be discussed in the following section on side-chain systems. There have been relatively few other reports on acetylenic systems; however, substituent constants for calculating ^{13}C chemical shifts in compounds of the form RC≡CR' have recently been derived.[185]

G. Side-chain systems

Substituent effects at carbons on the side-chains of aromatic series have been reviewed very recently,[8] but for continuity, the main trends to emerge are described briefly here. Ring-substituted styrenes have been extensively studied,[33] and provide a useful framework in which to illustrate many of the features observed in other unsaturated side-chain series. In *para*-substituted derivatives, larger ^{13}C SCS values are observed at C-β compared with C-α, even though the former site is further from the substituent. This can be attributed to the greater influence of resonance effects at C-β, evidenced, for example, by the through-conjugation shown in [39].

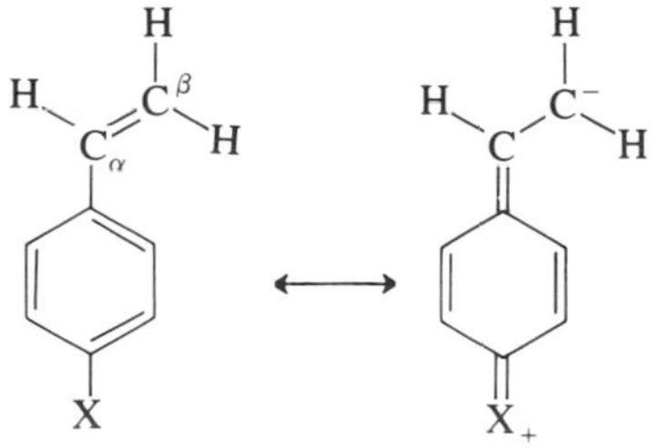

[39]

Calculated electron densities confirm that C-α remains relatively unperturbed by resonance effects of X, whereas C-β obtains excess electron density when X is a donor, and becomes deficient when X is an acceptor.[33] The C-β SCS values correlate very well with calculated *ab initio* π electron densities with a slope of 193 ppm per electron (Fig. 7).[73] DSP analyses of

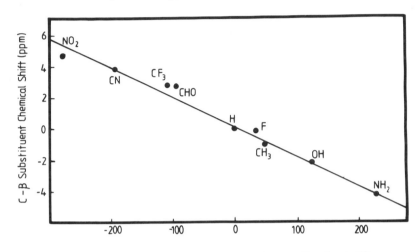

FIG. 7. Plot of C-β SCS versus Δq_π in *para*-substituted styrenes. Chemical shifts are taken from reference 33a and the *ab initio* π electron densities from reference 73. For the fluoro substituent, the SCS datum from reference 189 is used. The OH point in the graph corresponds to the SCS value for OCH_3, whilst that for NH_2 corresponds to the SCS value for NMe_2. Slope = 193 ppm per electron.[73]

the ^{13}C SCS values[33] for both side-chain carbon sites show the relatively greater contribution from resonance effects at C-β compared to C-α:

$$C\text{-}\alpha\ SCS = -2.4\sigma_I - 0.4\sigma_R^0 \qquad (25)$$

$$C\text{-}\beta\ SCS = +5.0\sigma_I + 8.9\sigma_R^0 \qquad (26)$$

These equations also indicate a reversal in the sign of ρ_I values at the C-α and C-β sites. This is consistent with the polar component of the shifts being controlled largely by π-polarization. Polar substituents bring about a shift of electron density from C-β to C-α. The magnitude of the effect is quite large, e.g. the ρ_I value in equation (26) is comparable to that obtained at the *para* position in monosubstituted benzenes. The fact that ρ_I at C-α has a magnitude half of that at C-β may be rationalized in terms of an extended polarization component.[33,177,186] Such effects are unique to conjugated π systems and are not observed in the aliphatic alkenes examined in the previous section. In early studies of styrenes[33] it was noted that a substituent dipole may bring about a polarization of electron density in each of the separate π systems (vinyl and phenyl) or it may polarize the whole π system as a unit. Theoretical studies[177] have confirmed this concept and the two effects have been named *localized polarization* and *extended polarization* respectively.[186] Since localized polarization of the vinyl π system would bring about equal and opposite changes in electron density at C-α and C-β, larger observed effects (monitored by SCS values) at C-β reflect an important contribution at this site from extended polarization of the conjugated phenyl/vinyl π system.

The styrene framework has also proved useful for the evaluation of correlation procedures. For example, Reynolds and coworkers[34] have recently used SCS data in this series to demonstrate the general validity of the DSP approach. Previous suggestions[187] that single parameter correlations are sufficient for the analysis of ^{13}C SCS data are disputed.[34] Similar controversy has been reported from studies of SCS data on a number of structurally related side-chain derivatives.[46,196] However, the necessity of dual parameter methods now seems unequivocal.

Numerous other styrene derivatives have been investigated.[188–209] These include series with fixed CN, NO_2, CO_2H, CO_2^-, H, or CH_3 substituents at the β position, as well as the usual range of ring substituents. Happer and coworkers[188,189] have made extensive investigations of these series, and have used C-β SCS values to estimate a variety of substituent constants. SCS data on these series have been used for the derivation of a multi-parameter correlation equation involving a logarithmic dependence of resonance effects on electron demand.[189] The fundamental basis of this equation and the DSP-NLR equation is similar in that the aim is to describe the non-linear response of a substituent to electron demand at a distant centre.

Styrene derivatives also provide good model systems for extending the treatment of shift–charge ratio (SCR) effects noted earlier. For example, C-β SCS values induced by ring substituents become smaller when a fluoro or nitro group is present at the β position, compared with the case when there is no β substituent.[42,195] The F and NO_2 groups may thus bring about a reduced SCR at the adjacent carbon, consistent with their effects in *para*-disubstituted benzenes.[42] Confirmation of this preliminary proposal[42,195] awaits further studies of chemical shifts and electron densities in β-substituted styrenes.

Styrene derivatives containing a fixed α-substituent as well as a range of ring substituents are also of interest. Since the resonance effect of the ring substituent plays an important role in determining C-β SCS values, loss of conjugation between the ring and the vinyl group, caused by steric interaction with a substituent at C-α, reduces resonance effects of ring substituents on side-chain chemical shifts. This can be seen from studies of α-methyl styrenes,[33] α-*t*-butyl styrenes[33] and α-methoxy styrenes.[194]

In another study closely related to styrenes the relative abilities of O, S and Se bridging groups to transmit substituent effects in 4-substituted phenylvinyl ethers, sulphides and selenides ([40]; G = O, S and Se) have been investigated.[209]

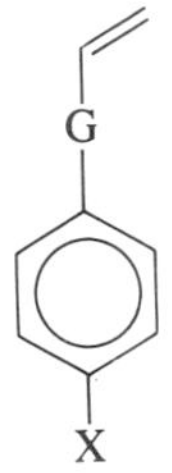

[40]

Taking into account the ability of the bridging group to modify the sensitivity of adjacent ^{13}C shifts to substituent effects of X (i.e. the SCR effect) it is found that the ability of the bridging group to *transmit* substituent effects follows the order − > O > S > Se ("−" means no bridging group is present).[209]

Substituent effects in aromatic carbonyl compounds have been extensively studied, especially those having the general structure [41]. For a given series, Y remains constant and X is varied over a range of electron-donating and electron-withdrawing substituents.

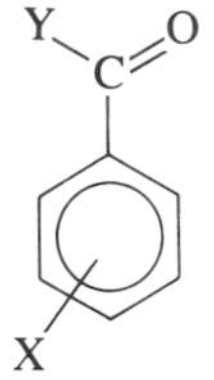

[41]

The range of SCS values at C-α is generally small (< 3 ppm); however, good DSP fits are obtained (Table 18).[186] This shows that SCS values at this site are sensitive to the electronic properties of the X substituent. As in the styrenes discussed above, a reverse polar component is observed at C-α in these series. A comparison of DSP correlations for *para* derivatives in Table 18 shows that polar effects at C-α are similar ($\rho_I = -2.7 \pm 0.3$) for all series. The Y group apparently does not have a large effect in modifying π-polarization of the adjacent carbonyl group.[186] Calculated electron densities (*ab initio*, STO-3G) have confirmed (for the series Y = F, i.e. benzoyl fluorides)[8,111] the importance of electronic effects in controlling

TABLE 18

DSP analysis[a] of carbonyl side-chain SCS data.

Side-chain	Para series				Meta series			
	ρ_I	ρ_R	Scale	f	ρ_I	ρ_R	Scale	f
$CONH_2$	−2.4	−0.4	$\sigma_R{}^{BA}$	0.10	−3.3	−0.6	$\sigma_R{}^+$	0.13
COF	−2.5	−1.3	$\sigma_R{}^0$	0.12	−3.1	−0.7	$\sigma_R{}^+$	0.12
COOEt	−2.6	−1.1	$\sigma_R{}^0$	0.08	−3.2	−0.4	$\sigma_R{}^+$	0.13
COOH	−2.3	−0.5	$\sigma_R{}^{BA}$	0.13	−2.8	−0.5	$\sigma_R{}^+$	0.15
COMe	−2.6	+0.8	$\sigma_R{}^+$	0.14	−3.5	−0.6	$\sigma_R{}^+$	0.14
COH	−3.0	+1.0	$\sigma_R{}^+$	0.15	−3.8	−0.6	$\sigma_R{}^+$	0.11
CH_2COOH	−1.2	−1.0	$\sigma_R{}^{BA}$	0.17	−1.0	−0.1	$\sigma_R{}^+$	0.17
OCOMe	−1.4	−1.7	$\sigma_R{}^0$	0.12	−1.2	−0.3	$\sigma_R{}^{BA}$	0.15
$CSNH_2$	−2.8	+1.1	$\sigma_R{}^+$	0.17	−4.1	−0.8	$\sigma_R{}^+$	0.16
$COMe(H^+)$	5.3	10.9	$\sigma_R{}^+$	0.16	2.8	0.9	$\sigma_R{}^-$	0.31
$COMe(BCl_3)$	2.0	6.8	$\sigma_R{}^+$	0.18				
$COMe(TiCl_4)$	−2.6	2.1	$\sigma_R{}^+$	0.12	−4.1	−3.1	$\sigma_R{}^{BA}$	0.12

[a] Reference 35.

the ^{13}C shifts. π electron densities are consistent with the reverse SCS direction, whereas total electron densities are not.[8]

Further evidence that SCS effects in carbonyl derivatives are controlled by polarization of the π electrons can be seen from studies of carbonyl series where the C=O bond is protonated, or perturbed by a Lewis acid complexing agent.[186] Table 18 shows that the reverse polar component of SCS effects at C-α in para-substituted acetophenones ($\rho_I = -2.6$) changes to a normal effect ($\rho_I = +5.3$) when the carbonyl π bond is removed by protonation.

Table 18 also shows that polar effects are larger in meta derivatives of series [41] than in para derivatives. This shows that π-polarization increases with a decrease in the X/C=O separation (although orientational effects also change and it is difficult to quantify contributions from both effects). The larger ρ_I values in the meta series confirm that conjugation of the substituent and carbonyl group is not required for the transmission of π-polarization effects.

Many other carbonyl side-chain series have been investigated,[210–215] with applications ranging from the investigation of solvent effects[211] and rotational barriers,[212,213] to studies of linear free energy relationships in dissociated and undissociated benzoic acids.[214]

Substituted benzonitriles [42] are one of the first series in which reverse C-α SCS values were observed.[36] Derived ρ_I values for meta- and para-substituted derivatives (equations (27) and (28))[36,110] are consistent with

[42]

π-polarization. The larger ρ_I value in the *meta* series reflects the shorter distance between the substituent and the polarizable C≡N bond.

$$C\text{-}\alpha \ SCS \ (para) = -2.7\sigma_I - 1.1\sigma_R^+ \tag{27}$$

$$C\text{-}\alpha \ SCS \ (meta) = -3.4\sigma_I - 0.6\sigma_R^+ \tag{28}$$

Ab initio MO calculations in this series[8,110] have shown that total electron densities do not predict the correct direction of SCS effects at C-α whereas π electron densities do. Thus, substituent effects in series like benzonitriles and benzoyl fluorides show that unsaturated side-chain sites adjacent to a benzene ring (i.e. C-α) are useful for differentiating between π and total electron densities in terms of relationships to ^{13}C chemical shifts. This differentiation is not achieved at C-*para* and C-β, since π and total components of electron density follow the same direction at these sites.

Phenylacetylenes[37,77,216,217] are subject to similar (reverse) substituent effects at C-α as those in the derivatives discussed above. The similarity in ρ_I values in all of these series shows that the polar components of SCS values in C≡C and C≡N triple-bonded systems are influenced by a distant dipole in a similar manner to those in double-bonded C=C and C=O systems. This is somewhat surprising since changes in electron populations (related to polarizability of the unsaturated bonds) and changes in SCR values might be expected. Compensatory changes in both factors may account for observed similar polar contributions to SCS values.

Several authors have performed theoretical calculations in *para*-substituted phenylacetylenes. While π electron densities calculated by both semi-empirical[37] and *ab initio*[77] methods accurately reproduce trends in SCS values at C-β, the corresponding correlations at C-α are more scattered. Nevertheless, the correct reverse SCS direction at this site is reproduced by π electron density calculations.[37,77] Slightly improved correlations with CNDO π electron densities are obtained when C-α shifts are corrected for ring current and diamagnetic susceptibility effects.[216]

A wide range of other side-chain series have been investigated;[218–246] however, space limitations prevent a full description of all of these. Series that deserve specific comment are those containing side-chain phenyl groups.[235-247] Many of these can be represented by the general structure shown in [43].

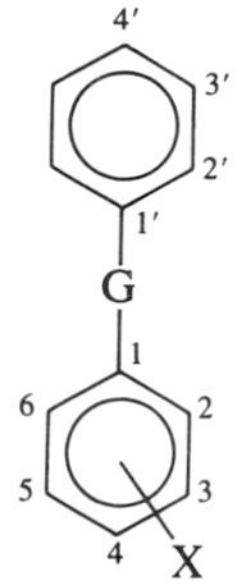

[43]

On the basis of the π-polarization mechanism it is expected that polar substituents should bring about high frequency shifts at C-4' and low frequency shifts at C-1'. A consideration of ρ_I values derived from DSP correlations in a variety of these series (Table 19) confirms this general expectation. The fact that the transmission of π-polarization effects does

TABLE 19

DSP analyses of ^{13}C SCS data for side-chain phenyl groups in series [43].

G	Position	ρ_I	ρ_R	Scale	f	Reference
$-(CO)CH{=}CH-$	C-1	-0.8	-0.4	$\sigma_R{}^{BA}$	0.21	222
	C-4	0.3	0.6	$\sigma_R{}^0$	0.27	222
$-CH{=}CH(CO)-$	C-1'	-0.7	-0.5	$\sigma_R{}^+$	0.07	222
	C-4'	0.8	0.8	$\sigma_R{}^{BA}$	0.14	222
$-CH_2{-}N{=}CH-$	C-1'	-0.6	-0.4	$\sigma_R{}^0$	0.19	226
	C-4'	0.6	0.3	$\sigma_R{}^+$	0.19	226
$-CH{=}N{-}CH_2-$	C-1	-1.1	-1.0	$\sigma_R{}^{BA}$	0.09	226
	C-4	0.4	0.3	$\sigma_R{}^{BA}$	0.03	226
Biphenyl	C-1'	-3.0	-1.2	$\sigma_R{}^-$	0.15	227
	C-4'	2.0	3.1	$\sigma_R{}^0$	0.11	227
$-CH_2-$	C-1'	-1.3	-1.7	$\sigma_R{}^-$	0.29	227
	C-4'	1.2	0.3	$\sigma_R{}^+$	0.30	227
$-CH{=}N-$	C-1	-1.4	-1.3	$\sigma_R{}^0$	0.27	230
	C-4	1.5	1.8	$\sigma_R{}^0$	0.15	230
$-N{=}CH-$	C-1'	-1.0	-1.0	$\sigma_R{}^{BA}$	0.05	230
	C-4'	1.6	1.7	$\sigma_R{}^0$	0.12	230
$-(CO)-$	C-1'	-1.6	-1.9	$\sigma_R{}^{BA}$	0.10	231
	C-4'	1.3	2.0	$\sigma_R{}^0$	0.10	231
$-\overset{+}{C}(OH)-$	C-4'	3.1	3.6	$\sigma_R{}^+$	0.03	234
$-CH^+-$	C-4'	10.4	10.9	$\sigma_R{}^+$	0.17	135
$-1,4{-}Ph-$	C-1	-0.8	-0.4	$\sigma_R{}^-$	—	242
	C-4	0.5	0.7	$\sigma_R{}^0$	—	242

not require the presence of an intermediate π system can be confirmed from an examination of SCS data in series [44].

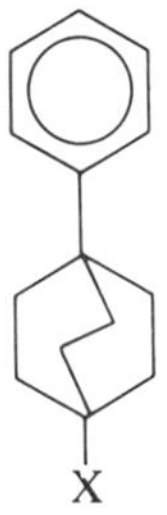

[44]

Substituents are found to influence significantly the ^{13}C chemical shifts in the distant phenyl ring in this series; however, apart from possible contributions from hyperconjugative interactions, resonance effects are expected to be negligible. Indeed, DSP analyses[243,244] reveal that the observed SCS values in the phenyl ring can be attributed solely to substituent polar effects (ρ_R contributions are found to be negligible).[243] The induced polarization pattern is shown in [45] and is consistent with that expected from a π-polarization effect transmitted through space.

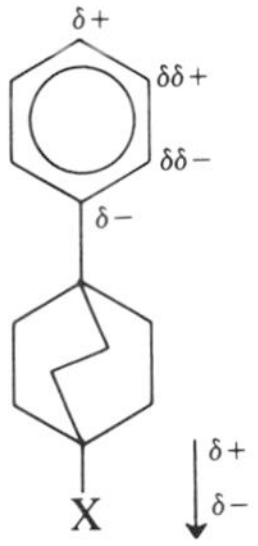

[45]

In contrast to series [44], significant resonance effects are observed (Table 19) in series of general form [43], where G is an unsaturated bridging group. Indeed, the transmission of resonance effects from the substituted phenyl ring to the unsubstituted ring depends strongly on the nature of the bridging group G. To illustrate the types of mechanistic information that can be obtained, it is useful to discuss series [46] and [47] as typical examples.[225] The smaller range of substituent effects observed at C-4' in the latter series, [47], compared with [46][225] can be attributed to a greater loss of conjugation between the two rings in [47] relative to [46] due to steric hindrance of the ortho protons with the methyl group.[225]

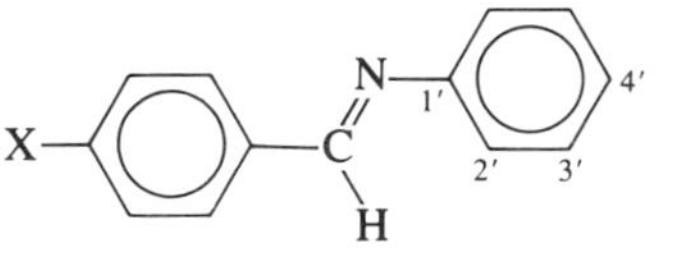

[46] 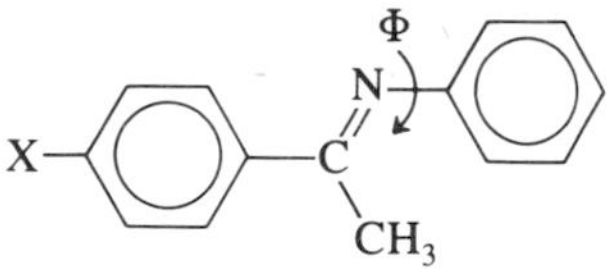   [47]

Indeed, the magnitudes of C-4′ SCS effects in both series are rather small (<1.6 ppm) indicating that both series are non-planar. The non-planarity appears to arise from a rotation about the N—Ph bond rather than the C—Ph bond, because ^{15}N SCS values[247] show that the C-aryl ring is highly conjugated with the imine bond (the ^{15}N SCS for 4-NO_2 is 12.7 ppm and for 4-OCH_3 it is -8.5 ppm).[247] This indicates the important contribution of resonance forms such as [48].

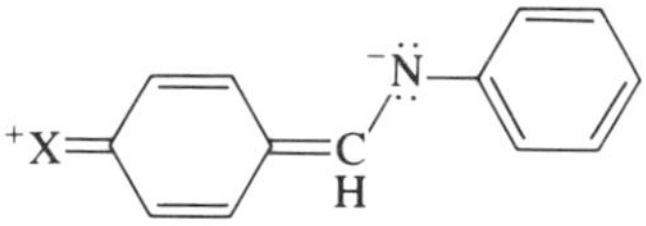

[48]

Buchanan and Dawson[225] suggested that changes in the shifts of C-2′ and C-4′ could be used to monitor varying amounts of electron release from the nitrogen lone pair into the adjacent ring. As the angle Φ approaches 90°, overlap of the phenyl ring orbitals with the nitrogen lone pair becomes more efficient and C-2′ and C-4′ are shielded accordingly. Greater shielding of these carbons is seen for series [47] compared with series [46], suggesting a greater out-of-plane twist in series [47]. This is confirmed from other ^{13}C NMR data[246] and X-ray data[248] which suggest that the angle of twist is $\sim 55°$ in series [46] (X = H) and $\sim 90°$ in series [47] (X = H).

A number of derivatives in Table 19 contain bridging groups not expected to be efficient at transmitting resonance effects, e.g. CH_2, yet significant resonance effects are still observed in the distant phenyl ring. This can be attributed to secondary resonance effects, in which the delocalized charge in the phenyl ring attached to the substituent acts as a dipole which polarizes (through space) the second, more distant phenyl ring as shown in [49].

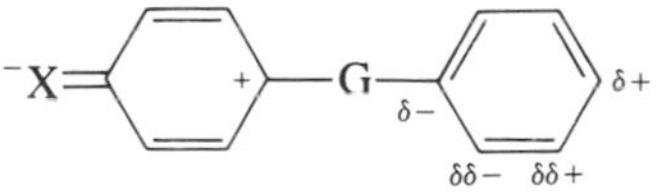

[49]

Since the charge delocalization, and hence the resultant dipole, in the first ring is related to σ_{R^o} of the substituent, a resonance component is present in the induced polarization of the second phenyl ring. Such effects will always be present in systems where the substituent is attached to an unsaturated centre, but are normally much smaller than direct resonance interactions that occur when the pathway between the substituent and probe is fully unsaturated.

Other DSP correlations given in Table 19 include cases when there is no bridging group (corresponding to biphenyls [50]) and cases when the bridging group is a 1,4-phenyl linkage (corresponding to terphenyls [51]). SCS data for the former series has been available for some time,[227] but results for the latter series have been reported only very recently.[242]

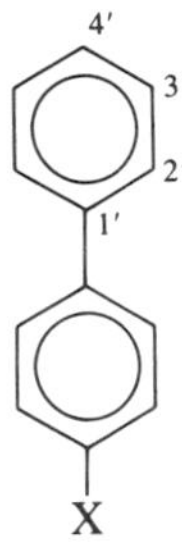

[50]

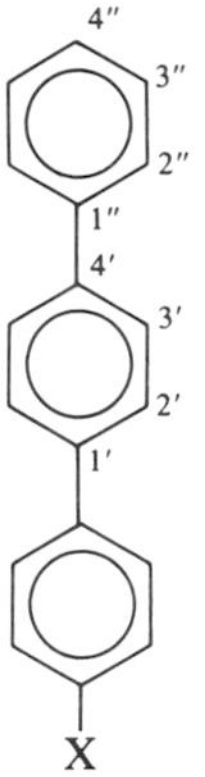

[51]

Wilson and Zehr[242] have carried out an extensive comparison of single parameter, DSP, and DSP-NLR correlation procedures in series [51], and generally confirmed the superiority of the two latter approaches. The DSP-NLR method is found to give better fits at the C-1' and C-4' positions in the central ring, whereas DSP fits are better at all other sites except C-2',6' and C-2'',6'' where single parameter treatments yield better statistical correlations. Despite the considerable distance between the substituent and the C-4'' probe, significant polar and resonance effects are observed at this site.[242] Successive ρ_I values at the *para* sites (C-4, C-4' and C-4'') are 4.43, 2.03 and 0.45, whereas successive ρ_R values are 20.3, 2.44 and 0.77. The decrease in resonance effects from the first ring to the second ring is more dramatic than the corresponding decrease in polar effects, but from the second to the third ring, the attenuation of both effects is approximately equal. Note that reverse resonance and polar effects are observed in both distant phenyl rings at the C-1' and C-1'' sites, in agreement with observations in other side-chain phenyl series.

Most of the discussion of side-chain series so far has centred on the effects of ring substituents on side-chain probe sites. In previous sections, the opposite effects have been examined, i.e. the effect of side-chains (which may be regarded as substituents) on ring shifts in substituted benzenes have been noted. Bradamante and Pagani[249] have recently examined a third alternative in a study aimed at monitoring interactions of contiguous functionalities in the side-chains of benzene derivatives. In these studies, series of general structure [52] are examined.

MON—⟨ ⟩—GX

[52]

Here X is a substituent group, G is some contiguous functionality (e.g. CH_2, PhCH, NH or O) and MON is a probe used to monitor the interactions of X and G. *Para* ^{13}C or ^{19}F chemical shifts are found to be the most sensitive monitoring groups.[249] Linear relationships are found between contiguous interactions of X and G (series [52], G = O, NH), as monitored by *para* SCS values, and longer range interactions in X—Ph—GH derivatives, monitored by ^1H SCS values of the side-chain GH group. In other observations[249] it is found that SCS values of the *para* ^{13}C monitor in Ph_2CHX (i.e. G = CHPh) are 0.9 times as large as those in $PhCH_2X$ (i.e. G = CH_2), indicating that there is negligible partitioning of substituent effects of X into the two phenyl rings. This is consistent with transmission of polar effects by π-polarization, but not by σ-inductive effects.[249] Finally, for substituents which are capable of only polar/inductive interactions with G = NH, the sensitivity of *para* SCS values in the series PhNHX is found to be twice that in $PhCH_2X$. It is suggested[249] that polar/inductive effects of X are able to modulate resonance delocalization of the NH group.

The examples given above in this section are illustrative of the types of effects that can be studied using ^{13}C SCS values in the side-chains of aromatic derivatives. SCS effects in saturated and cationic side-chains are not discussed in detail here as they have been fully covered elsewhere very recently.[8] It suffices to say that SCS effects in neutral aliphatic side-chains are generally small, and less well understood than those in unsaturated derivatives. On the other hand, SCS effects in cationic species are generally *very* large[135,250,251] and have been used extensively for the generation of new enhanced substituent constants.[251]

H. SCS/electron density correlations as assignment aids

From the electron density/SCS correlations discussed so far, it is clear that SCS values cannot be used as a precise experimental probe of electron densities, but often they can be used to derive *relative* changes in electron densities. In many cases a knowledge of relative electron densities can be used for assignment purposes. For example, Bangov[252] has recently discussed a method for the automated assignment of ^{13}C NMR spectra based on chemical shift–charge density relationships. For a proton-decoupled ^{13}C spectrum containing N peaks, there are formally as many as $N!$ ways of assigning those peaks. If a relationship between ^{13}C chemical shifts and charge densities of the form

$$\delta^{13}C = a\,\Delta qC + b \qquad\qquad (29)$$

is assumed, then the fit to this equation will be most precise when the assignment of each peak in the ^{13}C spectrum to its corresponding atomic position (and hence electron density) is correct. In the ideal situation of rigorous validity of equation (29) for all carbons in a molecule then, except for fortuitous degeneracies, there should be only one solution, or permutation, of the N peaks in the spectrum to the N carbons of the molecule which gives an ideal fit to this equation. In practice, a computer is used to perform a regression analysis of a set of N chemical shifts and N electron densities for each of the $N!$ possible permutations. The permutation giving the best correlation coefficient (or other fitting criterion) is considered to be the correct assignment. The number of possible permutations which must be considered can be reduced by taking into account signal multiplicity, or fixing certain assignments on the basis of additional chemical information.

The method has been tested[252] for only a limited number of aromatic derivatives and further testing is clearly in order before judgements on the general applicability of this approach can be made. However, as has already been pointed out,[252] there are several inherent limitations:

(a) Equation (29) is only an approximation, and often not a very good one, especially for carbons in different environments or having different hybridizations. In addition, even carbons in similar compounds may have different SCR values (slopes of equation (29)).

(b) Quantum chemical methods are not yet sufficiently refined to produce meaningful absolute electron densities (although trends are often reproduced well). Different MO methods often produce different absolute electron densities.

(c) Solvent is not taken into account in the calculations, and the assumption in this method is that solvent-induced shifts affect each carbon similarly.

(d) Geometrical information must be arbitrarily assigned in many cases before a calculation of electron densities can be done.

These difficulties appear formidable, but since the method relies on picking the best result from a number of alternatives, and does not necessarily need an exact absolute fit, it is possible that it will be a useful assignment aid in some cases. General application of the method appears somewhat limited, however.

I. Heterocyclic compounds

^{13}C chemical shifts in 2-, 3- or 4-substituted pyridines[253] show generally similar substituent effects to those in corresponding monosubstituted benzenes. For example, in series [53] good correlations are found[253] between C-6 SCS values and either C_p SCS values in monosubstituted benzenes or σ_p values. This occurs in spite of the fact that the monitoring site is adjacent to an electronegative nitrogen atom.

[53]

Since most pyridine studies were carried out many years ago, for concentrated solutions, it is more than likely that systematic second-order effects arising from substituent–ring-nitrogen interactions are being masked by intermolecular effects. A modern high dilution study, including an investigation of solvent effects, would therefore seem well worthwhile.

One related system which has been measured recently is the pyridine N-oxide series.[254] Carbons C-2, C-4 and C-6 show a large shielding increase (~ 10 ppm) relative to pyridines, reflecting increased ring electron densities in the oxides.[254] Substituent effects on ^{13}C shifts for a range of these derivatives are shown in Table 20, where pyridine SCS values are also given

TABLE 20

^{13}C SCS values in pyridine N-oxides and (pyridines).^{a}

X	C-2	C-3	C-4	C-5	C-6
4-NMe$_2$	0.4 (−0.2)	−17.9 (−17.1)	25.7 (18.3)	−17.9 (−17.1)	0.4 (−0.2)
4-OCH$_3$	1.6 (0.6)	−13.7 (−14.2)	33.3 (29.3)	−13.7 (−14.2)	1.6 (0.6)
4-Cl	1.7 (0.9)	1.2 (0.5)	8.9 (8.2)	1.2 (0.5)	1.7 (0.9)
4-CH$_3$	−1.0 (−0.5)	0.3 (0.7)	11.4 (10.8)	0.3 (0.7)	−1.0 (−0.5)
4-COCH$_3$	1.0 (1.0)	−0.5 (−2.6)	7.2 (6.7)	−0.5 (−2.6)	1.0 (1.0)
4-NO$_2$	1.9 (−1.3)	−4.6 (−3.0)	17.2 (4.1)	−4.6 (−3.0)	1.9 (−1.3)
3-Cl	0.4 (0.1)	7.9 (8.6)	0.8 (0.0)	0.5 (0.7)	−0.6 (−2.2)
3-Br	2.2 (1.2)	−5.2 (−2.7)	4.1 (2.8)	0.5 (1.1)	−0.5 (−2.0)
3-COCH$_3$	1.0 (0.1)	10.2 (8.7)	1.1 (−0.4)	−0.9 (0.0)	4.0 (3.7)
3-CN	3.4 (2.6)	−12.5 (−13.6)	4.6 (3.4)	1.4 (0.0)	4.8 (3.1)
2-OCH$_3$	19.7 (14.5)	−17.2 (−12.6)	5.2 (2.7)	−8.2 (−6.9)	1.4 (−2.8)
2-Cl	2.9 (1.7)	0.1 (0.8)	1.9 (2.9)	−1.7 (−1.3)	1.8 (0.0)
2-CH$_3$	10.3 (8.6)	−0.3 (−0.4)	1.3 (0.4)	−2.2 (−3.0)	0.6 (−0.6)
2-COCH$_3$	8.3 (3.8)	0.8 (−2.1)	1.6 (1.0)	2.5 (3.4)	2.2 (−0.8)

^{a} Reference 254. Values in parentheses are for substituted pyridines.

for comparison. The most striking differences are for *ipso* SCS values of
4-N(CH$_3$)$_2$, 4-OCH$_3$ and especially 4-NO$_2$, where much larger deshielding
of the adjacent site is noted in pyridine oxides compared with pyridines.
(Note, however, that the *ipso* SCS of a 4-NO$_2$ group in pyridine is particularly small relative to its effect in benzene, and *ipso* SCS effects of the
NO$_2$ group are similar in benzene and pyridine *N*-oxide.)

^{13}C chemical shifts in a range of pyrimidines [54] containing 2-, 4- or
5-substituents have also been recently measured.[255,256]

[54]

In 2-substituted derivatives, substituents such as NH$_2$, OCH$_3$ and F exert
smaller deshielding effects at the *ipso* position relative to pyridine and
benzene. For example, the fluoro group causes a deshielding of ~ 35 ppm
in fluorobenzene, ~ 15 ppm in 2-fluoropyridine and ~ 4 ppm in 2-
fluoropyrimidine.[255] This trend is apparently due to increased electron
demand from the competing ring nitrogens. Where the electron demand is
not so great, e.g. at C-5 in 5-substituted pyrimidines, the *ipso* SCS values
are quite similar to those in monosubstituted benzenes.[256] Moderately good
correlations between semi-empirical π electron densities and ^{13}C shifts are
obtained for a number of pyrimidine derivatives.[255,256]

In another recent study of 6-membered ring heterocycles, ^{13}C shift
increments are derived for methyl and styryl groups in a number of
diazines.[257] Other recent reports relating to nitrogen heterocycles include
studies of aminopyrazoles,[258] *N*-aryl-2-pyrazolines,[259] benzamida-
zoles,[260-262] 1,3-diazaazulenes,[262,263] quinolines[264-266] and isoquinolines.[267]
Of these series, benzimidazoles [55] provide a particularly interesting
framework in which to study substituent effects. Larina *et al.*[260] have recently
examined series [55] and [56] using both ^{1}H and ^{13}C NMR, for a range of
10 substituents in a number of solvents (CH$_3$OH, CH$_3$CN, (CH$_3$)$_2$SO).

[55]   [56]   [57]

In preliminary ^{1}H studies[261] it was found that the transmission of substituent effects of X changes from series [55] to series [56]. The ^{1}H shifts

were also found to be substantially affected by the solvent. Therefore, to eliminate effects due to the acid–base equilibrium

$$[55]+H^+ \rightleftharpoons [56]$$

^{13}C and ^{1}H studies of [57] were initiated. In 1,3-dimethylbenzimidazolium perchlorate salts [57], this prototropic equilibrium is blocked and the methyl groups do not significantly affect chemical shifts relative to series [56]. In contrast to the earlier results,[261] substituent effects on ^{13}C and ^{1}H chemical shifts are very similar in [55] and [57].[260] This is surprising since it would be expected that the interaction of the substituent X with the nearby positive charge would change the electronic properties of X and hence it should affect the distant ring sites differently. DSP analyses of the ^{13}C SCS values for ring carbons 4–7 are given in Table 21 and show similar resonance and

TABLE 21

DSP transmission coefficients in benzimidazole derivatives.[a]

	Series [55]		Series [57]	
Carbon	ρ_I	ρ_R	ρ_I	ρ_R
C-4,7	0.7	6.9	0.7	6.6
C-5,6	2.7	6.6	2.4	7.5

[a] Reference 260.

polar effects at comparable sites in the two series. The observation that the polar influence of the substituent on the benzene ring carbons is largely unaffected by the nature of the intermediate framework is consistent with a through-space transmission mode for these effects. The smaller ρ_I value for C-4,7 compared with C-5,6 reflects the expected polarization pattern shown in [58]. This is confirmed by noting that polar effects at C-8 and C-9 are small.[260]

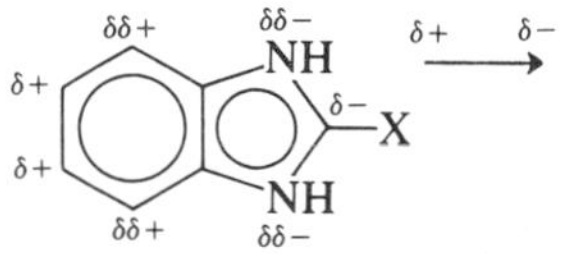

[58]

On the other hand, the similarity of ρ_R values in the two series probably reflects the fact that resonance effects are transmitted efficiently via the double-bonded nitrogen in both the neutral and cationic species. The similarity of the magnitude of resonance effects for C-4,7 and C-5,6 arises

because both sites can conjugate with the substituent as shown in [59a] and
[59b].

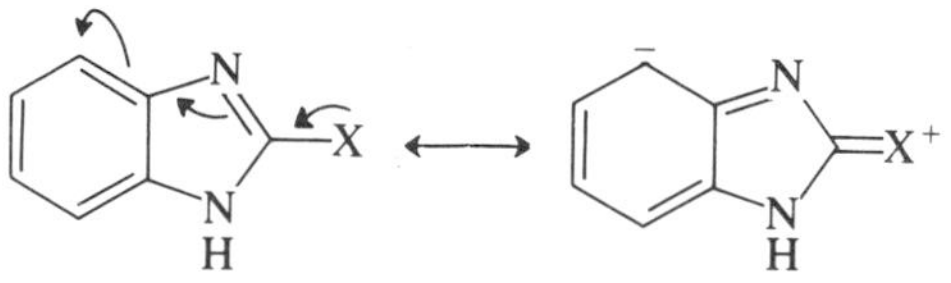

[59a]

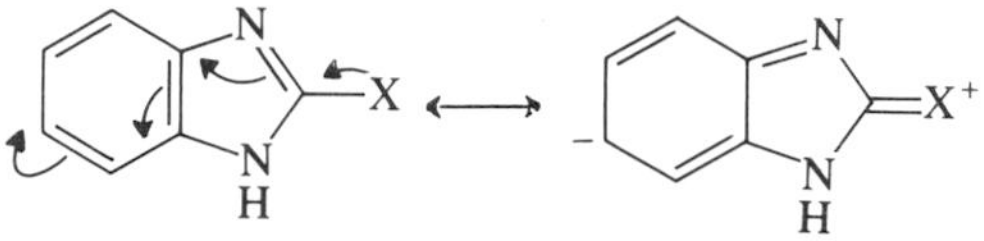

[59b]

   The series discussed above display quite different SCS effects from those
in monosubstituted benzenes, which is to be expected, since the substituent
is attached to an imidazole ring. On the other hand, benzimidazoles substituted in the benzene ring are expected to display trends more akin to
those in monosubstituted benzenes. Mathias and Overberger[262] examined
5-substituted benzimidazoles and found reasonable correlations of C-8 and
C-2 SCS values with Hammett-type σ parameters. However, the lack of a
very precise correlation at C-8 (a site *para* to the substituent) is taken as
evidence that the σ constants employed do not contain the correct relative
proportions of field and resonance effects. This again illustrates the
inadequacy of single parameter treatments and clearly demonstrates the
need for a DSP approach.
   Mathias and Overberger[262] have measured ^{13}C shifts for two 2-substituted
1,3-diazaazulenes [60].

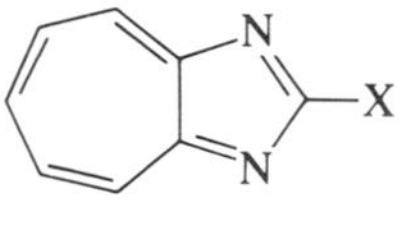

[60]

   Takeshita and Mametsuka[263] have made a more extensive investigation
of this series, and made revisions to an earlier assignment.[262] Shifts are
measured for a range of 16 substituents in several solvents; however, a
DSP analysis of the SCS data would not be meaningful because no π acceptor
substituents are included. As was observed in 2-azulenes,[108] substituent
effects at C-5,7 are negligible because of the non-conjugating nature of
these sites.
   Other fused ring nitrogen heterocycles recently investigated include 5-
quinoxalines,[264] 2- and 4-quinolones[265] and a number of quinolines and

isoquinolines.[266,267] The most extensive of these in terms of substituent electronic effects is the study[264] on 5-quinoxalines [61].

Chemical shifts in this series are found to be similar to those in phenazines [62] and 1-X-naphthalenes[144] (except for C-9 and C-10).

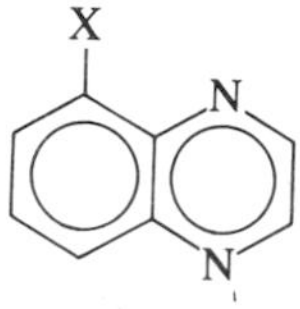

[61] 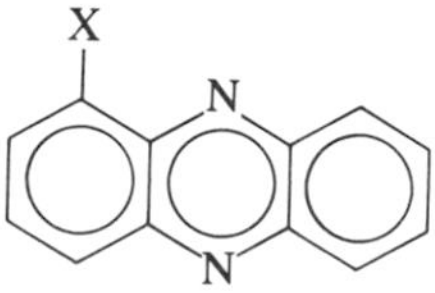   [62]

Finally, to complete the discussion of heterocyclic systems examined here we note that there have been a number of studies of series containing two different heterocyclic atoms, e.g. 4-thiazoline-2-thiones,[269] isothiazoles,[270] benzoxazoles,[271] thiazoles[272] and benzothiazoles.[273,274] The last series is of most interest because of its structural similarity to the benzimidazoles discussed earlier. Faure and coworkers[273] measured ^{13}C shifts for 43 benzothiazoles and discussed the results in terms of equation (23). The effects of ring annelation in benzoheterocyclic series and prototropic tautomerism in benzothiazolic series are also discussed. Sawhney and Boykin[274] examined a larger range of compounds, shown below:

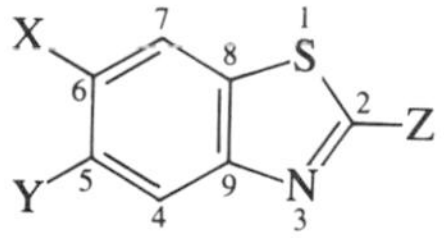

[63] X varies; Y = H; Z = NH$_2$
[64] X varies; Y = H; Z = CH$_3$
[65] X varies; Y = H; Z = H
[66] X = H;   Y varies; Z = CH$_3$
[67] X = H;   Y = H; Z varies

These series are selected to examine the effects of substituents at carbons 5 and 6 on the chemical shifts at C-2. In particular, the question of the relative transmission via sulphur and nitrogen is examined. Series [67] is included to determine the effects of transmission in the reverse direction from the other series. The results from DSP analyses are given in Table 22. Resonance transmission from a substituent at C-6 to the C-2 site (via the nitrogen) is clearly very efficient (see, for example, ρ_R values in Table 22 for series [63]–[65]), whereas much smaller effects are seen for substituents at C-5 (series [66]). The greater localization of substituent effects in the benzene ring in the latter series can be seen from the larger ρ_R value at C-8 in series [66] compared with ρ_R values at C-9 in series [63]–[65].

TABLE 22

DSP transmission coefficients in benzothiazoles.[a]

Series	Observed carbon	ρ_I	ρ_R
63	2	5.2	6.1
63	9	4.8	12.1
64	2	8.4	9.6
64	9	3.3	10.8
65	2	8.0	9.2
65	9	4.3	9.9
66	2	6.2	1.4
66	8	7.2	16.7
67	5	3.0	1.4
67	6	4.6	5.6

[a] Reference 274. Correlations with σ_I and $\sigma_R{}^{BA}$.

These results show that the transmission of substituent effects through the nitrogen atom is very much more efficient than via the sulphur atom.

The range in chemical shifts at C-6 in series [67] (as the substituent at C-2 is varied) is ~ 8 ppm, whereas the range for C-2 in series [65] (as the substituent at C-6 is varied) is ~ 14 ppm.[274] This shows that the transmission of substituent effects in the two directions is not equal (or that SCR values at the C-2 and C-6 are different).

J. Organometallics

^{13}C NMR has been useful as a structural probe in a variety of organometallic compounds; however, there have not been a large number of substituent effect studies. One recent trend has been the use of the metal nucleus itself as a NMR probe, and some of these studies are discussed in later sections. A few illustrative examples of the role of ^{13}C NMR in organometallic species are given here.

Chromium carbonyl complexes of monosubstituted benzenes [68] have been studied and the *para* SCS values[275,276] are closely related to those in the corresponding uncomplexed species.

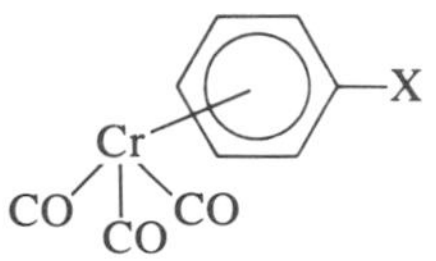

[68]

However, a larger chemical shift range is observed for the *ipso* SCS values in the complexed species and this may be a reflection of a withdrawal of σ electron density from the ring framework by $Cr(CO)_3$.[275] In an extension of these early studies, Fedorov and coworkers[277] have recently examined a number of other π-arenechromium carbonyl complexes [69].

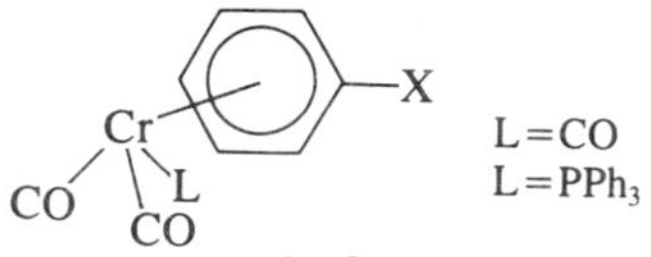

L = CO
L = PPh$_3$

[69]

For complexes with L = CO, the previously observed trend[275] that C-*ipso* shifts are larger than in uncomplexed derivatives is confirmed.[277] At the *ortho* site, some differences between SCS values in complexed and uncomplexed species are observed but a rationalization of effects is not attempted. At the *meta* site, complexation brings about a significant increase in the magnitude of SCS values. At this position, an increase in the donor ability of substituents in the order COR, H, Hal, OR, R_2N is associated with a *decrease* in shielding; however, a good correlation with the corresponding shifts in uncomplexed species is not obtained. Finally, at C-*para* a satisfactory linear correlation (with unity slope) is obtained for the shifts in complexed and uncomplexed species. Similar trends are observed for the case when L = PPh$_3$.[277]

The main conclusion to be derived from studies of these systems is that the transfer of π electron density to the *para* position remains relatively unchanged by complexation, but effects at the *meta* site become considerably enlarged, and indeed follow a reverse direction (i.e. donors induce downfield shifts). Theoretical studies predicting such an effect at *meta* sites in *uncomplexed* species have been made[278] but the range of effects is generally too small to confirm unequivocally. Complexation apparently amplifies these effects.

In related studies, Adcock and Aldous[279] have examined ^{13}C and ^{19}F SCS values in phenyl- and fluorophenyl[2.2.2]octyltricarbonylchromium(0) derivatives. Only the non-fluorinated species are considered here [70], and comparison is made with previously reported data[243] for the uncomplexed derivatives.

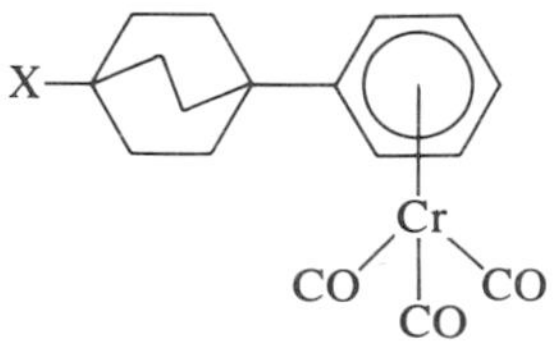

[70]

Table 23 shows DSP correlations of ^{13}C SCS values for complexed and uncomplexed derivatives. In these series it is expected that only polar effects contribute to the shifts, and hence negligible ρ_R terms are expected (and

TABLE 23

DSP correlations of ^{13}C SCS values in $Cr(CO)_3$ complexes of bridgehead substituted phenylbicyclo[2.2.2]octanes.[a]

	ρ_I	$\rho_R{}^b$	f	n
Uncomplexed				
C-1	−5.3	1.3	0.08	11
C-2	−0.6	0.1	0.22	11
C-3	0.4	−0.1	0.22	11
C-4	1.2	−0.05	0.11	11
Complexed				
C-1	−6.5	0.5	0.14	7
C-2	−1.3	−0.1	0.21	7
C-3	−0.3	0.1	0.72	7
C-4	0.2	0.05	0.65	7

[a] Reference 279. Correlations with σ_I and $\sigma_R{}^{BA}$.
[b] The contribution from σ_R is expected to be statistically negligible, and reflects minor extraneous influences.[279]

observed). As is pointed out,[279] the fits for C-3 and C-4 in the complexed species are unexpectedly poor. This can be explained in terms of a substituent-induced change in the conformation of the $Cr(CO)_3$ group relative to the phenyl ring.[279] Nevertheless, trends in the ρ_I values in Table 23, or trends in SCS values for axially symmetric halogen substituents, show that the polarization pattern in complexed and uncomplexed derivatives may be represented as shown in [71].[279]

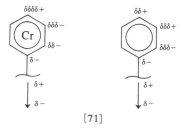

[71]

These series thus provide a means of determining the effect of complexation on π-polarization in the absence of resonance effects. The main trend on complexation is a change in sign of the effect at C-3. This corresponds to a shift in the nodal point of the polarization pattern. An attempt has been made to determine the effect of complexation on π-polarization in monosubstituted benzenes; however, the results must be regarded as somewhat tentative because of contributions from conformational changes.[279]

[72]

[13]C shifts in aryl-substituted ferrocenes [72] have been shown to be sensitive to substituent effects.[280] The C-1' shifts in this series are linearly related to those at the corresponding position in biphenyls, but are 66% larger. Shifts at C-2' and C-3' are also crudely related to shifts at a number of positions in the unsubstituted ring of biphenyls, but no correlations are obtained between C-4' shifts in phenylferrocenes and those at any position in biphenyls.[280] Correlations with Swain–Lupton substituent constants reveal a variety of sensitivities to polar and resonance effects at different sites in the complexed rings.[280]

The final discussion of organometallic series here centres on benzene derivatives with attached selenium or tellurium side-chains.[281–286] In a number of these systems, substantial non-additivity of ^{13}C shifts is observed at the site adjacent to the metallo-substituent.[281–284] In some series, conformational changes may contribute to the non-additivity;[281–283] however, the ability of Se or Te to modify the sensitivity factor (SCR) of the adjacent carbon is also important.[284] In compounds of the form X—Ph—G—CH$_3$, the effective SCR at the site *para* to X (adjacent to G) increases in the order O < S < Se.[284] Non-additivity of ^{13}C shifts is also observed at the C-1 position in phenyllead derivatives.[287,288]

K. Aliphatic compounds

In this section, most of the discussion relates to the effects of substituents on saturated aliphatic *probe sites*. Many other series have been examined in which the substituent is attached to a saturated section of a molecule, yet the probe site is unsaturated. Most notable in this category are the bridgehead substituted phenylbicyclo[2.2.2]octanes discussed earlier. In these types of compounds, SCS effects are well understood because of the importance of π-polarization of the unsaturated probe by the substituent dipole. In contrast, series in which the substituent is located at an unsaturated centre, and the probe is a saturated site, are difficult to treat. This is because of possible secondary resonance effects arising from substituent delocaliz-ation into the adjacent π system. The resultant change in the electron distribution in the unsaturated fragment of the molecule near the substituent can cause a related effect at the probe site. This section thus focuses mainly on series where both the substitution site and the probe site are saturated.

Although substituent effects in saturated systems are not as well under-stood as those in unsaturated systems, chemical shifts at saturated carbons are often markedly affected by distant substituents. Electric field effects play an important role. For example, changes in ^{13}C shifts arising from protonation of carboxylate ions or amines,[29,30] and substituent effects in cyclohexanes have been interpreted in terms of electric field effects.[289] The complexity of substituent effects at saturated carbons has been emphasized in a recent discussion of possible contributions to the *syn/anti* anisochrony in *N,N*-dialkylamides.[290] In this case, although electric field effects are found not to be the primary cause of this effect, no definitive conclusions as to other possible contributions are drawn.[290] Orientational effects of sub-stituents have also been observed in *N,N*-diisopropylamides.[291]

Recent studies of substituent effects in aliphatic compounds have been concerned mostly with carbons in a γ position with respect to the substituent. The magnitude and sign of substituent effects at these positions depend strongly on the stereochemistry of the system investigated.[292,293] γ-SCS effects in molecular systems in which the X—C-α/C-β—C-γ dihedral angle is 60–90° are referred to as γ-*gauche* effects, [73a], whereas those where the angle is in the 150–180° range are termed γ-*anti* effects, [73b].

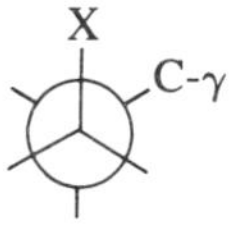

[73a]    [73b]

The former effects generally produce low frequency shifts whereas the latter shifts may be either high or low frequency. A classic example concerns the effect of a γ methyl group,[294] which causes an increase in shielding of about 5 ppm (relative to hydrogen as a substituent) on a *gauche* methylene carbon, compared to a negligible effect on a similar *anti* carbon [74].

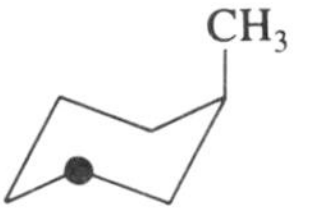

SCS = −5.4 ppm   SCS ~ 0 ppm
(a)   (b)

[74]

The γ-*gauche* effect was originally attributed to a steric (van der Waals) interaction between groups at the substituent and probe sites.[294] However, a variety of other mechanisms have now been proposed to account for this effect, including electric field effects,[295] bond angle changes,[296] or shielding properties of the hydrogen that is replaced by the γ substituent.[297,298] Lambert and Vagenas[298] have summarized much of the available evidence. In structurally well defined series, a reasonable linear stereochemical dependence of γ-SCS values is established.[298] The γ-*anti* effect has also been suggested to be dependent on electric field effects[295] or on hyperconjugative interactions of the substituent lone pair.[299] In closely related series, where the substituent is part of a ring system, electronegativity effects appear to be important.[293,300] The general conclusion that can be derived from all of these studies is that both steric and electronic effects contribute significantly.

γ-*anti* SCS values vary considerably from one substituent to another. Forrest and Webb[300] attribute this largely to substituent electronegativity. In a related study, Duddek and Islam[301] have recently examined the effects of additional fixed substituents at the α and γ carbons in a number of adamantane derivatives [75].

[75]

γ-*anti* SCS effects of hetero substituents are increased by between +2 and +5 ppm if the α hydrogen is replaced by any group or atom other than hydrogen. Thus, when Y = H, γ-*anti* SCS values of a number of X groups are as follows: OH = −1.1 ppm, OCH_3 = −1.2 ppm, F = −2.0 ppm, Cl = +0.5 ppm and Br = +1.0 ppm. OH, OCH_3 and F are thus shielding, whereas Cl and Br are slightly deshielding.[301] However, when Y = CH_3, the γ-*anti* SCS values all become deshielding, e.g. OH = +3.7 ppm and Br = +4.5 ppm.[301] A similar trend is found when the axial γ-hydrogen is replaced with CH_3, OH or F, but low frequency SCS values are seen when this hydrogen is replaced with Cl or Br groups.[301] The importance of 1,3 diaxial interactions between α and γ hydrogens has been clearly established by all of the recent studies. Removal of these interactions (e.g. by α or γ substitution) markedly changes γ-SCS effects.

Other series examined include adamantanes,[302-304] diamantanes,[302] triamantanes,[302] tricyclenes,[305] halomethanes[306] and a range of other substituted alkane derivatives.[307-310] In many series, empirical rules have been developed for making assignments.[311] The application and extension of a number of these predictive schemes have recently been discussed.[312,313]

While many of these relationships are of considerable practical utility, they do not in themselves provide an explanation of nuclear shielding. As Wiberg, Pratt and Bailey[310] have pointed out, "the successful prediction of a chemical shift neither requires nor implies an understanding of the origin of the chemical shift". Because of the lack of a uniform theoretical treatment describing ^{13}C shifts at saturated carbons there has been a recent trend to utilize a more statistical approach. Wiberg et al.[310] have, for example, recently applied factor analysis[314] to the study of substituent effects on ^{13}C chemical shifts in a large range of aliphatic halides. Since most positions examined are α, β or γ to the halogen substituent, this study strictly falls outside the scope of the present review of electronic substituent effects, but a brief discussion is given here to illustrate the method, variations of which may be useful for extension to other more distant sites.[34,221]

The fundamental difference between factor analysis and the usual regression analyses applied to SCS data is that in the latter case the number of independent variables to which the data are fitted must be defined before the analysis can be carried out. In factor analysis the initial $m \times n$ data matrix (n is the number of positions examined and m is the number of substituents), denoted $\mathbf{D}$, is transformed via a simple matrix manipulation[310,314] into the product of two new matrices,

$$\mathbf{D} = \mathbf{HM} \tag{30}$$

where $\mathbf{M}$ is a $n \times n$ matrix that is associated with the molecular positions and $\mathbf{H}$ is a $m \times n$ matrix associated with the substituents.[310] Clearly, the input data are reproduced *exactly* by multiplying $\mathbf{H}$ by $\mathbf{M}$; however, it is sometimes possible to delete some rows of $\mathbf{M}$ and columns of $\mathbf{H}$ without leading to large errors in the calculated data. The number of rows which are retained corresponds to the number of independent factors required to reproduce the experimental data within a specified error limit. Each row can in fact be assigned a relative importance (derived from eigenvalues, λ, which are obtained from the matrix manipulation noted above). For the analysis of 62 sets of SCS data for substituents (F, Cl, Br and I) it was found[310] that three independent substituent factors are required:

$$SCS_{Xi} = a_{X1}b_{1i} + a_{X2}b_{2i} + a_{X3}b_{3i} \tag{31}$$

Here SCS_{Xi} is the SCS of substituent X at site i, the a_{Xk} terms may be regarded as substituent parameters representing the halogens X = F, Cl, Br, I, while the b_{Xk} values reflect the sensitivity of substituent effects at a given molecular position. From an analysis of the derived a and b values,[310] it is concluded that halogen substituents affect ^{13}C chemical shifts in three independent ways, related to (a) the a_1 factor which is constant for all halogens, (b) the a_2 factor which increases in the order F < Cl < Br < I, and represents the availability of the lone pair electrons, and (c) the less significant a_3 factor which exerts its effect mainly at the site of direct substitution. The b parameters are interpreted in terms of transmission modes of substituent effects.[310]

From this discussion, it can be seen that the final result of a factor analysis is an equation which can be interpreted in a similar manner to more familiar correlation equations (e.g. the DSP equation). However, in factor analysis, the independent variables (i.e. effective σ constants) are not predetermined. For cases where large, uniform data sets are available, the method will probably be extremely useful; however, it is obviously inappropriate for the analysis of limited data sets.

To conclude the discussion of substituent effects in aliphatic derivatives, a number of aliphatic carbonyl series are examined. In these systems, reverse carbonyl SCS effects consistent with π-polarization are routinely observed.[315-323] Examples of typical systems are shown in structures [76] and [77].

[76]

[77]

In the 4-substituted camphors, [76], direct steric interactions between the substituent and carbonyl group are minimized, and the carbonyl SCS values correlate well with substituent σ_I values.[322] The shifts are moderately large, e.g. $SCS(NO_2) = -8.0$ ppm and $SCS(CH_3) = +0.3$ ppm.[322] In *trans* bicyclo[4.4.0]decan-3-ones [77][323] the substituent and probe group are closer, and may interact sterically. Here larger SCS values are observed. For example, an equatorial NO_2 group causes a low frequency shift of 10.63 ppm while an equatorial methyl causes a high frequency shift of 1.22 ppm. In this series, the axial/equational arrangement of a given substituent is found to have a significant effect on carbonyl chemical shifts.[323] Carbonyl-substituent non-bonded interactions have been used to account in part for the observed trends.[323]

IX. ^{15}N CHEMICAL SHIFTS

The ^{15}N nucleus is one of the most difficult common nuclei to study via NMR; however, the tremendous importance of nitrogen sites in chemical and biological molecules has prompted NMR spectroscopists to develop improved instrumentation and methods optimized for the study of this nucleus. ^{15}N chemical shifts are an extremely sensitive monitor of the electronic environment of the nitrogen atom.[324–326] This is due, in part, to the presence of the lone pair of electrons on this atom, and series in which substituents are able to interact with the lone pair generally have larger ^{15}N SCS values. For example, the range of shifts in *para*-substituted anilines is of the order of 20 ppm.[327] Factors which reduce the ability of the lone pair to interact with substituents reduce the range of the ^{15}N shifts, as can be seen from a comparison of ^{15}N SCS data for *para*-substituted anilines and anilinium ions.[328] In anilines, donation of electrons from the NH_2 lone pair to the ring is enhanced by *para* acceptor substituents, as shown in [78].

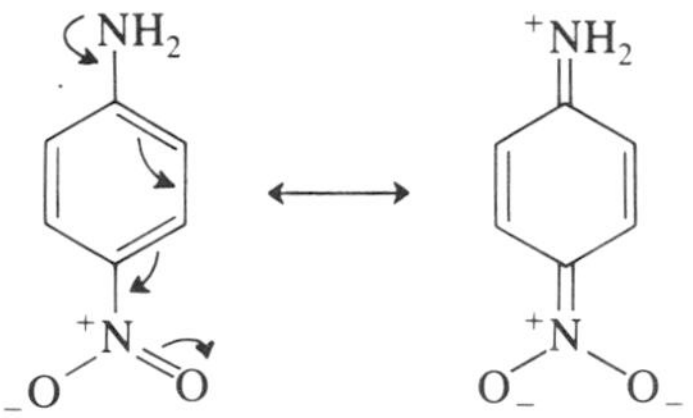

[78]

Here, the *para* nitro substituent induces a high frequency shift of 14.9 ppm for the amine resonance. In anilinium ions, interactions such as shown in [78] are not possible, and a *para* nitro group brings about a high frequency shift of only 1.1 ppm.[328]

^{15}N SCS values of side-chain $N(CH_3)_2$ probe groups are very sensitive to the electronic nature of *para* substituents[329] (and to solvent effects, as are most ^{15}N chemical shifts). A reasonable linear relationship has been established between ^{15}N shifts in this series and those in anilines:[329]

$$SCS(X—Ph—N(CH_3)_2) = 0.98 \times SCS(X—Ph—NH_2) - 20.6 \quad (32)$$

In these studies, Martin and coworkers[329] have confirmed that ^{15}N spectroscopy is a useful means of studying electron delocalization in N—C bonds. Earlier, Jones and Wilkins[213] had shown that rotational barriers of *para*-substituted N,N-dimethylbenzamides[330] correlate well with ^{15}N SCS values of corresponding benzamides.[331]

Series having the general structure [79] can be used to illustrate the importance of lone pair availability in determining ^{15}N SCS values.[247,332,333]

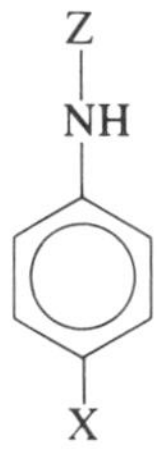

[79]

DSP analyses[8] of the ^{15}N SCS values[332] show that ρ_I and ρ_R values decrease in the following order for Z: $H > CH=C(Ph)CH_3 > SO_2Ph > COCH_3$. That the largest shifts occur in anilines (Z = H) presumably reflects the greater ability of the X substituent to interact with the nitrogen lone pair in this series. In the other series, the nitrogen lone pair is delocalized to some extent with the Z group as shown in [80].

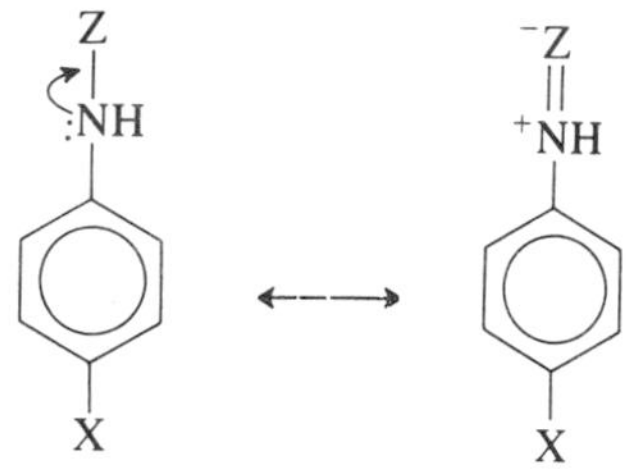

[80]

An increased π electron-withdrawing ability of the Z group causes increasing delocalization of the nitrogen lone pair, which means that its interactions with *para* X substituents are reduced, and this is reflected in reduced ^{15}N SCS values.

In related series [81], substituent effects (of X) on ^{15}N chemical shifts have been measured for the cases when $Z = c\text{-}C_6H_{12}$, Ph, OH and NHPh.[247]

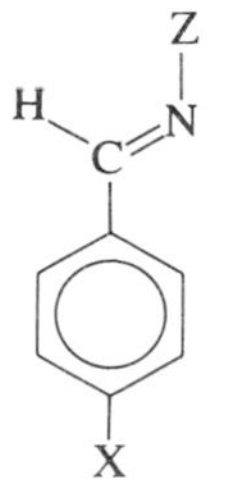

[81]

When Z = OH, the range of shifts is ~ 25 ppm as X is varied from NMe$_2$ to NO$_2$. A plot of the ^{15}N chemical shifts versus the corresponding ^{13}C shifts for C-β in styrenes has a slope of 2.3, indicating that the ^{15}N shifts are more than twice as sensitive to substituent effects than are the ^{13}C shifts.[247] When Z = cyclohexane, protonation of the nitrogen has very little effect on the transmission of substituent effects of the *para* group X to the nitrogen.[334] This is because the nitrogen lone pair is not involved in interactions with the substituent, and hence its removal by protonation does not significantly influence existing transmission modes in the unprotonated series (although the absolute ^{15}N shifts are changed considerably).

There have been relatively few reports pertaining to substituent effects on ^{15}N chemical shifts of nitro groups. This reflects the especially low sensitivity of the NO_2 group to ^{15}N NMR detection. However, SCS data from a number of studies suggest that the N—O bonds of nitro groups undergo π-polarization. In early studies of 4-X-nitrobenzenes[335] and in a more recent study of 8-X-1-nitronaphthalenes,[336] the SCS values can be seen to follow a reverse direction (i.e. donors induce high frequency shifts). A very recent investigation of ^{15}N and ^{17}O shifts in *meta-* and *para-* substituted nitrobenzenes, measured at low concentration, has confirmed that this is the case and has demonstrated that the shifts can be treated quantitatively.[337] Table 24 shows that the ^{15}N SCS values are relatively

TABLE 24

^{15}N and ^{17}O substituent chemical shifts[a] (ppm) in *meta-* and *para-*substituted nitrobenzenes.[337]

	Para		Meta	
X	^{15}N SCS[b]	^{17}O SCS[c]	^{15}N SCS[b]	^{17}O SCS[c]
NH_2	−0.4	−25.6	0.8	−1.5
OCH_3	−1.0	−9.8		
OH		−6.7	0.1	−0.2
F	−2.6	−1.3	−3.0	2.9
Cl	−2.5	0.6		
Br	−2.1		−3.3	4.0
CH_3	−0.1	−3.3	0.3	−0.4
H	0.0	0.0	0.0	0.0
CHO	−1.9	8.1	−2.3	1.9
CF_3	−3.1	7.7	−3.8	2.7
CN	−3.5	7.5	−4.2	3.5
NO_2	−4.0	11.5	−5.0	5.0

[a] Positive shifts are to high frequency.
[b] Error ±0.1 ppm.
[c] Error ±2 ppm.

small, however; the following equations indicate the electronic nature of the shifts:

$$^{15}\text{N SCS}\,(para) = -6.0\sigma_I - 1.0\sigma_{R^0} \qquad f = 0.09 \qquad (33)$$

$$^{15}\text{N SCS}\,(meta) = -7.5\sigma_I - 2.5\sigma_{R^0} \qquad f = 0.14 \qquad (34)$$

These DSP correlations show that the shifts are dominated by polar substituent effects. As is observed for ^{13}C shifts in aromatic carbonyl series, ρ_I increases as the substituent is moved from the *para* to the *meta* position. *Ab initio* MO calculations[337] show that π electron densities correctly reproduce the reverse ^{15}N SCS trend; however, the correlation between Δq_π and

[15]N SCS values in the *para* series is somewhat scattered. This is largely because the nitrogen is at a non-conjugating site, and the ranges of shifts and electron densities are rather small. Figure 8 shows this correlation, as well as the related plot for [17]O SCS values.

A number of other side-chain series can be used to demonstrate the sensitivity of ^{15}N chemical shifts to substituent effects. SCS values for a range of *para*-substituted benzenediazonium [82] salts are given in Table 25.[338]

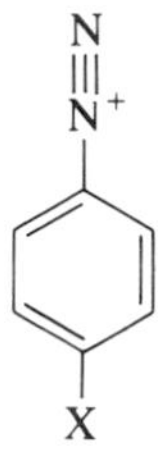

[82] 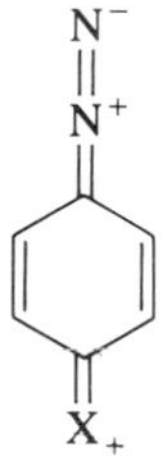   [83]

TABLE 25

[15]N SCS values (ppm) in *para*-substituted benzenediazonium tetrafluoroborates.[a]

X	N-1	N-2
O$^-$	33.1	47.6
OCH$_3$	1.7	4.0
CH$_3$	0.8	0.3
H	0.0	0.0
NO$_2$	−2.0	0.1

[a] Reference 338. Positive shifts are to high frequency. N-1 is the nitrogen closest to the benzene ring.

Donor substituents increase the diazo character, [83], of the system and deshield both nitrogens.[338,339] On the other hand, withdrawing substituents such as NO$_2$ have a relatively small effect, reflecting the difficulty of interaction between acceptor substituents and the powerful electron-withdrawing diazonium group. In contrast to this series, ^{15}N SCS values in *para*-substituted benzonitriles follow a normal direction.[331] Donors induce low frequency shifts (e.g. −2.7 ppm for *p*-OCH$_3$) and acceptors high frequency ones (e.g. +8.8 ppm for *p*-NO$_2$). These trends are consistent with expected resonance interactions between the substituent and the terminal nitrogen atom.

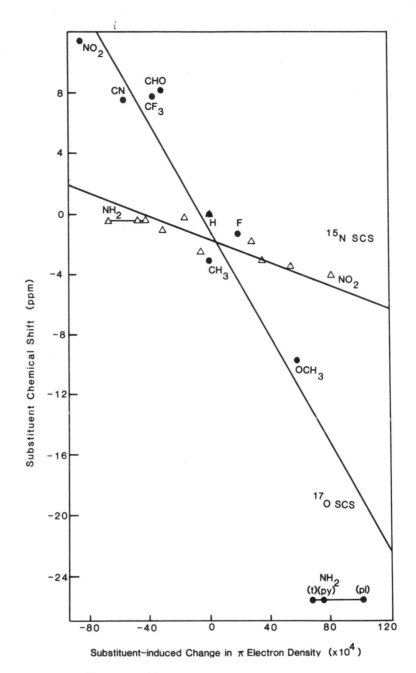

FIG. 8. Plot of ^{15}N ($\triangle$) and ^{17}O ($\bullet$) substituent chemical shifts (ppm) versus substituent-induced changes in electron density ($\Delta q_\pi \times 10^4$) for *para*-substituted nitrobenzenes. A positive SCS denotes a high frequency shift and a positive Δq_π indicates an increase in electron density. Electron densities for three conformations of the NH_2 substituent were calculated corresponding to a planar (pl), pyramidal (py) and tetrahedral (t) arrangement of bonds about the nitrogen atom.

Substituents exert significant effects on ^{15}N shifts in pyridines, as can be seen from a recent tabulation and discussion of these effects.[341] As expected, conjugative interactions of 4-substituents play an important role, with donor substituents producing low frequency shifts, and acceptors high frequency ones. Gust and Neal[342] have also recently shown that ^{15}N SCS values of the porphyrin nitrogen in complexes formed by zinc *meso*-tetraphenylporphyrin and substituted pyridines correlate well with the electronic nature of the substituted pyridines.

[84]   [85]

Quinolines [84][328] and indoles [85][343] are two other common nitrogen heterocycles in which ^{15}N SCS values have been reported. In both series, the nitrogen atom is at a similar distance from the substituent as it is in anilines. However, in series [84] the lone pair is confined to the plane of the ring and is hence unable to interact electronically with the substituent. This results in a reduced range of ^{15}N SCS values. For example, ^{15}N SCS values for CH_3 and NO_2 groups are -1.3 ppm and $+2.2$ ppm respectively. By contrast, in 5-substituted indoles, [85], the substituent interacts more strongly with the nitrogen lone pair. Here, SCS values for CH_3 and NO_2 are -0.7 ppm and $+6.9$ ppm. Although Roberts and coworkers[343] report that the shifts do not correlate well with known substituent parameters, the DSP correlation shown below confirms the systematic nature of the shifts:

$$^{15}N\ SCS = 8.2\sigma_I + 8.4\sigma_{R^0} \qquad f = 0.12 \qquad (35)$$

X. ^{17}O CHEMICAL SHIFTS

^{17}O NMR has found extensive application in inorganic chemistry, largely in structural and mechanistic studies. More recently, there has been an increasing number of reports[343-348] dealing with relationships between ^{17}O and ^{13}C chemical shifts in organic molecules. The general parallel observed suggests that ^{17}O chemical shifts will play a more important role in future studies of electronic structure. In their pioneering work, Christ and coworkers[349] demonstrated the large chemical shift range of the ^{17}O nucleus in organic molecules, and showed that different functional groups have ^{17}O resonances falling in characteristic regions. The broad spectral lines associated with this nucleus are thus to some extent offset by a large chemical shift range.

Systematic studies of *substituent* effects on ^{17}O chemical shifts in organic molecules are just beginning to appear. In 1969, Sardella and Stothers[350] examined substituent effects on ^{17}O shifts in enriched acetophenones and demonstrated the electronic origin of these effects. This work has been followed more recently by that of Fiat and coworkers,[351] who extended the range of substituents studied. Shifts in *para*-substituted acetophenones are spread over 60 ppm while those in the *meta* series are restricted to a 10 ppm range. Similar trends are noted in substituted benzaldehydes.[351] The different ranges of the *meta* and *para* shifts are of course typical of observations in other chemical shift studies, and reflect the dominance of resonance effects in *para* compared with *meta* series. The contributing structures, [86], show the importance of resonance effects in the *para* series.

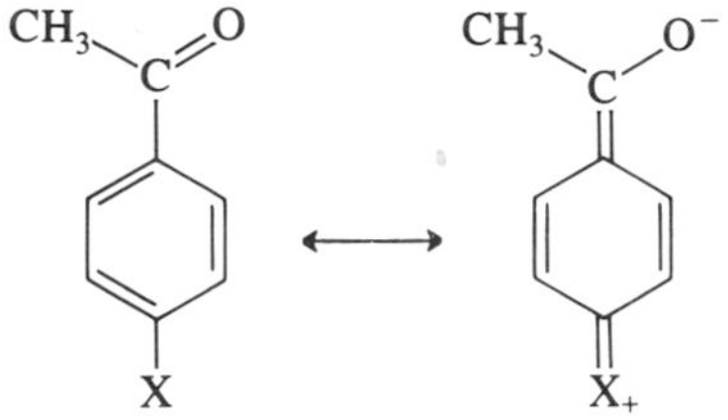

[86]

Significant polar effects are also observed, as can be seen from a DSP analysis[8] of the ^{17}O shift data:[351]

$$^{17}O \text{ SCS}(p\text{-acetophenones}) = 18.6\sigma_I + 24.0\sigma_{R^+} \qquad f = 0.14 \qquad (36)$$

The polar component of this equation monitors the effect of π-polarization of the $C{=}O$ bond. This provides complementary support for similar observations in ^{13}C SCS data discussed earlier.

A recent report has shown that ^{17}O SCS values in substituted nitrobenzenes (Table 24) are similar to those in acetophenones. This can also be seen from a comparison of equation (36) and equation (37)[337]

$$^{17}O \text{ SCS}(p\text{-nitrobenzenes}) = 13.5\sigma_I + 15.6\sigma_{R^+} \qquad f = 0.15 \qquad (37)$$

The equivalent equation for *meta*-substituted nitrobenzenes[337] is

$$^{17}O \text{ SCS}(m\text{-nitrobenzenes}) = 7.1\sigma_I + 1.4\sigma_{R^+} \qquad f = 0.23 \qquad (38)$$

and shows that resonance effects are negligible. Polar effects decrease by a factor of two in going from the *para* to the *meta* series, in line with related trends for ^{13}C shifts at the C-β carbon in substituted styrenes.[33,34]

Figure 8 shows that there is a moderately good correlation between ^{17}O SCS values and calculated π electron densities in *para*-substituted nitrobenzenes. The slope is 1600 ± 300 ppm per electron ($r = 0.962$).[337]

Substituent effects on ^{17}O chemical shifts in anisoles[352] are also very large and systematic trends are readily discerned. A DSP analysis of the ^{17}O SCS values shows a dominant resonance contribution in the *para* series:

$$^{17}O \ SCS(p\text{-anisoles}) = 11.2\sigma_I + 23.8\sigma_{R^-} \qquad f = 0.22 \qquad (39)$$

Unfortunately the correlation for the *meta* series is not very well defined due to the limited set of substituents studied:

$$^{17}O \ SCS(m\text{-anisoles}) = 15.3\sigma_I + 9.3\sigma_{R^-} \qquad f = 0.33 \qquad (40)$$

However, as expected, resonance effects are clearly decreased, compared with the *para* series. Interestingly, polar effects increase in going from the *para* to the *meta* series. This is consistent with ^{13}C data for C-α in carbonyl series (see, for example, Table 18). There is thus a general pattern that a change of substitution from the *para* to the *meta* position brings about an increase in ρ_I at the first atom of a side-chain, but a decrease at the second atom.

Changes in ^{17}O chemical shifts in the anisole series may be explained on the basis of increased electron donation from the oxygen lone pair to the ring with the introduction of electron-withdrawing substituents. The fact that good correlations are obtained between ^{17}O SCS values and oxygen π electron densities (slope = 2040 ppm per electron, $r = 0.993$)[352] supports this view.

XI. ^{19}F CHEMICAL SHIFTS

Fluorine was one of the earliest nuclei studied by NMR, reflecting its high natural abundance and sensitivity to NMR detection. Along with ^{13}C, this nucleus has provided chemists with the most useful information relating to the transmission of substituent effects. Since many of the developments and controversies[166,353] concerning the use of ^{19}F SCS values as electronic probes have been reviewed recently,[23,63] a selective, rather than exhaustive coverage is given here.

Early studies of ^{19}F SCS values in *meta*- and *para*-substituted fluoroben-zenes clearly established the importance of the fluorine probe.[4] Larger observed SCS values in *para* compared with *meta* derivatives suggested an electronic origin for the shifts. Initial semi-empirical,[4,109] and later *ab initio*[63] MO calculations confirmed the electronic nature of ^{19}F chemical shift perturbations in these series. For *para*-substituted fluorobenzenes, the relationship between ^{19}F SCS values and *ab initio* π electron densities has a slope of 2100 ppm per electron.[63]

Benzoyl fluorides [87] are another useful series in which ^{19}F shifts have been studied.[354]

[87]

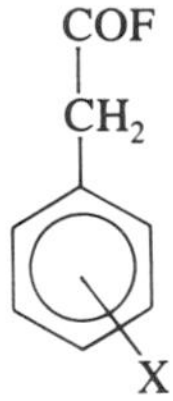

[88]

π electron densities on the side-chain fluorine atom in this series accurately reproduce the trends in ^{19}F SCS values. This relationship has a slope of 5000 ppm per electron.[354] ^{19}F SCS values become successively smaller as the COF probe group is moved further from the substituent, as has been demonstrated in a study of *para*-substituted phenylacetyl fluorides [88].[355] The DSP analyses in Table 26 show that both polar and resonance effects are decreased by the imposition of the bridging CH_2 group. In the latter series, the shifts cover a range of less than 2 ppm. However, in series where there is a strong conjugative pathway between the substituent and probe, moderately large ^{19}F SCS values are observed, even at distant sites. This can be seen in 4-substituted β-difluorostyrenes[195,356] [89].

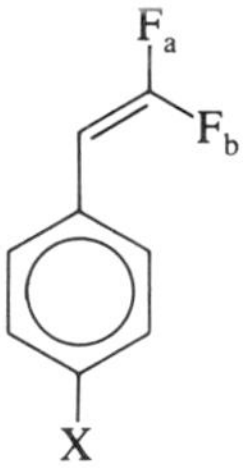

[89]

TABLE 26

DSP analyses of ^{19}F SCS values in series [87]–[89].[a]

Side-chain	ρ_I	ρ_R	Reference
CO*F*	3.3	6.0	354
CH$_2$CO*F*	1.0	2.3	355
CH=C*F*$_a$F$_b$	5.8	8.2	195
CH=CF$_a$*F*$_b$	4.9	7.9	195

[a] All correlations use the σ_R^0 resonance scale.

^{19}F SCS values for both fluorines in this series are similar in sign, with values for F_a being slightly larger in magnitude. For example, an NO_2 group induces an SCS of 5.07 ppm for F_a and 4.47 for F_b. DSP analyses[195] (Table 26) of the SCS data for all substituents studied reveal that this trend reflects differences mainly in the polar component of the shifts of the two fluorines. The internal shift difference, $\delta F_a - \delta F_b$, has therefore been used as a probe of the direct electrostatic field contribution to C—F bond polarization.[195] Resonance effects, and π-polarization effects which perturb the adjacent carbon π system and are presumed to bring about complementary changes in fluorine electron density,[195] are expected to be equal for both fluorines, and hence are cancelled out in the subtraction process. The term $\delta F_a - \delta F_b$ thus reflects the differential field component arising from the different orientations of the two C—F bonds with respect to the substituent. This is analogous to the procedure applied for the analysis of ^{1}H chemical shifts in styrenes[33] described earlier.

^{19}F shifts for both fluorines in series [89] correlate well with *ab initio* π electron densities, yielding a single correlation line, with a slope of 1500 ppm per electron.[195]

Adcock and coworkers have also made extensive investigations of substituent effects on ^{19}F chemical shifts. Series [90] provides an ideal framework in which to examine the transmission of polar substituent effects.[243] Studies using ^{13}C probes in the non-fluorinated analogue [44] were described previously.

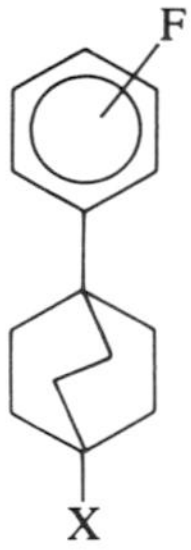

[90]

Resonance effects are absent in this series and good correlations are obtained[243] between ^{19}F SCS values and the polar substituent parameter σ_I. Relationships between ^{19}F and ^{13}C SCS values and calculated π electron densities have been discussed.[243,244] Attempts have also been made to determine the relative proportion of the ^{19}F SCS values arising indirectly from π-polarization of the adjacent benzene ring and from direct C—F bond polarization.[243]

In more recent studies,[357] the effects of shorter range substituent-induced perturbations have been examined in series [91].

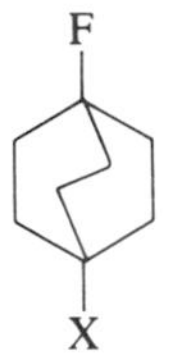

F

X

[91]

F

X

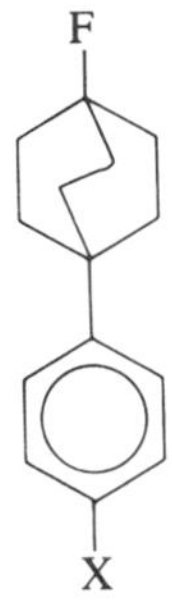

[92]

Here, polar substituents induce significant low frequency shifts (8.39 ppm for NO_2 and 6.40 for OCH_3) of the ^{19}F resonance. A variety of statistical correlations with σ_I, σ_{R^o} and the electronegativity parameter of Inamoto,[130] ι, were attempted and a dual parameter treatment incorporating σ_I and ι was found to yield the most significant fit.[357] Since resonance effects are apparently not transmitted through the bicyclo[2.2.2]octyl skeleton of [91], it is suggested that the ^{19}F shifts in this series are determined by perturbations of fluorine σ orbitals.[357]

In contrast, the ^{19}F shifts in series [92] show a definite dependence on resonance effects of X.[357,358] This can be seen from the following DSP analysis:[357]

$$^{19}F \; SCS = -1.0\sigma_I - 0.6\sigma_{R^o} \qquad f = 0.12 \qquad (41)$$

The resonance component probably arises from secondary effects of the type discussed earlier.

The most interesting feature of equation (41) is the negative sign of the ρ_I and ρ_R values, indicative of a reverse SCS effect, despite the favourable alignment of the C—F dipole with the C—X dipole and the expectation that polar substituents should induce a high rather than a low frequency shift. Similar comments apply to the anomalous SCS data for series [91],[357] and yet it is noted that the same relative alignment of C—X and C—F dipoles in substituted p-fluorophenylbicyclo[2.2.2]octanes [90] is associated with a normal direction of SCS effects.[243] This difference appears to be related to the fact that the fluorine probe group is bonded to an sp^2 centre in series [90], but to an sp^3 centre in [91] and [92]. Previous observations of large reverse ^{19}F SCS effects in substituted benzyl fluorides [93][359] fit this pattern; however, these shifts also reflect an appreciable contribution from conformational changes.

This has been confirmed from variable temperature (VT) ^{19}F studies[360] and from studies of conformationally rigid model systems [94].[361,362] It has been shown[360] that in the benzyl fluoride series, para donor substituents

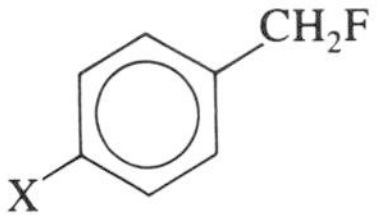

[93]

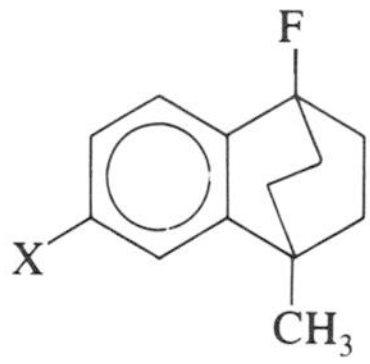

[94]

bring about increased populations of conformers having the C—F bond out of the plane of the benzene ring, whereas *para* acceptors tend to favour in-plane conformations of the C—F bond. Since out-of-plane conformations of the C—F bond are apparently associated with high frequency shifts, the anomalous SCS effects arise more from a substituent-induced conformational change than from direct electronic interactions. In support of this view is the fact that ^{19}F SCS data for model system [94], which is structurally related to series [93], except that the C—F bond is constrained to the nodal plane of the benzene ring, are extremely small. This shows that substituent effects for a fixed in-plane conformation of the C—F bond cannot alone be responsible for the large reverse SCS effect in *para*-substituted benzyl fluorides [93]. It is interesting to note that reverse effects are also observed in benzal fluorides and benzotrifluorides, although the range of ^{19}F SCS values decreases with increasing fluorine substitution on the probe group.[363]

Model systems similar to series [94] have also been utilized in a number of other studies.[165,364] For example, series [17] previously described is formally analogous, except that the variable substituent X is located on the saturated ring instead of the benzene ring. The 6- and 7-fluoro derivatives of this series have been used to probe electronic effects of substituted alkyl groups.[165] This provides another example of the use of ^{19}F NMR to investigate electronic properties of substituent groups. More specifically, the use of ^{19}F SCS values to derive substituent constants[40] has been a very important practical tool, and applications continue.[161–164,365–367]

Another continuing interest has been the use of ^{19}F NMR to determine the effects of the intermediate molecular framework on the transmission of substituent effects. Resonance effects are clearly expected to be sensitive to the nature of the framework between substituent and probe, but the sensitivity of polar effects to the intermediate cavity has been the subject of considerable controversy.[353] In an attempt to examine these effects, Taft and coworkers[368,369] made extensive investigations of series [95].

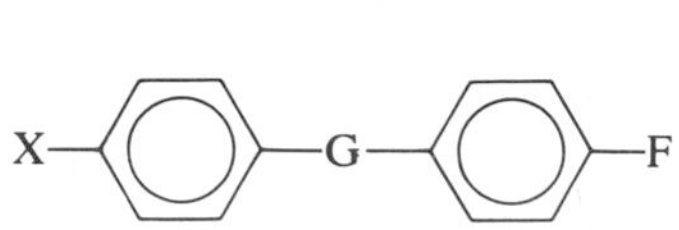

[95]

Here the group G is varied to encompass a range of electronic properties, and the DSP method is used to dissect out the polar component of the observed ^{19}F SCS values. These studies demonstrate that the electronic nature of the intervening π system strongly influences the transmission of polar substituent effects on ^{19}F chemical shifts. This presumably reflects the sensitivity of ^{19}F shifts to π-polarization of the adjacent π framework.[353] Similar methods have recently been applied to study the transmission of ^{19}F SCS effects through CH, Si, Sn, Bi, N and P bridging groups in triarylphosphines and related derivatives.[370]

XII. [31]P CHEMICAL SHIFTS

Despite the high sensitivity of this nucleus, relatively few substituent effect studies have been reported. A number of isolated examples are discussed below, with the first being related to ^{31}P shifts in phenylphosphonic acids [96]. Observed reverse SCS effects in this series,[371] and in related phenylphosphonic acid difluorides,[372] are clearly analogous to those seen for ^{13}C shifts in side-chain carbonyl derivatives.[35,373]

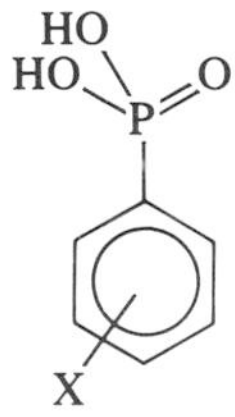

[96] 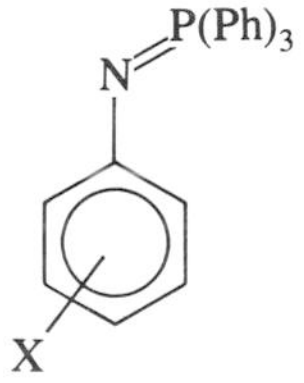   [97] 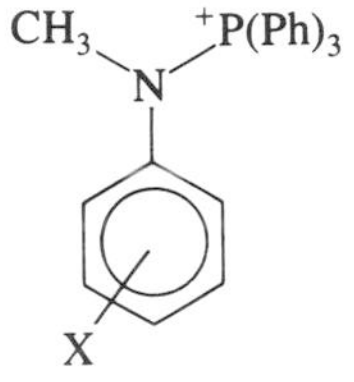   [98]

Other more recent reports have been concerned with substituent effects on ^{31}P shifts in N=P bonds[374–376] and in —PH$_2$ groups.[377] [31]P SCS values in series [97] have been reported by two groups[374,375] and are found to follow a normal direction, with a range of just over 5 ppm. DSP analyses for *para* and *meta* derivatives show that the shifts are sensitive to both polar and resonance effects:

$$^{31}\text{P SCS}(para) = 3.5\sigma_I + 4.4\sigma_{R^-} \qquad f = 0.12 \qquad (42)$$

$$^{31}\text{P SCS}(meta) = 2.3\sigma_I + 2.1\sigma_{R^o} \qquad f = 0.23 \qquad (43)$$

The ρ_I values reflect contributions from π-polarization of the N=P bond, and, as for ^{13}C shifts at the C-β position in substituted styrenes,[33] the effect is smaller in the *meta* than in the *para* series. The ρ_R value in the *para* series arises from the contribution of resonance forms shown in [99].

Brown and coworkers[375] have shown that substituents have an almost negligible effect on ^{31}P chemical shifts in the closely related series [98].

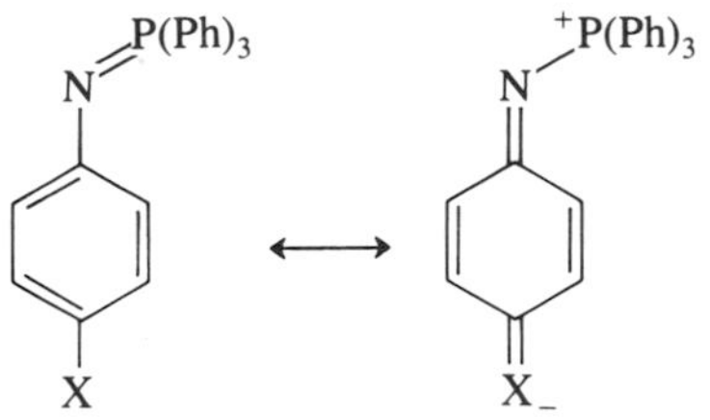

[99]

This can be attributed to two factors: (a) the double-bond character of the N—P bond has been substantially reduced compared with series [97] and hence π-polarization is no longer a significant transmission mechanism; (b) resonance interactions of the substituent with the phosphorus atom in series [98] are limited because such effects (for acceptor substituents) would lead to an unfavourable interaction between adjacent positive charges. Both effects can be seen from a consideration of the canonical forms shown in [100].

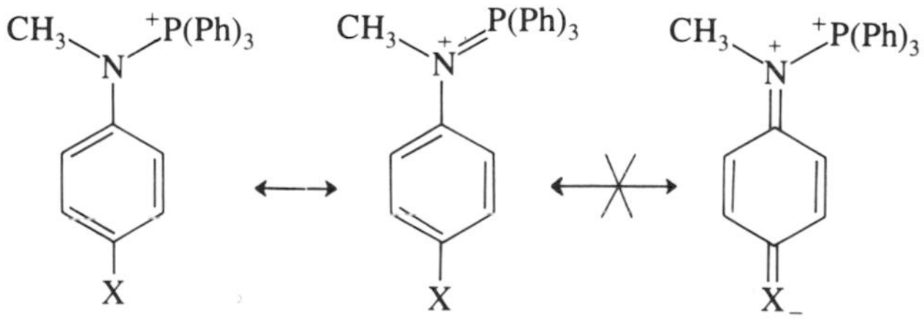

[100]

XIII. ^{33}S CHEMICAL SHIFTS

^{33}S NMR has recently assumed a new interest because of the possibility of using this technique to obtain structural information from compounds in petroleum products. Since ^{33}S is a quadrupolar nucleus, its NMR linewidths depend on the electric field gradient at the nucleus and only compounds having a high symmetry yield relatively sharp ^{33}S lines. Because of the tetrahedral atomic arrangement about the sulphur, sulphones represent one such class of compounds. Faure et al.[378] have recently investigated a series of substituted sulphones and sulphonic acids and preliminary estimates of ^{33}S SCS values can be derived from this data.

In series [101] the substituents CH_3, OH and NH_2 cause low frequency shifts of 5, 6 and 9 ppm respectively, relative to the parent derivative (X = H).

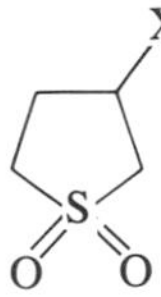

[101]

XIV. ^{77}Se CHEMICAL SHIFTS

Selenium-77 NMR is another technique that is becoming increasingly popular. Baiwir and coworkers have reported ^{77}Se chemical shifts for a large number of mono- and disubstituted benzo(b)selenophenenes [102].[379] For 2- and 3-substituted derivatives, the shifts correspond closely with those in the corresponding selenophenes [103].[380]

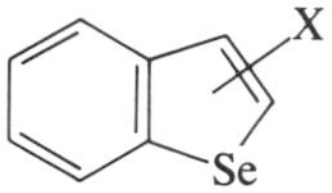

[102] 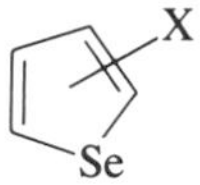　　　　　　　　　　　　[103]

A good correlation is also obtained between ^{77}Se shifts in 2-substituted benzo(b)selenophenes and ^{125}Te shifts of the corresponding benzo(b)tellurophenes.[381] The slope of this relationship indicates that the ^{125}Te shifts are more than twice as sensitive to substituent effects as are ^{77}Se shifts. A similar trend holds true in substituted selenol and tellurol esters.[382] In general, both nuclei are also significantly more sensitive to substituent effects than are ^{13}C shifts in related series.

Analyses of ^{77}Se shifts in terms of calculated electron densities have been moderately successful in some series,[383] but much work remains to be done.

Several series containing side-chain selenium groups have been studied and large SCS ranges are observed. These include aryl- and benzylseleninic acids,[384] 4,4′-disubstituted diphenylselenides,[286] and a range of substituted selenoanisoles X—Ph—SeOH.[385] Shifts in the latter series are particularly large, and cover a range of about 50 ppm as X is varied from p-N(CH$_3$)$_2$ to p-NO$_2$. This is much larger than is seen for ^{13}C chemical shifts in any side-chain system. The direction of the shifts is normal, and DSP analyses reveal substantial contributions from both polar and resonance effects for the *para* series:

$$^{77}\text{Se SCS}(p\text{-selenoanisoles}) = 32\sigma_I + 52\sigma_R^0 \qquad f = 0.17 \qquad (44)$$

XV. ^{95}Mo CHEMICAL SHIFTS

Studies of ^{95}Mo are relatively rare, although there has been one recent report on the effects of substituents on ^{95}Mo chemical shifts in cyclopentadienyl molybdenum tricarbonyl benzyl compounds [104].[386]

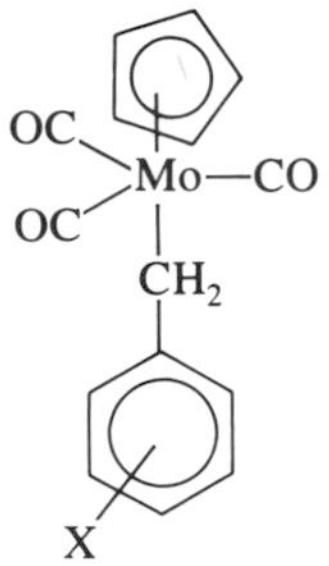

[104]

The large sensitivity of the ^{95}Mo chemical shifts to both polar and resonance effects can be seen from the following DSP analyses for *para* and *meta* derivatives of series [104]:[386]

$$^{95}\text{Mo SCS}(p) = 45\sigma_I + 49\sigma_{R^-} \qquad f = 0.08 \qquad (45)$$

$$^{95}\text{Mo SCS}(m) = 45\sigma_I + 17\sigma_{R^-} \qquad f = 0.10 \qquad (46)$$

XVI. ^{199}Hg CHEMICAL SHIFTS

Mercury is another nucleus that has become readily accessible to the NMR spectroscopist in recent years. ^{199}Hg chemical shifts are extremely sensitive to the electronic environment of this nucleus and hence offer great potential as an electronic probe.[387,388] However, this high sensitivity means that care must be taken to ensure that uniform conditions are used for the measurement of a series of spectra, since the shifts are highly solvent dependent.[389] Wells and Hawker[387] have recently reported ^{199}Hg chemical shifts for several series containing mercury adjacent to a substituted benzene ring. The ^{199}Hg chemical shifts of series [105] change by more than 76 ppm as X is changed from CF_3 to OCH_3. Further selected data for series [105]–[107] are given in Table 27. In these series the shifts follow a reverse direction, i.e. resonance donors such as OCH_3, F, Cl and CH_3 produce high frequency shifts while the polar and resonance withdrawing CF_3 substituent produces low frequency shifts.

Overall, resonance effects appear to be more important than polar effects in determining these shifts. Note that almost exact additivity of the ^{199}Hg shifts is observed, as SCS values in [106] are twice the magnitude of those

TABLE 27

[199]Hg substituent chemical shifts (ppm) in *para*-substituted phenylmercury derivatives.[a]

| I | Series | | |
	[105][b]	[106][c]	[107][c]
OCH$_3$	52.4	81.1	40.8
F	22.8		
Cl	13.8	−19.2	−9.9
Br	18.5		
CH$_3$	27.0	46.7	23.4
H	0.0	0.0	0.0
CF$_3$	−23.8	−88.0	−44.4

[a] Positive shifts are to high frequency. Data from reference 387.
[b] Solvent 1:1 CDCl$_3$–CH$_2$Cl$_2$.
[c] Solvent 1:1 CDCl$_3$–DMSO.

[105]

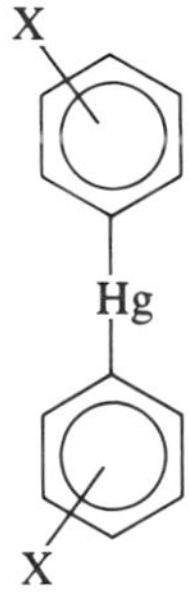

[106]

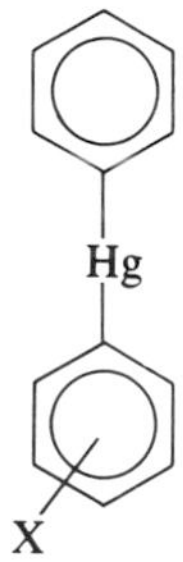

[107]

in [107]. Finally, it is noted that [199]Hg shifts in series such as those above are significantly more sensitive to substituent effects than those of other heavy metal nuclei such as [119]Sn and [207]Pb in related series.[390,391]

XVII. CONCLUSIONS

SCS values for many nuclei have now been investigated and have provided much useful information relating to the transmission of substituent effects. The ability of SCS values to prove local π electron densities has been clearly established in many systems; however, the slopes of Δq_π–SCS correlations vary markedly from one system to another. Variations from additivity of substituent effects in polysubstituted aromatics are now well established and, although further work needs to be done to understand fully the reasons for this non-additivity, and for variations in shift–charge ratios, good empirical descriptions are currently available. ^{13}C and ^{19}F remain the most useful nuclei for substituent effect studies, although many others display a remarkable sensitivity to substituent effects. This sensitivity, especially for many metal nuclei, means that uniform conditions must be adhered to in the measurement of SCS values, which can be strongly solvent dependent.

ACKNOWLEDGMENTS

Support from a CSIRO Postdoctoral Studentship during part of this work is gratefully acknowledged. I thank Dr R. T. C. Brownlee for providing material prior to publication. Helpful discussions with Drs G. K. Hamer, I. R. Peat and Professor W. F. Reynolds are also acknowledged, as is financial support from Professor G. C. Levy. The valuable assistance of Robyn Craik in the preparation of the manuscript was also much appreciated.

REFERENCES

1. J. A. Pople, W. G. Schneider and H. J. Bernstein, *High Resolution Nuclear Magnetic Resonance*, McGraw-Hill, New York, 1959, Chap. 11.
2. M. T. Tribble and J. G. Traynham, in *Advances in Linear Free Energy Relationships* (N. B. Chapman and J. Shorter, eds), Plenum, New York, 1972, Chap. 4.
3. H. S. Gutowsky, D. W. McCall, B. R. McGarvey and L. H. Meyer, *J. Amer. Chem. Soc.*, 1952, **74**, 4809.
4. (a) R. W. Taft, E. Price, I. R. Fox, I. C. Lewis, K. K. Andersen and G. T. Davis, *J. Amer. Chem. Soc.*, 1963, **85**, 3146; (b) R. W. Taft, E. Price, I. R. Fox, I. C. Lewis, K. K. Andersen and G. T. Davis, *J. Amer. Chem. Soc.*, 1963, **85**, 709; (c) R. W. Taft, F. Prosser, L. Goodman and G. T. Davis, *J. Chem. Phys.*, 1963, **38**, 380.
5. P. C. Lauterbur, *J. Chem. Phys.*, 1957, **26**, 217.
6. C. H. Holm, *J. Chem. Phys.*, 1957, **26**, 707.
7. G. E. Maciel, in *Topics in Carbon-13 NMR Spectroscopy*, Vol. 1 (G. C. Levy, ed.), Wiley-Interscience, New York, 1974, p. 53.
8. D. J. Craik and R. T. C. Brownlee, *Prog. Phys. Org. Chem.*, 1983, **14**, 1.
9. N. F. Ramsey, *Phys. Rev.*, 1951, **78**, 699.
10. A. Saika and C. P. Slichter, *J. Chem. Phys.*, 1954, **22**, 26.
11. G. J. Martin, M. L. Martin and S. Odiot, *Org. Magn. Reson.*, 1975, **7**, 2.
12. (a) M. Karplus and T. P. Das, *J. Chem. Phys.*, 1961, **34**, 1683;
    (b) M. Karplus and J. A. Pople, *J. Chem. Phys.*, 1963, **38**, 2803.

13. K. A. K. Ebraheem and G. A. Webb, *Prog. NMR Spectrosc.*, 1977, **11**, 149.
14. (a) J. S. Griffith and L. E Orgel, *Trans. Faraday Soc.*, 1957, **53**, 601; (b) R. Freeman, G. R. Murray and R. E. Richards, *Proc. Roy. Soc. A*, 1957, **242**, 455; (c) N. S. Biradar and M. A. Pujar, *Current Sci.*, 1966, **35**, 385; (d) N. S. Biradar and M. A. Pujar, *Z. Anorg. Allg. Chem.*, 1970, **379**, 88; (e) R. L. Martin and A. H. White, *Nature (Lond.)*, 1969, **223**, 394; (f) S. Fujiwara, F. Yajima and A. Yamasaki, *J. Magn. Reson.*, 1969, **1**, 203.
15. M. Witanowski, L. Stefaniak, B. Na Lamphun and G. A. Webb, *Org. Magn. Reson.*, 1981, **16**, 57.
16. M. Witanowski, L. Stefaniak and G. A. Webb, in *Annual Reports on NMR Spectroscopy*, Vol. 7 (G. A. Webb, ed.), Academic Press, London, 1977, p. 117.
17. M. Witanowski, L. Stefaniak and G. A. Webb, in *Anual Reports on NMR Spectroscopy*, Vol. 11B (G. A. Webb, ed.), Academic Press, London, 1981.
18. (a) B. Tiffon and J. P. Doucet, *Canad. J. Chem.*, 1976, **54**, 2045; (b) B. Tiffon and J. E. Dubois, *Org. Magn. Reson.*, 1978, **11**, 295; (c) B. Tiffon and J. E. Dubois, *Org. Magn. Reson.*, 1979, **12**, 24.
19. I. Ando, A. Nishioka and M. Kondo, *J. Magn. Reson.*, 1976, **21**, 429.
20. M. Jallali-Heravi and G. A. Webb, *Org. Magn. Reson.*, 1980, **13**, 116.
21. B. Tiffon and B. Ancian, *Org. Magn. Reson.*, 1981, **16**, 247.
22. R. D. Topsom, *Prog. Phys. Org. Chem.*, 1976, **12**, 1.
23. W. F. Reynolds, *Prog. Phys. Org. Chem.*, 1983, **14**, 165.
24. W. F. Reynolds, *J. Chem. Soc. Perkin Trans. II*, 1980, **1980**, 985.
25. M. Godfrey, in *Correlation Analysis in Chemistry: Recent Advances* (N. B. Chapman and J. Shorter, eds), Plenum, New York, 1978, chap. 3, p. 85.
26. Godfrey has recently proposed an alternative approach. For example, see reference 25 and M. Godfrey, *J. Chem. Soc. Perkin Trans. II*, 1977, **1977**, 769.
27. (a) A. R. Katritzky and R. D. Topsom, *Agnew. Chem. (Int. Ed.)*, 1970, **9**, 87; (b) A. R. Katritzky and R. D. Topsom, *J. Chem. Educ.*, 1971, **48**, 427.
28. (a) J. A. Pople and M. Gordon, *J. Amer. Chem. Soc.*, 1967, **89**, 4253; (b) N. C. Baird, M. J. S. Dewar and R. S. Sustmann, *J. Chem. Phys.*, 1969, **50**, 1275; (c) G. R. Howe, *J. Chem. Soc. B*, 1971, **1971**, 984.
29. W. J. Horsley and H. Sternlicht, *J. Amer. Chem. Soc.*, 1968, **90**, 3738.
30. J. G. Batchelor, J. Feeney and G. C. K. Roberts, *J. Magn. Reson.*, 1975, **20**, 19.
31. A. D. Buckingham, *Canad. J. Chem.*, 1960, **38**, 300.
32. W. F. Reynolds, I. R. Peat, M. H. Freedman and J. R. Lyerla, *Canad. J. Chem.*, 1973, **51**, 1857.
33. G. K. Hamer, I. R. Peat and W. F. Reynolds, *Canad. J. Chem.*, 1973, **51**, 897, 915.
34. W. F. Reynolds, P. Dais, D. W. MacIntyre, G. K. Hamer and I. R. Peat, *J. Magn. Reson.*, 1981, **43**, 81.
35. J. Bromilow, R. T. C. Brownlee, D. J. Craik, P. R. Fiske, J. E. Rowe and M. Sadek, *J. Chem. Soc. Perkin Trans. II*, 1981, **1981**, 753.
36. J. Bromilow and R. T. C. Brownlee, *Tetrahedron Lett.*, 1975, **1975**, 2113.
37. D. A. Dawson and W. F. Reynolds, *Canad. J. Chem.*, 1975, **53**, 373.
38. J. Shorter, in *Correlation Analysis in Chemistry: Recent Advances* (J. Shorter and N. B. Chapman, eds), Plenum, New York, 1978, Chap. 4.
39. D. F. Ewing, in *Correlation Analysis in Chemistry: Recent Advances* (J. Shorter and N. B. Chapman, eds), Plenum, New York, 1978, Chap. 8.
40. P. R. Wells, S. Ehrenson and R. W. Taft, *Prog. Phys. Org. Chem.*, 1968, **6**, 147.
41. S. Ehrenson, R. T. C. Brownlee and R. W. Taft, *Prog. Phys. Org. Chem.*, 1973, **10**, 1.
42. J. Bromilow, R. T. C. Brownlee, D. J. Craik, M. Sadek and R. W. Taft, *J. Org. Chem.*, 1980, **45**, 2429.
43. J. Bromilow and R. T. C. Brownlee, *J. Org. Chem.*, 1979, **44**, 1261.

44. C. G. Swain and E. C. Lupton, *J. Amer. Chem. Soc.*, 1968, **90**, 4328.

45. Y. Yukawa and Y. Tsuno, *Bull. Chem. Soc. Japan*, 1959, **32**, 971.

46. D. J. Craik, R. T. C. Brownlee and M. Sadek, *J. Org. Chem.*, 1982, **47**, 657.

47. J. W. Blunt and D. A. R. Happer, *J. Chem. Educ.*, 1979, **56**, 56.

48. G. L. Nelson and E. A. Williams, *Prog. Phys. Org. Chem.*, 1976, **12**, 229.

49. D. W. Jones, in *Specialist Periodical Reports on NMR*, Vol. 7 (R. J. Abraham, ed.), The Chemical Society, London, 1978, Chap. 2.

50. L. M. Jackman and S. Sternhell, *Applications of Nuclear Magnetic Resonance Spectroscopy in Organic Chemistry*, 3rd edn, Pergamon, New York, 1969, Chaps 3–6.

51. D. G. Farnum, *Adv. Phys. Org. Chem.*, 1975, **11**, 123.

52. D. W. Beistel, H. E. Chen and P. J. Fryatt, *J. Amer. Chem. Soc.*, 1973, **95**, 5455.

53. (a) P. L. Corio and B. P. Dailey, *J. Amer. Chem. Soc.*, 1956, **78**, 3043; (b) P. Diehl, *Helv. Chim. Acta*, 1961, **44**, 829; (c) S. Martin and B. P. Dailey, *J. Chem. Phys.*, 1963, **39**, 1722; (d) G. W. Smith, *J. Mol. Spectrosc.*, 1964, **12**, 146; (e) J. J. R. Reed, *Anal. Chem.*, 1967, **39**, 1586.

54. (a) M. Holik, *Org. Magn. Reson.*, 1977, **9**, 491; (b) D. Bruck and M. Rabinovitz, *Org. Magn. Reson.*, 1978, **11**, 587; (c) J. M. Hachey, *Canad. J. Spectrosc.*, 1978, **23**, 147.

55. (a) K. Hayamizu and O. Yamamoto, *J. Mol. Spectroscopy*, 1968, **28**, 89; (b) *ibid.*, 1969, **29**, 183.

56. B. M. Lynch, *Org. Magn. Reson.*, 1974, **6**, 190.

57. (a) Y. Yukawa, Y. Tsuno and N. Shimizu, *Bull. Chem. Soc. Japan*, 1971, **44**, 2843; (b) H. Yamada, Y. Tsuno and Y. Yukawa, *Bull. Chem. Soc. Japan*, 1970, **43**, 1459.

58. M. Charton, *J. Org. Chem.*, 1971, **36**, 266.

59. (a) W. B. Smith, A. M. Ihrig and J. L. Roark, *J. Phys. Chem.*, 1970, **74**, 812; (b) W. B. Smith, D. L. Deavenport and A. M. Ihrig, *J. Amer. Chem. Soc.*, 1972, **94**, 1959.

60. M. Zanger, *Org. Magn. Reson.*, 1972, **4**, 1.

61. J. Beevy, S. Sternhell, T. Hoffmann-Ostenhof, E. Pretsch and W. Simon, *Anal. Chem.*, 1973, **45**, 1571.

62. (a) B. D. Batts, S. J. Pasaribu and L. R. Williams, *Org. Magn. Reson.*, 1977, **9**, 210; (b) B. D. Batts and G. Pallos, *Org. Magn. Reson.*, 1980, **13**, 349.

63. W. J. Hehre, R. W. Taft and R. D. Topsom, *Prog. Phys. Org. Chem.*, 1976, **12**, 159.

64. V. Lucchini and P. R. Wells, *Org. Magn. Reson.*, 1976, **8**, 137.

65. R. M. Letcher and K.-M. Wong, *J. Chem. Soc. Perkin Trans. II*, 1978, **1978**, 739.

66. S. Akiyama, M. Iyoda, K. Yoshitsugu, M. Fukuoka and M. Nakagawa, *Bull. Chem. Soc. Japan*, 1978, **51**, 3351.

67. (a) M. Bremond, G. J. Martin and M. Cariou, *Org. Magn. Reson.*, 1978, **11**, 433; (b) M. Mikolajczyk, S. Grzejszczak and A. Zatorski, *Tetrahedron*, 1979, **35**, 1019.

68. (a) V. A. Lopyrev, L. I. Larina, T. I. Vakul'skaya, M. F. Larin, O. B. Nefedova, E. F. Shibanova and M. G. Voronkov, *Org. Magn. Reson.*, 1981, **15**, 219; (b) E. Smakula Hand and W. W. Paudler, *Org. Magn. Reson.*, 1980, **14**, 52; (c) R. H. Contreras and D. G. de Kowalewski, *J. Mol. Struct.*, 1974, **23**, 209; (d) H. Ranganathan, D. Ramaswamy, T. Ramasami and M. Santappa, *Chem. Lett.*, 1979, **1979**, 1201; (e) M. S. A. Abd-El-Mottaleb, A. M. Kamel and M. S. Antonious, *Indian J. Chem. B*, 1978, **16**, 620; (f) E. E. Liepin'sh, A. V. Eremeev, D. A. Tikhomirov and R. S. El'kinson, *Khim. Geterots. Soedin.*, 1978, 338; (g) G. M. Sanders, M. van Dijk and A. van Veldhuizen, *J. Royal Netherlands Chem. Soc.*, 1978, **97**, 95; (h) O. P. Shkurko and V. P. Mamaev, *Khim. Geterots. Soedin.*, 1979, 1683; (i) O. P. Shkurko and V. P. Mamaev, *Khim. Geterots. Soedin.*, 1978, 526; (j) O. P. Shkurko and V. P. Mamaev, *Khim. Geterots. Soedin.*, 1978, 673; (k) N. O. Saldabol, Y. Y. Popelis and E. E. Liepin'sh, *Zh. Org. Khim.*, 1980, **16**, 1494.

69. T. Yokoyama, G. R. Wiley and S. I. Miller, *J. Org. Chem.*, 1969, **34**, 1859.

70. G. R. Wiley and S. I. Miller, *J. Org. Chem.*, 1972, **37**, 767.
71. A. B. Turner, R. E. Lutz, N. S. McFarlane and D. W. Boykin, *J. Org. Chem.*, 1971, **36**, 1107.
72. R. H. Kohler and W. F. Reynolds, *Canad. J. Chem.*, 1977, **55**, 530.
73. W. F. Reynolds, P. G. Mezey and G. K. Hamer, *Canad. J. Chem.*, 1977, **55**, 522.
74. C. Srinivasana and K. Pitchumani, *J. Magn. Reson.*, 1982, **46**, 134.
75. E. Solcaniova and S. Toma, *Org. Magn. Reson.*, 1980, **14**, 138.
76. C. D. Cook and S. S. Danyluk, *Tetrahedron*, 1963, **19**, 177.
77. R. T. C. Brownlee and D. J. Craik, unpublished data.
78. (a) Z. Bankowska and M. Jedrzejewska-Krawczyk, *Pol. J. Chem.*, 1980, **54**, 1473; (b) J. M. Hachey, *Canad. J. Spectrosc.*, 1978, **23**, 147; (c) F. Grambal, J. Lasovsky, V. Bekarek and V. Simanek, *Coll. Czech. Chem. Comm.*, 1978, **43**, 2008; (d) I. D. Sadekov, M. L. Cherkinskaya, V. L. Pavlova, V. A. Bren and V. I. Minkin, *Zh. Obshch. Khim.*, 1978, **48**, 390; (e) J. Csaszar, *Acta Phys. Chem.*, 1979, **25**, 137; (f) N. A. Barba, A. P. Gulya and K. F. Keptanaru, *Zh. Obshch. Khim.*, 1978, **48**, 1627; (g) V. Koleva, B. Galabov and D. Simov, *Org. Magn. Reson.*, 1978, **11**, 475; (h) M. Holik, M. Potacek and J. Svaricek, *Coll. Czech. Chem. Comm.*, 1978, **43**, 734; (i) M. Holik, J. Belusa and J. Brichacek, *Coll. Czech. Chem. Comm.*, 1978, **43**, 610; (j) J. Kuthan, A. Kurfurst, Z. Prosek and J. Palecek, *Coll. Czech. Chem. Comm.*, 1978, **43**, 1068; (k) J. P. Idoux, G. E. Kiefer, G. R. Baker, W. E. Puckett, F. J. Spence, Jr., K. S. Simmons, R. B. Constant, D. J. Watlock and S. L. Fuhrman, *J. Org. Chem.*, 1980, **45**, 441; (l) M. S. A. Abd-El Mottaleb and Z. H. Khalil, *J. Signalauf.*, 1980, **8**, 109; (m) R. J. Abraham and J. M. Bakke, *Org. Magn. Reson.*, 1980, **14**, 312.
79. H. O. Castaneda, R. H. Contreras and D. G. de Kowalewski, *Org. Magn. Reson.*, 1980, **13**, 308.
80. C. J. O'Connor, R. W. Martin and D. J. Calvert, *Aust. J. Chem.*, 1981, **34**, 2297.
81. D. J. Calvert and C. J. O'Connor, *Aust. J. Chem.*, 1979, **32**, 337.
82. (a) M. Rouillard, S. Geribaldi, J. Damiano-Gal and M. Azzaro, *Org. Magn. Reson.*, 1977, **10**, 5; (b) R. Escale, A. Khayat, J. P. Girard, J. C. Rossi and J. P. Chapat, *Org. Magn. Reson.*, 1981, **17**, 217; (c) M. S. Salakhov, N. F. Musaeva, S. N. Suleimanov and A. A. Bairamov, *Org. React. (Tartu)*, 1979, **16**, 75.
83. J. Gasteiger and M. Marsili, *Org. Magn. Reson.*, 1981, **15**, 353.
84. J. Gasteiger and M. Marsili, *Tetrahedron*, 1980, **36**, 3219.
85. (a) N. C. Baird and M. A. Whitehead, *Theoret. Chim. Acta*, 1966, **6**, 167; (b) H. Stahl-Lariviere, *Org. Magn. Reson.*, 1974, **6**, 170.
86. J. M. Sichel and M. A. Whitehead, *Theoret. Chim. Acta*, 1966, **5**, 35.
87. (a) R. Janoschek, *Z. Naturforsch.*, 1970, **25a**, 1716; (b) M. Sterk, W. Fabian, J. J. Suschnigg and R. Janoschek, *Org. Magn. Reson.*, 1977, **9**, 389.
88. R. K. Harris and B. E. Mann (eds), *NMR and the Periodic Table*, Academic Press, London, 1978.
89. (a) H. C. Beachell and D. W. Beistel, *Inorg. Chem.*, 1964, **3**, 1028 (b) A. R. Siedle and G. M. Bodner, *Inorg. Chem.*, 1972, **11**, 3108.
90. (a) P. M. Tucker, T. Onak and J. B. Leach, *Inorg. Chem.*, 1970, **9**, 1430; (b) C. J. Foret, K. R. Korzekwa and D. R. Martin, *J. Inorg. Nucl. Chem.*, 1980, **42**, 1223; (c) V. I. Stanko, V. V. Khrapov, T. V. Klimova, T. A. Babushkina and T. P. Klimova, *Zh. Obshch. Khim.*, 1978, **48**, 368.
91. J. T. Vandeberg, C. E. Moore and F. P. Cassaretto, *Org. Magn. Reson.*, 1973, **5**, 57.
92. (a) N. J. Fitzpatrick, and N. J. Mathews, *J. Organomet. Chem.*, 1975, **94**, 1; (b) J. Kroner and B. Wrackmeyer, *J. Chem. Soc. Faraday II*, 1976, **1976**, 2283.
93. B. Wrackmeyer, *Prog. NMR Spectrosc.*, 1979, **12**, 227.
94. G. C. Levy, R. L. Lichter and G. L. Nelson, *Carbon-13 Nuclear Magnetic Resonance Spectroscopy*, 2nd edn, Wiley–Interscience, New York, 1980.

95. J. B. Stothers, *Carbon-13 NMR Spectroscopy*, Academic, New York, 1972.
96. F. Wehrli and T. Wirthlin, *Interpretation of Carbon-13 NMR Spectra*, Heyden, London, 1976.
97. D. F. Ewing, *Org. Magn. Reson.*, 1979, **12**, 499.
98. G. L. Nelson, G. C. Levy and J. D. Cargioli, *J. Amer. Chem. Soc.*, 1972, **94**, 3089.
99. H. Spiesecke and W. G. Schneider, *J. Chem. Phys.*, 1961, **35**, 731.
100. P. C. Lauterbur, *Ann. New York Acad. Sci.*, 1958, **70**, 841. *J. Amer. Chem. Soc.*, 1961, **83**, 1846; *Tetrahedron Lett.*, 1961, 274.
101. G. E. Maciel and J. J. Natterstad, *J. Chem. Phys.*, 1965, **42**, 2427.
102. G. P. Syrova, V. F. Brystrov, V. V. Orda and L. M. Yagupol'skii, *Zh. Obshch. Khim.*, 1969, **39**, 1395.
103. P. C. Lauterbur, *J. Amer. Chem. Soc.*, 1961, **83**, 1846; *Tetrahedron Lett.*, 1961, **1961**, 274.
104. W. B. Smith and D. L. Deavonport, *J. Magn. Reson.*, 1972, **7**, 364.
105. J. E. Bloor and D. L. Breen, *J. Phys. Chem.*, 1968, **72**, 716.
106. G. A. Olah, P. W. Westerman and D. A. Forsyth, *J. Amer. Chem. Soc.*, 1975, **97**, 3419.
107. (a) A. Domenicano and A. Vacigao, *Acta Cryst. B*, 1979, **35**, 1382; (b) A. Domenicano, G. Schultz, M. Kolonits and I. Hargittai, *J. Mol. Struc.*, 1979, **53**, 197.
108. T. A. Holak, S. Sadigh-Esfandiary, F. R. Carter and D. J. Sardella, *J. Org. Chem.*, 1980, **45**, 2400.
109. R. T. C. Brownlee and R. W. Taft, *J. Amer. Chem. Soc.*, 1970, **92**, 7007.
110. J. Bromilow, PhD thesis, La Trobe University, Melbourne, Australia, 1977.
111. D. J. Craik, PhD thesis, La Trobe University, Melbourne, Australia, 1980.
112. H. M. Relles and R. W. Schluenz, *J. Org. Chem.*, 1972, **37**, 1742.
113. H. L. Retcofsky and C. E. Griffin, *Tetrahedron Lett.*, 1975, 1966.
114. B. G. Ramsey and K. Longmuir, *J. Org. Chem.*, 1980, **45**, 1322.
115. C. W. Fong, *Aust. J. Chem.*, 1980, **33**, 1291.
116. R. T. C. Brownlee and M. Sadek, *Aust. J. Chem.*, 1981, **34**, 1593.
117. D. Calvert, P. B. D. de la Mare and N. S. Isaacs, *J. Chem. Res. (S)*, 1978, **1978**, 156.
118. J. D. Odom, T. F. Moore, R. Goetze, H. Noth and B. Wrackmeyer, *J. Organomet. Chem.*, 1979, **173**, 15.
119. A. Ricci, F. Bernardi and J. H. Ridd, *Tetrahedron*, 1978, **34**, 193.
120. (a) J. Llinares, J. Elguero, R. Faure and E.-J. Vincent, *Org. Magn. Reson.*, 1980, **14**, 20; (b) J. Horyna, A. Lycka and D. Snobl, *Coll. Czech. Chem. Comm.*, 1980, **45**, 1575.
121. J. Bromilow, R. T. C. Brownlee, V. O. Lopez and R. W. Taft, *J. Org. Chem.*, 1979, **44**, 4766.
122. (a) M. J. Kamlet, J. L. Abboud and R. W. Taft, *J. Amer. Chem. Soc.*, 1977, **99**, 6027; (b) J. L. Abboud, M. L. Kamlet and R. W. Taft, *J. Amer. Chem. Soc.*, 1977, **99**, 8325.
123. J. Bromilow, R. T. C. Brownlee, R. C. Topsom and R. W. Taft, *J. Amer. Chem. Soc.*, 1976, **98**, 2020.
124. (a) R. T. C. Brownlee, S. K. Dayal, J. L. Lyle and R. W. Taft, *J. Amer. Chem. Soc.*, 1972, **94**, 7208; (b) S. K. Dayal and R. W. Taft, *J. Amer. Chem. Soc.*, 1973, **95**, 5595.
125. B. Chawla, S. K. Pollack, C. B. Lebrilla, M. J. Kamlet and R. W. Taft, *J. Amer. Chem. Soc.*, 1981, **103**, 6924.
126. B. M. Lynch, *Canad. J. Chem.*, 1977, **55**, 541.
127. H. M. Hugel, D. P. Kelly, R. J. Spear, J. Bromilow, R. T. C. Brownlee and D. J. Craik, *Aust. J. Chem.*, 1979, **32**, 1511.
128. A. J. Hoefnagel, M. A. Hoefnagel and B. M. Wepster, *J. Amer. Chem. Soc.*, 1976, **98**, 6194.
129. N. Inamoto, S. Masuda, K. Tori and Y. Yoshimura, *Tetrahedron Lett.*, 1978, **1978**, 4547.
130. N. Inamoto and S. Masuda, *Tetrahedron Lett.*, 1977, **1977**, 3287.
131. I. B. Afanas'ev and V. L. Trojanker, *Org. Magn. Reson.*, 1982, **18**, 22.
132. I. B. Afanas'ev, *Zh. Org. Khim.*, 1981, **17**, 449.

133. R. W. Taft, in *Steric Effects in Organic Chemistry* (M. S. Newman, ed.), Wiley, New York, 1956, p. 587.

134. G. A. Olah and D. A. Forsyth, *J. Amer. Chem. Soc.*, 1975, **97**, 3137.

135. (a) D. P. Kelly and R. J. Spear, *Aust. J. Chem.*, 1977, **30**, 1993; (b) D. P. Kelly and R. J. Spear, *Aust. J. Chem.*, 1978, **31**, 1209.

136. F. Membrey, B. Ancian and J.-P. Doucet, *Org. Magn. Reson.*, 1978, **11**, 580.

137. F. Membrey, L. Boutin and J.-P. Doucet, *Tetrahedron Lett.*, 1980, 823.

138. (a) B. I. Istomin, *Org. React.* (*Tartu*), 1978, **15**, 216; (b) Y. Sasaki, H. Takai and T. Tsujimoto, *Chem. Pharm. Bull.*, 1980, **28**, 667.

139. (a) R. Laatikainen, *Org. Magn. Reson.*, 1980, **14**, 366; (b) J. Knuutinen, R. Laatikainen and J. Paasivirta, *Org. Magn. Reson.*, 1980, **14**, 360; (c) M. Ilczyszyn, Z. Latajka and H. Ratajczak, *Org. Magn. Reson.*, 1980, **13**, 132.

140. M. Mishima, M. Fujio and Y. Tsuno, *Mem. Fac. Sci, Kyushu Univ.*, *Ser. C*, 1980, **12**, 219.

141. H. Nery, D. Canet, B. Azoui, L. Lalloz and P. Caubere, *Org. Magn. Reson.*, 1977, **10**, 240.

142. H. M. Relles, *J. Magn. Reson.*, 1980, **39**, 481.

143. P. E. Hansen, *Org. Magn. Reson.*, 1979, **12**, 109.

144. W. Kitching, M. Bullpitt, D. Gartshore, W. Adcock, T. C. Khor, D. Doddrell and I. D. Rae, *J. Org. Chem.*, 1977, **42**, 2411.

145. (a) M. Bullpitt, W. Kitching, D. Doddrell and W. Adcock, *J. Org. Chem.* 1976, **41**, 760; (b) W. Adcock, B. D. Gupta, T. C. Khor, D. Doddrell and W. Kitching, *J. Org. Chem.*, 1976, **41**, 751.

146. W. Adcock, B. D. Gupta and W. Kitching, *J. Org. Chem.*, 1976, **41**, 1498.

147. (a) P. R. Wells, D. P. Arnold and D. Doddrell, *J. Chem. Soc. Perkin Trans. II*, 1974, **1974**, 1745; (b) W. Kitching, M. Bullpitt, D. Doddrell and W. Adcock, *Org. Magn. Reson.*, 1974, **6**, 289.

148. B. Mechin, J. C. Richer and S. Odiot, *Org. Magn. Reson.*, 1980, **14**, 79.

149. L. Ernst, *Chem. Ber.*, 1975, **108**, 2030.

150. L. Ernst, *Z. Naturforsch. B*, 1975, **30**, 788, 794.

151. L. Ernst, *J. Magn. Reson.*, 1975, **20**, 544.

152. L. Ernst, *J. Magn. Reson.*, 1976, **22**, 279.

153. J. Seita, J. Sandstrom and T. Drakenberg, *Org. Magn. Reson.*, 1978, **11**, 239.

154. N. K. Wilson and J. B. Stothers, *J. Magn. Reson.*, 1974, **15**, 31.

155. H. Takai, A. Odani and Y. Sasaki, *Chem. Pharm. Bull.*, 1978, **26**, 1966.

156. W. Adcock, J. Alste, S. Q. A. Rizvi and M. Aurangzeb, *J. Amer. Chem. Soc.*, 1976, **98**, 1701.

157. W. Kitching, W. Adcock and G. Aldous, *J. Org. Chem.*, 1979, **44**, 2652.

158. G. A. Olah, P. W. Westerman and J. Nishimura, *J. Amer. Chem. Soc.*, 1974, **96**, 3548.

159. G. A. Olah and R. J. Spear, *J. Amer. Chem. Soc.*, 1975, **97**, 1539.

160. G. A. Olah and P. W. Westerman, *J. Amer. Chem. Soc.*, 1973, **95**, 7530.

161. W. Adcock and G. Aldous, *Tetrahedron Lett.*, 1978, **1978**, 3387.

162. W. Adcock, D. P. Cox and W. Kitching, *J. Organomet. Chem.*, 1977, **133**, 393.

163. W. Adcock and D. P. Cox, *J. Org. Chem.*, 1979, **44**, 3004.

164. W. Kitching, V. Alberts, W. Adcock and D. P. Cox, *J. Org. Chem.*, 1978, **43**, 4652.

165. W. Adcock and A. N. Abeywickrema, *J. Org. Chem.*, 1982, **47**, 779.

166. W. F. Reynolds and G. K. Hamer, *J. Amer. Chem. Soc.*, 1976, **98**, 7296.

167. M. J. Shapiro, *J. Org. Chem.*, 1978, **43**, 3769.

168. P. R. Wells, K. G. Penman and I. D. Rae, *Aust. J. Chem.*, 1980, **33**, 2221.

169. T. Drakenberg, J. Sandstrom and J. Seita, *Org. Magn. Reson.*, 1978, **11**, 246.

170. R. Faure, J.-P. Galy, E.-J. Vincent, J. Elguero, A.-M. Galy and J. Barbe, *Chem. Scripta*, 1980, **15**, 62.

171. W. B. Smith and T. W. Proulx, *Org. Magn. Reson.*, 1976, **8**, 567.
172. F. Hruska, H. M. Hutton and T. Schaefer, *Canad. J. Chem.*, 1965, **43**, 2392.
173. Y. Berger, M. Berger–Deguee and A. Castonguay, *Org. Magn. Reson.*, 1981, **15**, 244, 303.
174. A. Solladie–Cavallo and M. Hilbert, *Org. Magn. Reson.*, 1981, **17**, 227; 1981, **16**, 44.
175. J. F. Bagli and M. St-Jacques, *Canad. J. Chem.*, 1978, **56**, 578.
176. (a) G. E. Maciel, P. D. Ellis, J. J. Nattersand and G. B. Savitsky, *J. Magn. Reson.*, 1969, **1**, 589; (b) G. E. Maciel, *J. Phys. Chem.*, 1965, **69**, 1947; (c) K. Hatada, K. Nagata and H. Yuki, *Bull. Chem. Soc. Japan*, 1970, **43**, 3195, 3267; (d) G. Miyajima, K. Takahashi and K. Nishimoto, *Org. Magn. Reson.*, 1974, **6**, 413.
177. R. T. C. Brownlee and D. J. Craik, *J. Chem. Soc. Perkin Trans. II*, 1981, **1981**, 760.
178. K. Seidman and G. E. Maciel, *J. Amer. Chem. Soc.*, 1977, **99**, 3254.
179. E. A. Hill and H. E. Guenther, *Org. Magn. Reson.*, 1981, **16**, 177.
180. R. T. C. Brownlee, G. Butt, M. P. Chan and R. D. Topsom, *J. Chem. Soc. Perkin Trans. II*, 1976, **1976**, 1486.
181. M. J. Albright, J. N. St Denis and J. P. Oliver, *J. Organomet. Chem.*, 1977, **125**, 1.
182. (a) M. E. van Dommelen, J. W. de Haan and H. M. Buck, *Org. Magn. Reson.*, 1980, **14**, 497; (b) H. Henry and S. Fliszar, *J. Amer. Chem. Soc.*, 1978, **100**, 3312.
183. D. G. Morris and A. M. Murray, *J. Chem. Soc. Perkin Trans. II*, 1975, **1975**, 539.
184. J. W. de Hann, L. J. M. van de Ven, H. Vlems, M. M. E. Scheffers-Sap, H. Gillissen and H. M. Buck, *Tetrahedron*, 1980, **36**, 799.
185. (a) R. I. Kruglikova, B. K. Berestevich and T. V. Sotnichenko, *Izv. Vyssh. Uchebn. Zaved., Khim. Khim. Teckhnol.*, 1980, **23**, 667; (b) A. G. Proidakov, B. I. Istomin, G. A. Kalabin, V. I. Donskikh and V. M. Polonov, *Org. React. (Tartu)*, 1979, **16**, 539.
186. R. T. C. Brownlee and D. J. Craik, *J. Chem. Soc. Perkin Trans. II*, 1981, **1981**, 753.
187. U. Edlund and S. Wold, *J. Magn. Reson.*, 1980, **37**, 183.
188. (a) D. A. R. Happer, *Aust. J. Chem.*, 1976, **29**, 2607; (b) D. A. R. Happer, S. M. McKerrow and A. L. Wilkinson, *Aust. J. Chem.*, 1977, **30**, 1715.
189. D. A. R. Happer and G. J. Wright, *J. Chem. Soc. Perkin Trans. II*, 1979, **1979**, 694.
190. G. M. Loudon and C. Berke, *J. Amer. Chem. Soc.*, 1974, **96**, 4508.
191. (a) K. Izawa, T. Okuyama and T. Fueno, *Bull. Chem. Soc. Japan*, 1973, **46**, 2881; (b) K. Harada, K. Nagata and H. Yuki, *Bull. Chem. Soc. Japan*, 1970, **43**, 3195.
192. V. I. Glukhikh, O. G. Yaroshk N. G. Glukhihk, and G. A. Pensionerova, *Dokl. Akad. Nauk SSSR*, 1979, **247**, 1405.
193. G. Butt and R. D. Topsom, *Spectrochim. Acta*, 1980, **36A**, 811; 1982, **38A**, 301.
194. J. Huet, D. Zimmermann and J. Reisse, *Tetrahedron*, 1980, **36**, 1773.
195. W. F. Reynolds, V. G. Gibb and N. Plavac, *Canad. J. Chem.*, 1980, **58**, 839.
196. A. Cornelis, S. Lambert, P. Laszlo and P. Schaus, *J. Org. Chem.*, 1981, **46**, 2130.
197. W. F. Reynolds, G. K. Hamer and A. R. Bassindale, *J. Chem. Soc. Perkin Trans. II*, 1977, **1977**, 971.
198. H. O. Krabbenhoft, *J. Org. Chem.*, 1978, **43**, 1830.
199. A. Cornelis, S. Lambert and P. Laszlo, *J. Org. Chem.*, 1977, **42**, 381.
200. T. B. Posner, and C. D. Hall, *J. Chem. Soc. Perkin Trans. II*, 1976, **1976**, 729.
201. C. N. Robinson, C. D. Slater, J. S. Covington, C. R. Chang, L. S. Dewey, J. M. Franceschini, J. L. Fritzsche, J. E. Hamilton, C. C. Irving, J. M. Morris, D. W. Morris, L. E. Rodman, V. I. Smith, G. E. Stablein and F. C. Ward, *J. Magn. Reson.*, 1980, **41**, 293.
202. C. N. Robinson and C. C. Irving, Jr, *J. Heterocyclic Chem.*, 1979, **16**, 921.
203. F. Membrey and J. P. Doucet, *J. Chim. Phys.*, 1976, **73**, 1024.
204. E. Taskinen and L. Tuominen, *Finn. Chem. Lett.*, 1978, **1978**, 240.
205. E. Taskinen, *Tetrahedron*, 1978, **34**, 429.

206. R. E. Bilbo and D. W. Boykin, *J. Chem. Res.* (*S*), 1980, **1980**, 332.
207. E. Solcaniova, P. Hrnciar and T. Liptaj, *Org. Magn. Reson.*, 1982, **18**, 55.
208. Z. Bankowska and M. Jedrzejewska-Krawczyk, *Pol. J. Chem.*, 1980, **54**, 1473.
209. W. F. Reynolds and R. A. McClelland, *Canad. J. Chem.*, 1977, **55**, 536.
210. N. Mandava and H. Finegold, *Spectrosc. Lett.*, 1980, **13**, 59.
211. C. W. Fong and H. G. Grant, *Org. Magn. Reson.*, 1980, **14**, 147.
212. C. W. Fong and H. G. Grant, *Aust. J. Chem.*, 1981, **34**, 1205.
213. R. G. Jones and J. M. Wilkins, *Org. Magn. Reson.*, 1978, **11**, 20.
214. Y. Kosugi and Y. Furuya, *Tetrahedron*, 1980, **36**, 2741.
215. P. E. Hansen, O. K. Poulsen and A. Berg, *Org. Magn. Reson.*, 1977, **9**, 649.
216. J. Niwa, *Bull. Chem. Soc. Japan*, 1980, **53**, 2685.
217. K. Izawa, T. Okuyama and T. Fueno, *Bull. Chem. Soc. Japan*, 1973, **46**, 2881.
218. A. M. Kamalyutdinova-Silikhova, T. G. Mannafov, R. B. Khismatova, S. G. Vul'fosn and A. N. Vereschchagin, *Izv. Akad. Nauk SSR, Ser. Khim.*, 1979, **1979**, 1757.
219. G. A. Kalabin, I. Kushnarev, T. G. Mannafov and A. A. Retinskii, *Izv. Akad. Nauk SSR, Ser. Khim.*, 1978, **1978**, 2410.
220. Y. Kusuyama, K. Hoyo, Y. Takai and T. Ando, *Nakayama Daigaku Kyoikugakubu Kiyo, Shizen Kagaku*, 1979, **280**, 13.
221. G. Musumarra, S. Wold and S. Gronowitz, *Org. Magn. Reson.*, 1981, **17**, 118.
222. E. Sol'aniova, S. Toma and S. Gronowitaz, *Org. Magn. Reson.*, 1976, **8**, 439.
223. Y. Kusuyama, C. Dyllick-Brenzinger and J. D. Roberts, *Org. Magn. Reson.*, 1980, **13**, 372.
224. W. Adcock, W. Kitching, V. Alberts, G. Wickham, P. Barron and D. Doddrell, *Org. Magn. Reson.*, 1977, **10**, 47.
225. G. W. Buchanan and B. A. Dawson, *Org. Magn. Reson.*, 1980, **13**, 293.
226. J. E. Arrowsmith, M. J. Cook and D. J. Hardstone, *Org. Magn. Reson.*, 1978, **11**, 160.
227. E. M. Schulman, K. A. Christensen, D. M. Grant and C. Walling, *J. Org. Chem.*, 1974, **39**, 2686.
228. Y. Nakai, T. Takabayashi and F. Yamada, *Org. Magn. Reson.*, 1980, **13**, 94.
229. Y. Nakai and F. Yamada, *Org. Magn. Reson.*, 1978, **11**, 607.
230. N. Inamoto, K. Kushida, S. Masuda, H. Ohta, S. Satoh, Y. Tamura, K. Tokumara, K. Tori and M. Yoshida, *Tetrahedron Lett.*, 1974, **1974**, 3617.
231. M. J. Shapiro, *Tetrahedron*, 1977, **33**, 1091.
232. A. W. Frahm and H. F. Hambloch, *Org. Magn. Reson.*, 1980, **14**, 444.
233. F. Membrey and J. P. Doucet, *C. R. Acad. Sci. Paris C*, 1976, **282**, 149.
234. F. Membrey, B. Ancian and J.-P. Doucet, *J. Chem. Soc. Perkin Trans. II*, 1980, **1980**, 1399.
235. E. Grimley, D. H. Collum, E. G. Alley and B. Layton, *Org. Magn. Reson.*, 1981, **15**, 296.
236. R. Radeglia and F. G. Weber, *Acta. Chim. Acad. Sci. Hung.*, 1980, **105**, 93.
237. F. G. Weber, R. Radeglia and W. Altenburg, *J. Pract. Chem.*, 1980, **322**, 849.
238. F. G. Weber and R. Radeglia, *J. Pract. Chem.*, 1979, **321**, 935.
239. C. M. Buess, K. S. Narayanan, D. J. O'Donnell, P. Arjunan and K. D. Berlin, *Org. Magn. Reson.*, 1979, **12**, 691.
240. G. Kreze, M. Berger, P. K. Claus and W. Rieder, *Org. Magn. Reson.*, 1976, **8**, 170.
241. D. A. Brown, N. J. Fitzpatrick, I. J. King and N. J. Mathews, *J. Organomet. Chem.*, 1976, **104**, C9.
242. N. K. Wilson and R. D. Zehr, *J. Org. Chem.* 1982, **47**, 1184.
243. (a) W. Adcock and T. C. Khor, *Tetrahedron Lett.*, 1976, **1976**, 3063; (b) W. Adcock and T. C. Khor, *J. Amer. Chem. Soc.*, 1978, **100**, 7799.
244. (a) D. F. Ewing, S. Sotheeswaran and K. J. Toyne, *Tetrahedron Lett.*, 1977, **1977**, 2041; (b) D. F. Ewing and K. J. Toyne, *J. Chem. Soc. Perkin Trans. II*, 1979, **1979**, 243.

245. K. Fuji, T. Yamada and E. Fujita, *Org. Magn. Reson.*, 1981, **17**, 250.
246. A. Solladie–Cavallo and G. Solladie, *Org. Magn. Reson.*, 1977, **10**, 235.
247. P. W. Westerman, R. E. Botto and J. D. Roberts, *J. Org. Chem.*, 1978, **43**, 2590.
248. H. B. Burgi and J. D. Dunitz, *Helv. Chim. Acta*, 1970, **53**, 1747.
249. (a) S. Bradamante and G. A. Pagini, *J. Org. Chem.*, 1980, **45**, 105; (b) S. Bradamante and G. A Pagini, *J. Org. Chem.*, 1980, **45**, 114.
250. D. G. Farnum, R. E. Botto, W. T. Chambers and B. Lam, *J. Amer. Chem. Soc.*, 1978, **100**, 3847.
251. (a) H. C. Brown, M. Periasamy and K.-T. Liu, *J. Org. Chem.*, 1981, **46**, 1646; (b) D. P. Kelly, M. J. Jenkins and R. A. Mantello, *J. Org. Chem.*, 1981, **46**, 1650.
252. I. P. Bangov, *Org. Magn. Reson.*, 1981, **16**, 296.
253. (a) H. L. Retcofsky and R. A. Friedl, *J. Phys. Chem.*, 1967, **71**, 3592; (b) H. L. Retcofsky and R. A. Friedl, *J. Phys. Chem.*, 1968, **72**, 291; (c) H. L. Retcofsky and F. R. McDonald, *Tetrahedron Lett.*, 1968, **1968**, 2575; (d) H. L. Retcofsky and R. A. Friedl, *J. Phys. Chem.*, 1968, **72**, 2619; (e) G. Miyajima, Y. Sasaki and M. Suzuki, *Chem. Pharm. Bull.*, 1972, **20**, 429.
254. S. A. Sojka, F. J. Dinan and R. Kolarczyk, *J. Org. Chem.*, 1979, **44**, 307.
255. C. J. Turner and G. W. H. Cheeseman, *Org. Magn. Reson.*, 1976, **8**, 357.
256. G. W. Cheeseman, C. J. Turner and D. J. Brown, *Org. Magn. Reson.*, 1979, **12**, 212.
257. H.-P. Erb and T. Bluhm, *Org. Magn. Reson.*, 1980, **14**, 285.
258. E. Gonzalez, R. Faure, E.-J. Vincent, M. Espada and J. Elguero, *Org. Magn. Reson.*, 1979, **12**, 587.
259. R. Faure, J. Llinares, E.-J. Vincent and J. Elguero, *Org. Magn. Reson.*, 1979, **12**, 579.
260. L. I. Larina, T. I. Vakul'skaya, A. V. Filatov, B. I. Istomin, E. F. Shibanova, V. A. Lopyrev and M. G. Voronkov, *Org. Magn. Reson.*, 1981, **17**, 1.
261. V. A. Lopyrev, L. I. Larina, T. I. Vakul'skaya, M. F. Larin, O. B. Nefedova, E. F. Shibanova and M. G. Voronkov, *Org. Magn. Reson.*, 1981, **15**, 219.
262. L. J. Mathias and C. G. Overberger, *J. Org. Chem.*, 1978, **43**, 3526.
263. H. Takeshita and H. Mametsuka, *Heterocycles*, 1979, **12**, 653.
264. U. Hollstein and G. E. Krisov, *Org. Magn. Reson.*, 1980, **14**, 300.
265. G. M. Coppola, A. D. Kahle and M. J. Shapiro, *Org. Magn. Reson.*, 1981, **17**, 242.
266. (a) J.-A. Su, E. Siew, E. V. Brown and S. L. Smith, *Org. Magn. Reson.*, 1977, **10**, 122; (b) I. V. Zuika, Y. Y. Popelis, I. P. Sekatsis, Z. P. Bruvers and M. A. Tsirule, *Khim. Geterots. Soedin.*, 1979, **1979**, 1665.
267. A. Van Veldhuizen, M. Van Dijk and G. M. Sanders, *Org. Magn. Reson.*, 1980, **13**, 105.
268. E. Breitmaier and U. Hollstein, *J. Org. Chem.*, 1976, **41**, 2104.
269. I. W. J. Still, D. M. McKinnon and M. S. Chauhan, *Canad. J. Chem.*, 1976, **54**, 1660.
270. R. E. Wasylishen, T. R. Clem and E. D. Becker, *Canad. J. Chem.*, 1975, **53**, 596.
271. J. Llinares, J.-P. Galy, R. Faure, E.-J. Vincent and J. Elguero, *Canad. J. Chem.*, 1979, **57**, 937.
272. R. Faure, J.-P. Galy, E.-J. Vincent and J. Elguero, *Canad. J. Chem.*, 1978, **56**, 46.
273. R. Faure, J. Elguero, E.-J. Vincent and R. Lazaro, *Org. Magn. Reson.*, 1978, **11**, 617.
274. S. N. Sawhney and D. W. Boykin, *J. Org. Chem.*, 1979, **44**, 1136.
275. G. M. Bodner and L. J. Todd, *Inorg. Chem.*, 1974, **13**, 360.
276. B. P. Roques, C. Segard, S. Combrisson and F. Wehrli, *J. Organomet. Chem.*, 1974, **73**, 327.
277. L. A. Fedorov, P. V. Petrovskii, E. I. Fedin, G. A. Panosyan, A. A. Tsoi, N. K. Baranetskaya and V. N. Setkina, *J. Organomet. Chem.*, 1979, **182**, 499.
278. (a) K. T. Wu and B. P. Dailey, *J. Chem. Phys.*, 1964, **41**, 2796; (b) P. Lazzaretti and F. Taddei, *Org. Magn. Reson.*, 1971, **3**, 283; (c) W. F. Hehre, L. Radom and J. A. Pople, *J. Amer. Chem. Soc.*, 1972, **94**, 1496.

279. W. Adcock and G. L. Aldous, *J. Organomet. Chem.*, 1980, **201**, 411.

280. S. Gronowitz, I. Johnson, A. Maholanyiova, S. Toma and E. Solcaniova, *Org. Magn. Reson.*, 1975, **7**, 372.

281. R. K. Chadha and J. M. Miller, *Canad. J. Chem.*, 1982, **60**, 596.

282. G. A. Kalabin, F. F. Kushnarev, G. A. Chmutova and L. V. Kashurnikova, *Zh. Org. Khim.*, 1979, **14**, 24.

283. G. A. Kalabin, D. F. Kushnarev, L. M. Kataeva, L. V. Kashurnikova and R. I. Vinokurova, *Zh. Org. Khim.*, 1978, **14**, 2478.

284. G. A. Kalabin, D. F. Kushnarev, V. M. Bzesovsky and G. A. Tschmutova, *Org. Magn. Reson.*, 1979, **12**, 598.

285. G. A. Kalabin, D. F. Kushnarev and T. G. Mannafov, *Zh. Org. Khim.*, 1980, **16**, 505.

286. S. Gronowitz, A. Konar and A.-B. Hornfeldt, *Org. Magn. Reson.*, 1977, **9**, 213.

287. R. H. Cox, *J. Magn. Reson.*, 1979, **33**, 61.

288. D. de Vos and J. Wolters. *J. Roy. Netherlands Chem. Soc.*, 1978, **97**, 219.

289. H.-J. Schneider and W. Freitag, *J. Amer. Chem. Soc.*, 1977, **99**, 8363.

290. C. Piccini-Leopardi and J. Reisse, *J. Magn. Reson.*, 1981, **42**, 60.

291. H. Fritz, P. Hug, H. Sauter, T. Winkler, S.-O. Lawesson, B. S. Pedersen and S. Scheibye, *Org. Magn. Reson.*, 1981, **16**, 36.

292. N. K. Wilson and J. B. Stothers, in *Topics in Stereochemistry*, Vol 8 (E. Eliel and N. Allinger, eds), Wiley–Interscience, New York, 1973, p. 1.

293. J. B. Lambert and A. R. Vagenas, *Org. Magn. Reson.*, 1981, **17**, 265.

294. D. M. Grant and B. V. Cheney, *J. Amer. Chem. Soc.*, 1967, **89**, 5315.

295. H.-J. Schneider and V. Hoppen, *J. Org. Chem.*, 1978, **43**, 3866.

296. D. G. Gorenstein, *J. Amer. Chem. Soc.*, 1977, **99**, 2254.

297. H. Beierbeck and J. K. Saunders, *Canad. J. Chem.*, 1976, **54**, 2985.

298. J. B. Lambert and A. R. Vagenas, *Org. Magn. Reson.*, 1981, **17**, 270.

299. E. L. Eliel, W. F. Bailey, L. D. Kopp, R. L. Willer, D. M. Grant, R. Bertrand, K. A. Christensen, D. K. Dalling, M. W. Duch, E. Wenkert, F. M. Schell and D. W. Cochran, *J. Amer. Chem. Soc.*, 1975, **97**, 322.

300. T. P. Forrest and J. G. K. Webb, *Org. Magn. Reson.*, 1979, **12**, 371.

301. H. Duddeck and M. R. Islam, *Org. Magn. Reson.*, 1981, **16**, 32.

302. H. Duddeck, F. Hollowood, A. Karim and M. A. McKervey, *J. Chem. Soc. Perkin Trans. II*, 1979, **1979**, 360.

303. Z. Majerski, V. Vinkovic and Z. Meic, *Org. Magn. Reson.*, 1981, **17**, 169.

304. H. Duddeck, *Tetrahedron*, 1978, **34**, 247.

305. D. G. Morris and A. M. Murray, *J. Chem. Soc. Perkin Trans. II*, **1975**, 1975, 734.

306. G. R. Somayajulu, J. R. Kennedy, T. M. Vickrey and B. J. Zwolinski, *J. Magn. Reson.*, 1979, **33**, 559.

307. R. M. Schwarz and N. Rabjohn, *Org. Magn. Reson.*, 1980, **13**, 9.

308. M. Sjostrom and S. Wold, *J. Chem. Soc. Perkin Trans. II*, 1979, **1979**, 1274.

309. W. Freitag and H.-J. Schneider, *J. Chem. Soc. Perkin Trans. II*, 1979, **1979**, 1337.

310. K. B. Wiberg, W. E. Pratt and W. F. Bailey, *J. Org. Chem.*, 1980, **45**, 4936.

311. (a) D. M. Grant and E. G. Paul, *J. Amer. Chem. Soc.*, 1964, **86**, 2984; (b) L. P. Lindeman and J. Q. Adams, *Anal. Chem.*, 1971, **43**, 1245.

312. J. C. MacDonald, *J. Magn. Reson.*, 1979, **34**, 207.

313. A. Ejchart, *Org. Magn. Reson.*, 1980, **13**, 368.

314. R. B. Catell, *Factor Analysis*, Harper and Row, New York, 1952.

315. C. G. Andrieu, D. Debruyne and D. Paquer, *Org. Magn. Reson.*, 1978, **11**, 528.

316. H. Duddeck and H.-T. Feuerhelm, *Tetrahedron*, 1980, **36**, 3009.

317. M. Azzaro, J. F. Gal, S. Geribaldi and N. Novo-Kremer, *Spectrochim. Acta*, 1978, **34A**, 157.

318. F. C. Brown and D. G. Morris, *J. Chem. Soc. Perkin Trans. II*, 1977, **1977**, 125.
319. J. A. Hirsch and A. A. Jarmas, *J. Org. Chem.*, 1978, **43**, 4106.
320. F. C. Brown, D. G. Morris and A. M. Murray, *Tetrahedron*, 1978, **34**, 1845.
321. A. Heumann and H. Kolshorn, *J. Org. Chem.*, 1979, **44**, 1575.
322. D. G. Morris and A. M. Murray, *J. Chem. Soc. Perkin Trans. II*, 1976, **1976**, 1579.
323. P. Metzger, E. Casadevall, A. Casadevall and M.-J. Pouet, *Canad. J. Chem.*, 1980, **58**, 1503.
324. G. C. Levy and R. L. Lichter, *Nitrogen-15 Nuclear Magnetic Resonance Sepctroscopy*, Wiley, New York, 1979.
325. G. J. Martin, M. L. Martin and J.-P. Gouesnard, ^{15}N-*NMR Spectroscopy*, Vol. 18, Springer-Verlag, Berlin, 1981.
326. M. Witanowski and G. A. Webb (eds), *Nitrogen NMR*, Plenum, London, 1973. M. Witanowski, L. Stefaniak and G. A. Webb, in *Annual Reports on NMR Spectroscopy*, Vol. 11B (G. A. Webb, ed.), Academic Press, London, 1981.
327. (a) T. Axenrod, P. S. Pregosin, M. J. Wieder, E. D. Becker, R. B. Bradley and G. W. A. Milne, *J. Amer. Chem. Soc.*, 1971, **93**, 6536; (b) R. L. Lichter and J. D. Roberts, *Org. Magn. Reson.*, 1974, **6**, 636.
328. T. Axenrod and M. J. Wieder, *Org. Magn. Reson.*, 1976, **8**, 350.
329. J. Dorie, B. Mechin and G. Martin, *Org. Magn. Reson.*, 1979, **12**, 229.
330. P. K. Korver, K. Spaargaren, P. J. van der Haak and T. J. de Boer, *Org. Magn. Reson.*, 1970, **2**, 295.
331. P. S. Pregosin, E. W. Randall and A. I. White, *J. Chem. Soc. Perkin Trans. II*, 1972, **1972**, 513.
332. I. I. Schuster, S. H. Doss and J. D. Roberts, *J. Org. Chem.*, 1978, **43**, 4693.
333. W. Schowotzer and W. von Philipsborn, *Helv. Chim. Acta*, 1977, **60**, 1501.
334. R. E. Botto and J. D. Roberts, *J. Org. Chem.*, 1979, **44**, 140.
335. W. Bremser, J. I. Kroschwitz and J. D. Roberts, *J. Amer. Chem. Soc.*, 1969, **91**, 6189.
336. I. I. Schuster and J. D. Roberts, *J. Org. Chem.*, 1980, **45**, 284.
337. D. J. Craik, G. C. Levy and R. T. C. Brownlee, *J. Org. Chem.*, 1983, **48**, 1601.
338. R. O. Duthaler, H. G. Forster and J. D. Roberts, *J. Amer. Chem. Soc.*, 1978, **100**, 4974.
339. Reference 324, p. 95.
340. P. S. Pregosin, E. W. Randall and A. I. White, *J. Chem. Soc. Perkin Trans. II*, 1972, **1972**, 513.
341. Reference 324, pp. 77–82.
342. D. Gust and D. N. Neal, *J. Chem. Soc., Chem. Comm.*, 1978, **1978**, 682.
343. E. Rosenberg, K. L. Williamson and J. D. Roberts, *Org. Magn. Reson.*, 1976, **8**, 117.
344. G. A. Kalabin, D. F. Kushnarev, R. B. Valeyev, B. A. Trofimov and M. A. Fedotov, *Org. Magn. Reson.*, 1982, **18**, 1.
345. C. Delseth and J.-P. Kintzinger, *Helv. Chim. Acta*, 1976, **59**, 466.
346. J. K. Crandall, M. A. Centeno and S. Borresen, *J. Org. Chem.*, 1979, **44**, 1184.
347. C. Delseth, T.-T. Nguyen and J.-P. Kintzinger, *Helv. Chim. Acta*, 1980, **63**, 498.
348. E. L. Eliel, K. M. Pietrusiewicz and L. M. Jewell, *Tetrahedron Lett.*, 1979, **1979**, 3649.
349. H. A. Christ, P. Diehl, H. R. Schneider and H. Dahn, *Helv. Chim. Acta*, 1961, **44**, 865.
350. D. J. Sardella and J. B. Stothers, *Canad. J. Chem.*, 1969, **47**, 2089.
351. T. E. St Armour, M. I. Burgar, B. Valentine and D. Fiat, *J. Amer. Chem. Soc.*, 1981, **103**, 1128.
352. M. Katoh, T. Sugawara, Y. Kawada and H. Iwamura, *Bull. Chem. Soc. Japan*, 1977, **52**, 3475.
353. W. F. Reynolds, *Tetrahedron Lett.*, 1977, **1977**, 675.
354. R. T. C. Brownlee and D. J. Craik, *Org. Magn. Reson.*, 1980, **14**, 186.
355. R. T. C. Brownlee and D. J. Craik, *Org. Magn. Reson.*, 1981, **15**, 248.

356. I. D. Rae and L. K. Smith, *Aust. J. Chem.*, 1972, **25**, 1465.
357. W. Adcock and A. Abeywickrema, *Tetrahedron Lett.*, 1981, 1135.
358. W. Adcock and T. C. Khor, *J. Org. Chem.*, 1977, **42**, 218.
359. J. Bromilow, R. T. C. Brownlee and A. V. Page, *Tetrahedron Lett.*, 1976, **1976**, 3055.
360. R. T. C. Brownlee and D. J. Craik, *Tetrahdron Lett.*, 1980, **1980**, 1681.
361. W. Adcock and A. N. Abeywickrema, *Tetrahedron Lett.*, 1979, **1979**, 1809.
362. W. Adcock and A. N. Abeywickrema, *Aust. J. Chem.*, 1980, **33**, 181.
363. R. T. C. Brownlee and D. J. Craik, *Aust. J. Chem.*, 1980, **33**, 2555.
364. W. Kitching, G. Drew, W. Adcock and A. N. Abeywickrema, *J. Org. Chem.*, 1981, **46**, 2252.
365. O. P. Shkurko, E. P. Khmeleva, S. G. Baram, M. M. Shakirov and V. P. Mamaev, *Khim. Geterots. Soedin.*, 1978, **1978**, 996.
366. W. Adcock, G. L. Aldous and W. Kitching, *J. Organomet. Chem.*, 1980, **202**, 385.
367. J. M. Angelelli, M. A. Delmas and J. C. Maire, *J. Organomet. Chem.*, 1978, **154**, 79.
368. J. Fukunaga and R. W. Taft, *J. Amer. Chem. Soc.*, 1975, **97**, 1612.
369. S. K. Dayal, S. Ehrenson and R. W. Taft, *J. Amer. Chem. Soc.*, 1972, **94**, 9113.
370. S. I. Pombrik, V. F. Ivanov, A. S. Peregudov, D. N. Kravtsov, A. A. Fedrov and E. I. Fedin, *J. Organomet. Chem.*, 1978, **153**, 319.
371. C. C. Mitsch, L. D. Freedman and C. G. Moreland, *J. Magn. Reson.*, 1970, **3**, 446.
372. L. L. Szafraniec, *Org. Magn. Reson.*, 1974, **6**, 565.
373. J. Bromilow, R. T. C. Brownlee and D. J. Craik, *Aust. J. Chem.*, 1977, **30**, 351.
374. J. Bodeker, P. Kockritz, H. Koppel and R. Radeglia, *J. Prakt. Chem.*, 1980, **322**, 735.
375. E. M. Briggs, G. W. Brown, P. M. Cairns, J. Jiricny and M. F. Meidine, *Org. Magn. Reson*, 1980, **13**, 306.
376. J.-P. Gouesnard, J. Dorie and G. J. Martin, *Canad. J. Chem.*, 1980, **58**, 1295.
377. N. Inamoto, *C.A.*, **92**, 180512z.
378. R. Faure, E. J. Vincent, J. M. Ruiz and L. Lena, *Org. Magn. Reson.*, 1981, **15**, 401.
379. M. Baiwir, G. Llabres, L. Christiaens and J.-L. Piette, *Org. Magn.Reson.*, 1982, **18**, 33; 1981, **16**, 14; 1980, **14**, 293.
380. S. Gronowitz, I. Johnson and A. B. Hornfeldt, *Chem. Scripta*, 1975, **8**, 8.
381. T. Drakenberg, A, B. Hornfeldt, S. Gronowitz, J.-M. Talbot and J.-L. Piette, *Chem. Scripta*, 1978, **13**, 152.
382. B. Kohne, W. Lohner, K. Praefcke, H. J. Jakobsen and B. Villadsen, *J. Organomet. Chem.*, 1979, **166**, 373.
383. I. A. Abronin, A. Z. Djumanazarova, V. P. Litvinov and A. Konar, *Chem. Scripta*, 1982, **19**, 75.
384. A. Fredga, S. Gronowitz and A. B. Hornfeldt, *Chem. Scripta*, 1977, **11**, 37.
385. G. A. Kalabin, D. F. Kushnarev, V. M. Bzesovsky and G. A. Tschmutova, *Org. Magn. Reson.*, 1979, **12**, 598.
386. R. T. C. Brownlee, A. F. Masters, M. J. O'Connor, A. G. Wedd, H. A. Kamlin and J. D. Cotton, *Org. Magn. Reson.*, 1982, **20**, 73.
387. P. R. Wells and D. W. Hawker, *Org. Magn. Reson.*, 1981, **17**, 26.
388. (a) A. J. Canty, P. Barron and P. C. Healy, *J. Organomet. Chem.*, 1979, **179**, 447; (b) T. N. Mitchell and H. C. Marsmann, *J. Organomet. Chem.*, 1978, **150**, 171; (c) Y. A. Strelenko, Y. G. Bundel, F. H. Kasumov, V. I. Rozenberg, O. A. Reutov and Y. A. Ustnyuk, *J. Organomet. Chem.*, 1978, **159**, 131.
389. M. A. Sens, N. K. Wilson, P. D. Ellis and T. D. Odom, *J. Magn. Reson.*, 1975, **19**, 323.
390. P. J. Smith and A. P. Tupciauskas, in *Annual Reports on NMR Spectroscopy*, Vol. 8 (G. A. Webb, ed.), Academic Press, London, 1978, p. 312.
391. D. C. van Beelen, H. O. van der Koo and J. Wolters, *J. Organomet. Chem.*, 1979, **179**, 37.

Isotope Effects on Nuclear Shielding

POUL ERIK HANSEN

*Institut I, Life Sciences and Chemistry, Roskilde University Centre, PO Box 260,
4000 Roskilde, Denmark*

ANNUAL REPORTS ON NMR SPECTROSCOPY,
VOLUME 15   ISBN 0-12-505315-0

I. INTRODUCTION

Isotope effects have been known for a long time and an excellent review covering the early reports exists.[1] Some of the conclusions drawn by Batiz–Hernandez and Bernheim[1] can briefly be summarized: the magnitude of the isotope shift is generally dependent on how remote the isotope substitution is from the nucleus under observation; the magnitude of the shift is also a function of the shielding range of the resonant nucleus; the isotope shift is largest where the fractional change in mass upon substitution is largest; and, in general, the isotope nuclear shielding is approximately proportional to the number of atoms in the molecule which has been substituted by isotopes.

Brief reviews of isotope effects may also be found elsewhere.[2–7]

Theories explaining one-bond isotope effects have been proposed[8] and reviewed recently.[9–11] Some isotopes occur in natural abundance in such amounts that their presence may be detected without enrichment. These isotopes include ^{13}C, ^{119}Sn, ^{37}Cl and ^{11}B, whereas to observe effects due to D, T, ^{18}O and ^{15}N, enrichment is necessary. Easy access to some of the isotopes may suddenly make an isotope popular, as demonstrated by ^{18}O (Section XII). Furthermore, as many isotope effects are small, the advent of high field NMR instruments has facilitated the measurement of isotope effects. The NMR observation of nuclei other than ^1H and ^{13}C has also become much more frequent, leading to the observation of new isotope effects.

Both primary and secondary isotope effects are discussed, but only a few primary effects are reported in the literature.

Two distinct types of secondary isotope effect on nuclear shielding are considered, the intrinsic effects (also known as direct effects) caused by isotopic substitution, and equilibrium isotope effects, caused by conformational changes or shifts in equilibria as a consequence of isotopic substitution.

The data treated and listed in the tables are those appearing later than 1965 and not included by Batiz–Hernandez and Bernheim.[1]

Theoretical aspects are not considered in much detail both because the advances in this field have been covered by Jameson[9,10] and because the theoretical understanding of long-range isotope effects is still too scant to be of any help.

This review concentrates upon empirical trends and practical uses of mostly secondary isotope effects, both of the intrinsic and equilibrium types.

The text and the tables are arranged in the following fashion. The most "popular" isotope effect is treated first, deuterium isotope effects on ^{13}C nuclear shielding, followed by deuterium on ^1H nuclear shieldings, etc. Focus is thus on the isotopes *producing* the effect rather than on the nuclei *suffering* the effect.

After a brief treatment of each type of isotope effect, general trends are dealt with. Basic trends of intrinsic isotope effects such as additivity, solvent effects, temperature effects, steric effects, substituent effects and hyperconjugation are discussed. Uses of isotope effects for assignment purposes, in stereochemical studies, in hydrogen bonding and in isotopic tracer studies are dealt with. Kinetic studies, especially of phosphates, are frequently performed by utilizing isotope effects. In addition, equilibrium isotope effects are treated in great detail as these are felt to be new and very important and may lead to new uses of isotope effects. Techniques used to obtain isotope effects are briefly surveyed at the end of the chapter.

For the sake of completeness all data that the reviewer has come by are included in the tables. The accuracy has thus not been judged before inclusion. When presented with other similar data it is, however, evident that some of the earlier data are less precise, and these are not taken into account in the discussion, although this is not stated explicitly except in the case of reference 12: comparison with other published data on identical compounds shows great inconsistency of the data from reference 12 to those of other workers, so data from this reference are not usually considered.

II. NOMENCLATURE

Isotope effects are divided into one-bond, two-bond, etc., isotope effects.

The large number of different nuclei investigated and the high number of possible isotope effects call for a notation showing the observed nuclei, the isotope causing the isotope effect, the number of bonds between the observed nuclei and the position of isotopic substitution; the notation should, in addition, emphasize that isotope effects are differences in chemical shielding. A $^n\Delta X(Y)$ form of notation has been adopted, which, when exemplified by ^{13}C and D, gives $^n\Delta C(D)$ meaning that an effect is observed at the ^{13}C atom by replacement of a hydrogen atom by a deuterium atom. $^n\Delta C(D)$ is the difference between $\delta C(H)$ and $\delta C(D)$ (the shielding of the carbons bearing the lighter isotope minus the shielding of the carbon bearing the heavier isotope). n indicates the number of bonds separating the observed nuclei and the isotope in question. The isotope mentioned is the heavier of the two, e.g. D or possibly T, ^{37}Cl, ^{15}N, ^{13}C, etc. In cases with more than two relevant isotopes a special notation must be used. All isotope effects are listed in the tables according to this notation, although it may be given in the original paper with opposite sign. In this case the data are marked with a superscript a in the tables. The sign convention used by Bernheim[1] is thus conserved for reasons of continuity, although good reasons could be given for the reverse notation as this would be more similar to that employed for substituent effects.

Throughout the paper D is used for ^2H and T for ^3H, both for convenience and to reduce the number of superscripts employed.

When the words "isotope effects" are mentioned, they refer to isotope effects *on nuclear shielding*, but the last two words are often left out for convenience and brevity.

III. DEUTERIUM ISOTOPE EFFECTS ON ^{13}C NUCLEAR SHIELDING

$^n\Delta D(^{13}C)$ are by far the best documented isotope effects, as evidenced in Tables 1–20. In this section only intrinsic isotope effects are treated.

Equilibrium isotope effects which have been well investigated are mentioned in Section XVI.

Substitution with deuterium was first used for assignment purposes due to the fact that the substituted carbon signal more or less disappeared. With improved techniques the isotope effect on the *ipso*, *ortho* and *meta* carbons could be used for assignment purposes.

The data are presented in tables according to the hybridization of the carbon in question and the hybridization of the deuterium-substituted carbon; the discussion is arranged in much the same way.

A. Csp³DCsp³ effects

Deuterium isotope effects have been studied in alkanes, cycloalkanes and derivatives thereof, as well as in amino acids and carbohydrates.

1. $^1\Delta C(D)$

The isotope effects observed in deuteriated methanes show a shielding change of $+0.187$ ppm per D, and the effects are largely additive.[13] For straight-chain $-CD_2-$ groups an effect of 0.36 ppm per D is observed[14] (Table 1). Slightly larger values are reported in cyclohexanes.[15,16] These also show a small difference between D in equatorial and axial positions. With D in an equatorial position a smaller effect is observed than when it is in an axial position.[16,17] This feature has been used in the measurement of activation energies[16] (Section XVI. C.2). No such effect is observed in *t*-butylcyclohexane-4-D.[19] A small difference is observed between cyclohexane, 1- and 2-D-adamantane.[16]

A shielding increase, in going from cyclopropane to cyclohexane, is noted. For cycloheptane no further increase is observed[15] (Table 1). For cyclobutane[15] and cyclobutene[22] no marked difference is present.

The small value observed for methane is clearly exceptional, probably related to the symmetry of methane. Substituted alkanes make possible an estimation of the effect of substituents (carbohydrates are treated later). A comparison of only CD_3H $(+0.58\text{ ppm})$,[13] CD_3Br $(+0.43\text{ ppm})$,[12] and CD_3I $(+0.13\text{ ppm})$[21] could lead to the conclusion that the heavier the substituent, the smaller the isotope effect. This is, however disproven by the series $RCHDCl$ $(+0.30\text{ ppm})$, $RCHDBr$ $(+0.41\text{ ppm})$, and $RCHDI$ $(+0.11\text{ ppm})$.[22] If one compares the influence of the number of substituents, CD_2H_2 $(+0.39\text{ ppm})$, CD_2Cl_2 $(+0.41\text{ ppm})$, $CDCl_3$ $(+0.18\text{ ppm})$, one sees that chlorines have very little effect.[22]

Grishin *et al.*[21] suggested that the small $^1\Delta C(D)$ value observed for CD_3I compared to CD_3COCD_3 or CD_3OD is perhaps related to a greater $H\hat{C}H$ angle for methyl iodide $(111°50')$ compared to that of CH_3COCH_3 $(108°40')$ and CH_3OH $(109°28')$.

P. E. HANSEN

TABLE 1

Deuterium isotope effects on ^{13}C nuclear shieldings of aliphatic compounds of Csp^{3}DCsp^3 type (in ppm).

Csp^{3}DCsp^3	$^1\Delta$	$^2\Delta$	$^3\Delta$	Ref.
CH$_3$D	+0.187			13
CH$_2$D$_2$	+0.385			13
CHD$_3$	+0.579			13
CD$_4$	+0.774			13
CH$_3$CD$_2$CH$_3$	+0.72			92
CH$_3$(CH$_2$)$_5$CH$_2$D	+0.28	Unresolved		22
CH$_3$(CH$_2$)$_{16}$CH$_2$D	+0.28^a	+0.08^a	+0.02	14
CH$_3$OOC(CH$_2$)$_{16}$CD$_3$		+0.36^a	+0.16	14
CH$_3$OOC(CH$_2$)$_{15}$CHDCH$_3$	+0.44^a	+0.12^a (C-16)	+0.02	14
CH$_3$OOC(CH$_2$)$_{15}$CD$_2$CH$_3$		+0.24^a (C-16)	+0.08	14
		+0.28^a (C-18)		
CH$_3$OOC(CH$_2$)$_{14}$CD$_2$CH$_2$CD$_3$		+0.48^a (C-17)		14
		+0.20^a (C-15)		
CH$_3$OOCCD$_2$(CH$_2$)$_{13}$CH$_3$		+0.12^a	+0.04	14
Cyclohexane-D$_{12}$	+1.36			21
	+1.33			93
	+1.45			94
	+1.40^a			12
Cyclohexane-D$_{11}$^{b}	+0.42^c	+0.11^c	0.02(4)^c	17
	+0.40^d	+0.12^d	+0.03(9)^d	
	+0.45^e	+0.10^e	+0.01(2)^e	
Adamantane-1-D	+0.482^f	+0.127^g		18
Adamantane-2-D^k	+0.436^h	+0.099^i	+0.008^i	18
			+0.026^j	
Norbornane-1-D			−0.037 (C-4)	15
			+0.008 (C-3)	
Cyclopropane-1-D	+0.309	+0.064		15
Cyclobutane-1-D	+0.363	+0.147	+0.027	15
Cyclobutene-3-D	+0.323^a	+0.136^a (C-4)		20
Cyclopentane-1-D	+0.374	+0.103	−0.012	15

^{a} Sign reversed compared to original paper due to use of different definition.
^{b} Original data given relative to C$_6$D$_{12}$. In this table the data are transformed to give the effect of the addition of a deuterium to C$_6$D$_{11}$.
^{c} Average values.
^{d} D$_{eq}$.
^{e} D$_{ax}$.
^{f} Relative to $^3\Delta$.
^{g} Relative to $^4\Delta$.
^{h} $^4\Delta = -0.014$ ppm.
^{i} Relative to $^5\Delta$.
^{j} Relative to $^5\Delta$. Deuterium and carbon in *anti* position.
^{k} $^4\Delta = +0.003$ ppm (C-5).

TABLE 1 (*cont.*)

Csp^{3}DCsp^3	$^1\Delta$	$^2\Delta$	$^3\Delta$	Ref.
Cyclohexane-1-D	+0.418	+0.104	+0.025	15
	+0.395^n			16
	+0.445^r			
	+0.419^s	−0.104^s	+0.025^s	
	+0.412^t	+0.104^t	+0.025(5)^t	
Cyclohexane-1,1-D$_2$	+0.826(5)^m	+0.210^m	+0.050(5)^m	16
Cycloheptane-1-D^h	+0.413	+0.110	+0.0027	15
cis-t-Butylcyclo-hexane-4-D	+0.43^a	+0.09^a		19
trans-t-Butylcyclo-hexane-4-D	+0.42^a	+0.13^a	+0.09^a	19
trans-t-Butylcyclo-hexyl mesylate-1-D$_{ax}$		+0.13^a		19
4-Homoisotwistane-2-*exo*-D, 2-*endo*-D, 3-*exo*-D		~0.4–0.5	~0.06–0.12	95
4-Methylcamphor-11-D		+0.30	+0.08 (C-4)	33
4-Methylcamphor-3,3'-D$_2$	+0.96	+0.12 (C-4)	−0.01 (C-1) +0.07 (C-5)	33
Norcamphor-3-*exo*-D$_1$	+0.35	0.09 (C-4)	+0.01 (C-1) +0.04 (C-5) +0.02 (C-7)	33
Norcamphor-3,3'-D$_2$	+0.77	+0.18 (C-4)	−0.01 (C-1) +0.07 (C-5) +0.04 (C-7)	33
1-Phenylcyclopentane-3-D$_2$		+0.2 (C-2) +0.2 (C-4)		98
Scylatone		+0.3 (C-4)		31
4-Methyl-[2.2]-*para*-cyclophane-4'-D$_2^o$				31
4-Methyl-[2.2]-*para*-cyclophane-4'-D$_3^p$				31

^{l} Relative to $^4\Delta$(C-5).
^{m} Equilibrium isotope effect.
^{n} $^1\Delta$(C-D$_{eq}$) at −80 °C. Solvent CS$_2$–CD$_2$Cl$_2$ (4:1).
^{o} $^4\Delta$ = +0.006^a ppm (C-2); $^5\Delta$ = −0.010^a ppm (C-1).
^{p} $^4\Delta$ = +0.010^a ppm (C-2); $^5\Delta$ = −0.013^a ppm (C-1); $^6\Delta$ = −0.004^a ppm (C-9).
^{q} $^4\Delta$ = +0.006^a ppm (C-2); $^5\Delta$ = −0.010^a ppm (C-1).
^{r} $^1\Delta$(C-D$_{ax}$) at −80 °C. Solvent CS$_2$–CDCl$_2$ (4:1).
^{s} Solvent CS$_2$–CD$_2$Cl$_2$ (4:1). Temperature 20 °C.
^{t} Solvent CCl$_4$. Temperature 20 °C.
^{u} $^5\Delta$ = +0.012^a ppm.
^{v} $^4\Delta$ = +0.01^a ppm (C-5); $^5\Delta$ = −0.02^a ppm (C-6).
^{w} Splitting between the C-3$_{aq}$ and the C-3$_{ax}$ is 0.184 ppm.

P. E. HANSEN

TABLE 1 (*cont.*)

Csp^{3}DCsp^3	$^1\Delta$	$^2\Delta$	$^3\Delta$	Ref.
4-Methyl-[2$_4$](1,2,4,5)-cyclophane-4'-D$_2$^{a}				31
trans-1-Ethyl-4-methylcyclohexane-4'-D$_3$^{u}	+0.927^a (CH$_3$)	+0.237^a (C-4)	+0.076^a (C-3)	31
trans-1,3-Dimethyl-cyclohexane-3'-D$_3$^{m,v}	−0.026^a	+0.24^a	+0.06^a (C-2) +0.06^a (C-4)	29
cis-2,6-Dimethyl-cyclohexanone-6'-D$_3$^{m}		+0.22^a		96
trans-2,6-Dimethyl-cyclohexanone-6-D$_3$^{m}	+0.018^a	+0.17^a		96
trans-3,5-Dimethyl-cyclohexanone-5'-D$_3$^{m}	−0.012^a	+0.26^a	+0.07^a (C-4) +0.02 (C-3)	96
1,1,3,3-Tetramethyl-cyclohexane-1'-D$_3$^{w}				323
PhCCD$_3$CD$_2$C(CH$_3$)$_3$		+0.17		97
4-F-Ph−ĊHCH$_2$D	+0.29			97
4-F-PhĊHCHD$_2$	+0.60			97
4-F-PhĊHCD$_3$	+0.91			97
4-F-PhCH$_2$ClCH$_2$D	+0.29			97
4-F-PhCH$_2$ClCHD$_2$	+0.58			97
4-F-PhCH$_2$ClCD$_3$	+0.86			97

The effects of an oxygen atom may be studied in carbon-deuterated carbohydrates. Serianni and Barker[23] give $^1\Delta$ data for a number of carbohydrates substituted at C-1; $^1\Delta$ varies from +0.25 to +0.44 ppm (Table 3).

Isotope effects may also vary according to the position of the deuterium. RCDOHCH$_3$ (+0.52 ppm) and RCHDOH (+0.38 ppm) show a distinct difference. Substituents at a β-carbon have little effect, as can be seen from Table 2. Gratwohl *et al.*[24] have observed no isotope effect at C-α of cyclo(Gly-Pro-Gly-α-D$_2$)$_2$.

A significant difference between $^1\Delta$ isotope effects on NCD$_3$ groups sitting *cis* or *trans* to a C=O group in an amide is observed [1]. This effect may possibly be used for assignment purposes of carbons that are otherwise difficult to assign.

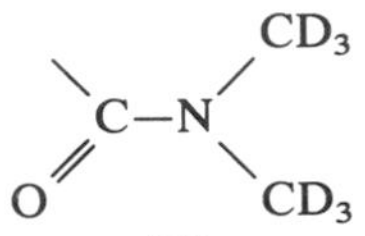

[1]

TABLE 2

Deuterium isotope effects on ^{13}C nuclear shieldings of substituted aliphatic compounds of Csp^3DCsp^3 type (in ppm).

Csp^3DCsp^3	$^1\Delta$	$^2\Delta$	$^3\Delta$	Ref.
$CDCl_3$	+0.18			21
	+0.20			93
	+0.32[a]			12
	+0.21			99
	+0.26			94
$CDBr_3$	+0.11[a]			12
$CH_3(CH_2)_5CHDCl$	+0.30	+0.11		22
$CH_3(CH_2)_5CHDBr$	+0.41	+0.11		22
$CH_3(CH_2)_5CHID$	+0.11	Unresolved		22
CD_2Cl_2	+0.65[a]			12
	+0.41			93
	+0.41			94
CD_3CD_2Br	+0.21[a] (C-1)			12
CD_2BrCD_2Br	+0.86[a]			12
CD_3Br	+0.43[a]			12
CD_3I	+0.13			21
	+0.22			12
CD_3CD_2Br	+0.53[a] (C-2)			12
$CDH_2C_6H_5$	+0.2755			30
$CD_2HC_6H_5$	+0.5508			30
$CD_3C_6H_5$	+0.8255			30
	+0.86[a]			100
	+1.08[a]			12
$CH_3CDHC_6H_4$	+0.3491	+0.0811		30
$CH_3CD_2C_6H_5$	+0.71[a]	+0.15[a]		100
$(CH_3)_2CDC_6H_5$	+0.44[a]	+0.10[a]		100
	+0.4344	+0.1094		30
CD_3CN	+0.45			93
	+0.54[a]			12
	+0.46			99
CD_3COCD_3	+0.87[a]			12
	+0.75			93
	+1.21			94
$CD_3COC_6H_5$	+0.24[d]			99
$CHD(COOC_2H_5)_2$	+0.26			99
CD_3COONa	+0.17			99
CD_3COOCD_3	+0.76 (CD_3)[a]			12
	+0.76 (OCD_3)[a]			
CD_3NO_2	+0.63[a]			12
	+0.69			93
CD_3SOCD_3	+0.89			21
	+0.90			93
	+0.92			94
	+1.18[a]			12

P. E. HANSEN

TABLE 2 (*cont.*)

Csp^3DCsp^3	$^1\Delta$	$^2\Delta$	$^3\Delta$	Ref.
$(CD_3)_2CON(CD_3)_2$	$+0.86^a$			12
$DCON(CD_3)_2$	$+0.86^a$			12
	$+0.81^a$			
$CD_3CON(CD_3)_2$	$+0.87^a$ (CD_3)			12
	$+0.91^a$ (NCD_3)			
	$+0.86^a$ (NCD_3)			
$(CD_3)_6N_3PO$	$+1.08^a$			12
$(CD_2)_4S$	$+0.30$ (CD_2)			12
	$+0.87$ (CD_2S)			
$(CD_3)_2CH_3COH$	$+0.9^a$	$+0.3^a$		35
$(CD_3)_3COH$	$+1.5^a$	$+0.4^a$		35
$CH_3CH_2C(CD_3)_2OH$	$+0.9^a$	$+0.2^a$		35
 cyclopentane CD_3, OH	$+0.9^a$	$+0.1^a$		35
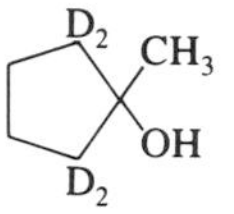 cyclopentane D_2, CH_3, OH, D_2		$+0.1^a$		35
$PhC(CD_3)_2OH$	$+0.9^a$	$+0.3^a$		35
(CH_3)-p-$PhCHCD_3OH$	$+0.9^a$	0		35
(CH_3O)-p-$PhCHCD_3OH$	$+0.8^a$	$+0.2^a$		35
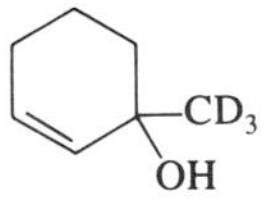 cyclohexene CD_3, OH	$+0.8^a$	$+0.1^a$		35
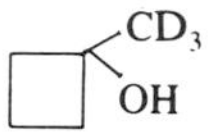 cyclobutane CD_3, OH	$+0.9^a$	$+0.1^a$		35
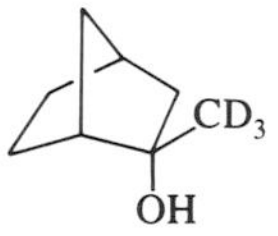 norbornane CD_3, OH	$+1.0^a$	$+0.3^a$		35
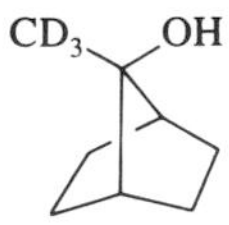 norbornane CD_3, OH	$+0.9^a$	$+0.1^a$		35
CD_3OD	$+0.86$			21
	$+0.94^a$			12
CD_3CD_2OD	$+1.19^a$ (C-2)			12
	$+1.05$ (C-2)			21
	$+0.97^a$ (C-1)			12
	$+0.88$ (C-1)			21

TABLE 2 (*cont.*)

Csp^3DCsp^3	$^1\Delta$	$^2\Delta$	$^3\Delta$	Ref.
$CD_3CDODCD_3$	$+1.18^a$ (C-2)			12
	$+0.54^a$ (C-1)			
CD_2ODCD_2OD	$+1.08$			12
$CH_3(CH_2)_5CHOHD$	$+0.38$	Unresolved		22
$CH_3OOC(CH_2)_{15}CDOHCH_3$	$+0.52^a$	$+0.12$		14
$CH_3OOC(CH_2)_{14}CD_2CHOHCD_3$		$+0.20$ (C-17)		14
		$+0.24$ (C-15)		
Methyl cholate-11,16-D_2	$+0.36$ (C-11)			84
	$+0.36$ (C-16)			
Methyl cholate-7,15,				
17,18,19,21-D_6	$+0.47$ (C-7)	$+0.22$ (C-13)		84
	$+0.34$ (C-15)	$+0.19$ (C-10)		
	$+0.47$ (C-17)			
	$+0.86$ (C-21)			
Methyl chenodeoxy-				
cholate-2,11,12,16,				
23-D_5	$+0.40$ (C-2)			84
	$+0.35$ (C-11)			
	$+0.36$ (C-12)			
	$+0.39$ (C-16)			
	$+0.23$ (C-23)			
Dihydrobotrydial				
ethyl ether-1,5-D_2	$+0.5$ (C-5)			90
	$+0.5$ (C-1)			
8-Hydroxyoctadeca-				
noate-11-D_2		$+0.21^a$ (C-10)	$+0.04^a$ (C-9)	28
Methyl 8-hydroxyocta-				
decanoate-5-D_2		$+0.20^a$ (C-6)	$+0.03$ (C-7)	28
2-Methylcyclohexa-				
none-2,6,6-D_3		$+0.14^a$ (CH_3)	$+0.05^a$ (C-4)	101
		$+0.13^a$ (C-5)		
		$+0.03^a$ (C-3)		
Methyl octadecanoate				
-2-D_2		$+0.14^a$ (C-3)	$+0.05^a$ (C-4)	27
-3-D_2		$+0.18^a$ (C-2)	$+0.06^a$ (C-5)	27
		$+0.20^a$ (C-4)		
-4-D_2		$+0.15^a$ (C-3)	$+0.05^a$ (C-2)	27
		$+0.21^a$ (C-5)	$+0.05^a$ (C-6)	
-5-D_2		$+0.19^a$ (C-4)	$+0.04^a$ (C-3)	27
		$+0.20^a$ (C-6)	$+0.05^a$ (C-7)	
-6-D_2		$+0.19^a$ (C-5)	$+0.04^a$ (C-4)	27
		$+0.20^a$ (C-7)		
-7-D_2		$+0.20^a$ (C-6)	$+0.05^a$ (C-6)	27
-8-D_2		$+0.19^a$ (C-7)	$+0.05^a$ (C-7)	27
-13-D_2			$+0.06^a$ (C-15)	27
-14-D_2		$+0.21^a$ (C-15)	$+0.06^a$ (C-16)	27
-15-D_2		$+0.21^a$ (C-16)	$+0.06^a$ (C-17)	27

P. E. HANSEN

TABLE 2 (*cont.*)

Csp^3DCsp^3	$^1\Delta$	$^2\Delta$	$^3\Delta$	Ref.
-16-D$_2$		+0.20^a (C-15)	+0.06^a (C-18)	27
-17-D$_2$		+0.20^a (C-16)	+0.06^a (C-15)	27
		+0.23^a (C-18)		
3,6-Epoxy-pentacyclo-[6.2.2.0.^{2,7}0.^{4,10}0.^{5,9}] dodecane-1,1'-D$_2$	+0.35^a	+0.11^a	+0.11^a	26
3,7-Epoxy-pentacyclo [7.2.1.0$^{2.8}$.0^{4,11}.0^{4,10}]dode- cane^c	+0.36^a (C-3)	+0.05^a (C-3)		26
	+0.44^a (C-4)	+0.11^a (C-4)		
	+0.38^a (C-3)	+0.12^a (C-3)		
	+0.34^a (C-1)	+0.11 (C-1)		
	+0.44^a (C-2)	+0.11 (C-2)		
	+0.43^a (C-12)	+0.11 (C-12)		
	+0.42^a (C-5)	+0.11 (C-5)		
Asperlactone-5,7,10-D$_3$	+0.37 (C-5)			89
	+0.28 (C-7)			
	+0.32 (C-10)			
Asperlactone-7,7-D$_2$	+0.54 (C-7)			89
Asperlactone-10,10-D$_2$	+0.55 (C-10)			89
Asperlactone-10,10,10-D$_3$	+0.72 (C-10)			89
$CD_3CD_2OCD_2CD_3$	+1.08^a (CD$_3$)			12
	+1.08^a (CD$_2$)			

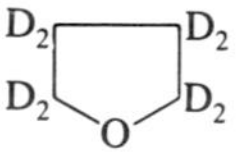

	+0.97^a (C-3)			12
	+0.86^a (C-2)			

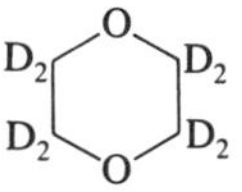

	+1.08^a			12

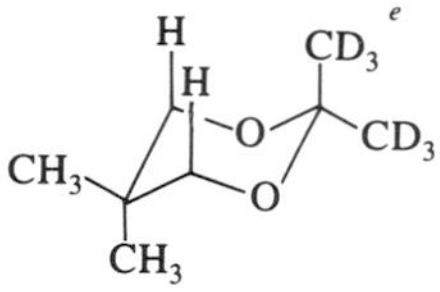

				102

1,3,5-Cyclohepta- triene-7-D	+0.32			115
2-Methyl-3-penta- none dimethylhydra- zone-5-D$_3$-1-D$_3$		+0.3 (C-4)		117
		+0.2 (C-2)		117
Dicyanocobyrinic acid heptanethyl ester-13-D$_1$^{b}		+0.082 (C-12)	+0.069 (12β-CH$_3$)	34
		+0.095 (e')	+0.025 (e'')	

TABLE 2 (*cont.*)

Csp^{3}DCsp^3	$^1\Delta$	$^2\Delta$	$^3\Delta$	Ref.
HOs$_3$(CO)$_{10}$CDH$_2$	~0.2			103
Sn(CH$_3$)$_3$CD$_3$	+0.777	+0.088		104
Sn(CH$_3$)$_2$(CD$_3$)$_2$	+0.876	+0.175		104
Sn(CH$_3$)(CD$_3$)$_3$	+0.964	+0.263		104
Sn(CD$_3$)$_4$	+1.052			104

[a] Sign reversed compared to original paper.
[b] $^4\Delta$(CH$_3$-15)=+0.019 ppm.
[c] Uniformly labelled. Numbering according to nomenclature. Some ambiguity exists in paper.
[d] Per deuterium.
[e] $^5\Delta$H-3(CD$_3$)=2.6 ppb. Believed to be a steric effect.

TABLE 3

Deuterium isotope effects on ^{13}C nuclear shieldings of carbohydrates with deuterium substitution at the ring carbons (in ppm).

Csp^{3}DCsp^3	$^1\Delta$	$^2\Delta$	$^3\Delta$	Ref.
α-D-Glucose-2-D		+0.05 (C-3)		25
		+0.01 (C-1)		
β-D-Glucose-2-D		+0.06 (C-3)		25
		+0.01 (C-1)		
β-D-Glucose-3-D		+0.05 (C-2)		25
		+0.06 (C-4)		
α-D-Glucose-5-D		+0.06 (C-4)		25
		+0.06 (C-5)		
β-D-Glucose-5-D		+0.06 (C-6)		25
α-D-Glucose-6-D$_2$[a]		+0.12 (C-5)		25
β-D-Glucose-6-D$_2$[a]		+0.12 (C-5)		25
α-D-Mannose-6-D$_2$		+0.13 (C-5)		25
β-D-Mannose-6-D$_2$		+0.11 (C-5)		25
α-D-Mannose-2-D		+0.06 (C-3)		25
		~+0.01 (C-1)		
β-D-Mannose-2-D		+0.06 (C-3)		25
		~+0.01 (C-1)		
α-D-Galactose-6-D		+0.12 (C-5)		25
β-D-Galactose-6-D		+0.10 (C-5)		25
α-D-Galactose-4-D		+0.05 (C-5)		25
		+0.10 (C-3)		
β-D-Galactose-4-D		+0.04 (C-5)		25
		+0.06 (C-3)		
α-D-Allose-3-D		+0.09 (C-2)		25
		+0.08 (C-4)		

TABLE 3 (*cont.*)

Csp^{3}DCsp^3	$^1\Delta$	$^2\Delta$	$^3\Delta$	Ref.
β-D-Allose-3-D		+0.09 (C-2)		25
		+0.08 (C-4)		
Methyl-2,3-di-*O*-methyl-				25
α-D-galactopyranoside-4-D		+0.04 (C-3)		
		+0.08 (C-5)		
Methyl-2,3-di-*O*-methyl-				25
α-D-glucopyranoside-4-D		+0.ᴜ4 (C-3)		
		+0.07 (C-5)		
α-D-Threose-1-D	+0.39			23
β-D-Threose-1-D	+0.34			23
D-1-Threose hydrate-1-D	+0.32			23
α-D-Erythrose-1-D	+0.34			23
β-D-Erythrose-1-D	+0.39			23
D-Erythrose hydrate-1-D	+0.34			23
Methyl-α-D-ribo-				23
furanoside-1-D	+0.34			
Methyl-β-D-ribo-				23
furanoside-1-D	+0.34			
α-D-Arabinopyranose-1-D	+0.44			23
β-D-Arabinopyranose-1-D	+0.39			23
α-D-Ribose 5-phosphate-1-D	+0.29			23
β-D-Ribose 5-phosphate-1-D	+0.29			23
D,L-Gluceraldehyde				23
3-phosphate	+0.34			

[a] An effect twice as large is reported in reference 359.

2. $^2\Delta C(D)$

The effect over two bonds is always smaller than the value of $^1\Delta$ and is usually smaller than 0.1 ppm. Fortunately, under high resolution conditions, $^2\Delta$ values may be observed. The line-broadening caused by unresolved carbon–deuterium coupling is in most cases negligible as $^2J(C—H)$ and hence $^2J(C—D)$ are small.

$^2\Delta$ values in open-chain hydrocarbons fall around +0.08 ppm. In cyclic hydrocarbons $^2\Delta$ varies in the following way: cyclopropane < cyclobutane ~ cyclobutene > cyclopentane ~ cyclohexane ~ cycloheptane.[15] In cyclohexane a small difference is observed between a D_{eq} and D_{ax}.[16] This difference is also observed between cyclohexane-D_{11} and cyclohexane-D_{12}.[17]

The effect of an oxygen atom is clearly demonstrated by a comparison of octadecanoic-18-D_3,16-D_2 acid and 17-hydroxyoctadecanoic-18-D_3,16-D_2 acid.[14] Oxygen at the carbon in question reduces the value of $^2\Delta$

dramatically. Effects of oxygen are also observed in carbohydrates. Carbon-deuterated hexoses show a shift to low frequency of $+0.01-0.10$ ppm (Table 3), which is less than for hydrocarbons.[25] No variation in the effect of an equatorial and an axial deuterium is found. An inverse relationship between the size of $^2\Delta$ and the deshielding of the ^{13}C nucleus has been established,[25] which can be expressed in other words as, the more electronegative the substituents at the β-carbon, the smaller is the value of $^2\Delta$; this is exemplified by glucose-2-D and mannose-2-D, in which $^2\Delta C\text{-}3(D) > {}^2\Delta C\text{-}2(D)$[25] (Table 3). Similar effects are observed in ethers.[26] The normal range is 0.04–0.06 ppm, but some unusual values are observed in α-D-galactose and α-D-allose[25] (Table 3).

Tulloch[27] found in octadecanoates that the $^2\Delta$ values of C-3 and C-4 in the 2,2'-D_2 derivative and of C-2 in the 3,3'-D_2 and of C-2 and C-3 of the 4,4'-D_2 derivative are smaller than other $^2\Delta$ and $^3\Delta$ values. They ascribe this to the presence of an ester group. In the investigation of the corresponding oxo derivatives, no effect of the oxo group is detected.

3. $^3\Delta C(D)$

The isotope effects over three bonds are small. They are not quoted in many papers as they are not resolved under the experimental conditions used. A splitting of 0.02 ppm is 0.5 Hz (at 25 MHz) and the lines may be broadened by unresolved C—D coupling.

Mazurek[14,27] found 0.02 ppm per D in long chain hydrocarbons. In cyclohexane an average value of 0.02 ppm is observed.[15,16] Temperature effects can be observed. In adamantane-2-D a difference dependent on the geometry is observed[18] [2] (see also Section III. K.2). In cyclobutane,

[2]

cyclohexane and cycloheptane the value of $^2\Delta$ is 0.027 ppm. However, in pentane a negative $^3\Delta$ is reported,[15] and in cycloheptane a negative $^4\Delta$ value is observed. The possibility of a conformational isotope effect was discussed, but considered less likely as a negative isotope effect is also observed in norbornane-1-D (a molecule with a rigid structure). In the latter C-4 shows a high frequency shift. The isotope effect is a sum of a $^3\Delta$ effect from the formally five-membered ring and a $^4\Delta$ effect from the six-membered ring ($^4\Delta$ is zero in cyclohexanes). The negative effect can, in consequence, be associated with the presence of a five-membered ring.[15]

Oxygen substitution may also diminish the value of $^3\Delta$, as shown by [3]. The possibility of observing $^3\Delta$ makes it possible to identify five carbons in long chains.[14,27,28]

$$C\underline{D}_2CHOH\underline{C} \qquad\qquad C\underline{D}_2CH_2\underline{C}$$
$$+0.03\ ppm \qquad [3] \qquad +0.05\ ppm$$

The value of $^4\Delta$ is usually too small to be observed. The effects of deuterium on dialkylcyclohexanes[29] are described in the section dealing with equilibrium effects. Likewise through-space effects are discussed separately.

B. Csp³DCsp² effects

Very few cases are reported. For cyclobutene values of $^2\Delta$ and $^3\Delta$ are given in Table 4. In toluene-4-D a long-range $^5\Delta = +0.0033$ ppm is reported.[30]

TABLE 4

Deuterium isotope effects on nuclear shieldings of sp³ hybridized carbons; deuterium at a sp² hybridized carbon (in ppm).

Csp³DCsp²	$^2\Delta$	$^3\Delta$	Ref.
Cyclobutene-1-D	+0.159[a] (C-1)	+0.053[a] (C-3)	20
Toluene-4-D[b]			30

[a] Sign reversed compared to original paper.
[b] $^5\Delta = +0.0033$ ppm.

C. Csp²DCsp³ effects

This combination is typically found in alkyl-substituted aromatics, in carbonyl derivatives and in carbonium ions.

1. $^2\Delta$

Deuteriated toluenes show a $^2\Delta$ effect of $+0.034$ ppm per deuterium and the effect is additive.[30] This isotope effect is much smaller than that observed in cyclobutene-3-D ($+0.078$ ppm)[20] (Table 5). The value of $^2\Delta$ observed in 4-methyl[2.2]paracyclophane-4′-D is similar,[31] but smaller than that of toluene. A comparison of methyl-, ethyl- and isopropylbenzene-α-D shows comparable effects (Table 5). A most unusual long-range effect is seen in ethylbenzene-β-D$_3$ (Table 5).

TABLE 5

Deuterium isotope effects on nuclear shieldings of sp^2 hybridized carbons; deuterium at sp^3 hybridized carbon (in ppm).

Csp^2DCsp^3	$^2\Delta$	$^3\Delta$	$^4\Delta$	Ref.
Cyclobutene-3-D	$+0.078^a$ (C-2)	-0.075^a (C-1)		20
Toluene-$D_1{}^c$	$+0.034$	-0.0002	0	30
Toluene-$D_2{}^d$	$+0.0692$	-0.0011	0	30
Toluene-$D_3{}^b$	$+0.10^a$	$+0.01^a$		100
Toluene-$D_3{}^e$	$+0.105$	-0.0016	0	30
Ethylbenzene-α-$D_1{}^f$	0.0296	$+0.002$	0	30
Ethylbenzene-β-$D_3{}^g$				30
Isopropylbenzene-α-$D_1{}^h$	0.0296	0.0143	0	30
4-Methyl[2.2]-*para*-cyclophane-4'-$D_1{}^i$	$+0.026^a$ (C-4)	-0.012^a (C-3)		31
4-Methyl[2.2]-*para*-cyclophane-4'-$D_2{}^j$	$+0.052^a$ (C-4)	-0.017^a (C-3)	$+0.01^a$ (C-8)	31
4-Methyl[2.2]-para-cyclophane-4'-$D_3{}^k$	$+0.080^a$ (C-4)	-0.026^a (C-3)	$+0.007^a$ (C-6) $+0.015^a$ (C-8)	31
4,7-Dimethyl[2.2]-*para*-cyclophane-4'-$D_2{}^l$	$+0.052$ (C-4)	-0.016^a (C-3) -0.006^a (C-5)	$+0.003^a$ (C-6) $+0.009^a$ (C-8)	31
4-Methyl[2_4](1,2,4,5)-cyclophane-4'-$D_2{}^m$	$+0.043^a$ (C-4)	-0.013^a (C-3) -0.013^a (C-5)		31
Indane-α,α-D_2		-0.037 -0.063		30
$CH_3\overset{+}{-}C(CD_3)_2$	-0.8^a			35
$(CD_3)_3\overset{+}{-}C$	-1.4^a			35
$CH_3CH_2\overset{+}{-}C(CD_3)_2$	-0.6^a			35
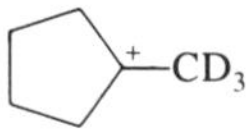 (cyclopentyl cation)$-CD_3$	-0.4^a			35
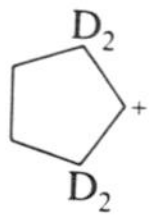 (cyclopentyl cation, D_2, D_2)	-0.5^a (C-1)			35
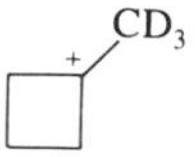 (cyclobutyl cation) CD_3	$+1.1^a$			35
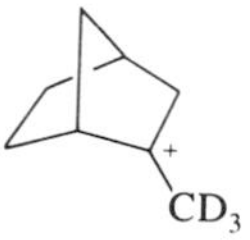 (norbornyl cation) CD_3	$+2.2^a$ (C-2)	$+0.3^a$ (C-3) -0.4^a (C-1)		35
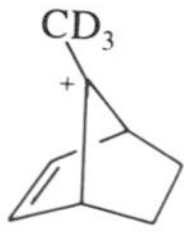 (norbornenyl cation) CD_3	$+0.8^a$			35

TABLE 5 (*cont.*)

Csp^2DCsp^3	$^2\Delta$	$^3\Delta$	$^4\Delta$	Ref.
$Ph\overset{+}{-}\!C(CD_3)_2{}^n$	0			35
	~0	0.05 (C-1')	−0.16 (C-2')	97
$CH_3PhC(CD_3)H$	0			35
$\triangleright\overset{+}{-}C(CD_3)H$	0			35
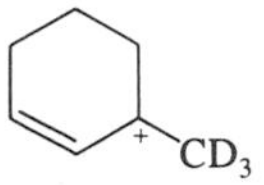 (cyclohexenyl) $\overset{+}{}CD_3$	0			35
$Ph\bar{C}CD_3CD_2C(CH_3)_3{}^o$	+0.20			97
Methyloctadecanoate-2-D_2	0 (COOH)			27
Methyloctadecanoate-3-D_2	0 (COOH)			27
CH_3COCH_3-D_6	−0.28^a			32
	−0.22^a			12
4-Methylcamphor-3,3'-D_2	−0.08 (C-2)			33
4-Chlorocamphor-3,3'-D_2	−0.01 (C-2)			33
4-Nitrocamphor-3,3'-D_2	−0.04 (C-2)			33
Norcamphor-3-D_1(exo)	+0.07			33
3,3'-D_2	−0.18			33
CH_3COOH-$D_4{}^p$	+0.43^a			12
$HCOOH$-$D_2{}^p$	+0.43^a			12
CH_3COOCH_3-D_6	0			12
$HCON(CH_3)_2$-$D_6{}^p$	+0.16			12
$(CD_3)_2C{=}NOH$	+0.08			105
1,3,5,Cycloheptatri-ene-7-D	+0.08			115
Dicyanocobyrimic acid heptanemethyl ester-13-D_1	+0.082			34

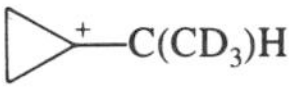

^{a} Sign reversed compared to original paper.
^{b} Recorded at 25.1 MHz.
^{c} $^5\Delta = -0.0039$ ppm.
^{d} $^5\Delta = -0.008$ ppm.
^{e} $^5\Delta = -0.0121$ ppm.
^{f} $^5\Delta = -0.0121$ ppm.
^{g} $^5\Delta = -0.0126$ ppm.
^{h} $^5\Delta = 0.0$ ppm.
^{i} $\Delta = -0.011$ ppm^a (C-15).
^{j} $\Delta = -0.023$ ppm^a (C-15).

TABLE 5—Footnotes (*Cont.*)

^{k} $^5\Delta = -0.010^a$ ppm (C-7); $\Delta = -0.035$ ppm^a (C-15).
^{l} $^5\Delta = -0.005$ ppm^a (C-5); $\Delta = -0.022$ ppm^a (C-15).
^{m} $^5\Delta = -0.012$ ppm^a (C-7); $\Delta = -0.030$ ppm^a (C-13).
^{n} $^5\Delta = \sim 0.00$ (C-3'); $^6\Delta = -0.22$ ppm (C-4').
^{o} $^6\Delta = 0.20$ ppm (C-4').
^{p} Combined effect.

The two-bond isotope effect of acetone was the first negative isotope effect to be reported.[32] The negative value has been ascribed to hyperconjugation.[30] Negative values, although smaller, have also been observed in substituted camphors.[33] In acids and acid derivatives the effects are either small, such as in methyloctadecanoates-2-D_2,[27] or in ethyl acetate-D_6,[12] or in dicyanocobyrinic acid heptamethyl ester.[34] Larger values are reported for formic acid derivatives,[12] but in these a $^1\Delta$ effect is present. In acetic acid the contribution from the acid deuterium should be subtracted (Section III.G.2) (Tables 5 and 16).

A large number of carbonium ions have been investigated.[35] Both positive and negative $^2\Delta$ values are found (Table 5). A negative value is taken as evidence for a classical carbonium ion, a zero effect indicates a classical delocalized ion and a positive effect as evidence for a non-classical σ- and π-bridged ion. The negative effects observed in the classical ions are related to hyperconjugation (Section XI.B). The effects observed in the other types are equilibrium effects to be discussed fully in Section XVI.

2. $^3\Delta$

Unusual isotope effects over three bonds are observed in cyclobutene-3-D^{20} and in the alkylbenzenes with deuterium at the α-carbon.[30,31] By unusual is meant a negative isotope effect. Until recently very few such effects had been reported, but they seem to me more common than originally anticipated (Section III.K.4 and XVI). The value of $^3\Delta$ for toluene-α-D, ethylbenzene-α-D, and isopropylbenzene-α-D increases from -0.2 ppb to $+14.3$ ppb. The conformations are given in [4]. Structure [4, B] is probably more highly populated in the latter which means that hyperconjugation becomes less favourable and hence a positive value is observed (Section XV.B). In indane-α,α'-D_2, a molecule with a fixed geometry, negative $^3\Delta$ values are observed.

A   B

[4]

D. Csp^2DCsp^2 effects

This combination occurs frequently in olefins and in aromatics, including heteroaromatics.

1. $^1\Delta$

In olefins the value of $^1\Delta$ falls close to 0.21 ppm. Similar values are derived from deuterated 1,4-benzoquinones,[36] whereas larger values are observed in 1,6-epoxybicyclo[4.3.0]nona-2,4-diene-4-D and in 1,6-methanobicyclo[4.3.0]nona-2,4-diene-4-D[37] (Table 6). Whether larger values will also be found in other 1,4-dienes remains to be seen.

TABLE 6

Deuterium isotope effects on ^{13}C nuclear shieldings of non-aromatic compounds with Csp^2 carbons with deuterium on Csp^2 (in ppm).

Csp^2DCsp^2	$^1\Delta$	$^2\Delta$	$^3\Delta$	Ref.
$CH_3(CH_2)_4CH{=}CHD^b$	+0.21	Unresolved		22
Cyclobutene-1-D	+0.245^a	+0.210^a		20
1,6-Epoxybicyclo - [4.3.0]nona-2,4-diene- 4-D^c	+0.35	+0.12 (C-3) +0.13 (C-5)	+0.04 (C-2)	37
1,6-Methanobicyclo- [4.3.0]nona-2,4-diene- 4-D^c	+0.30	+0.10 (C-3) +0.14 (C-5)	+0.01 (C-2)^d	37
Terrein	+0.30			88
1,4-Benzoquinone-2,5-D$_2$	+0.19^e			36
1,4-Benzoquinone-2,6-D$_2$	+0.21^e			36
Dicyanocobyrimic acid heptamethyl ester-D$_1$		+0.05 (C-11)		34

^{a} Sign reversed compared to original paper.
^{b} Probably a mixture of **E** and **Z** isomers.
^{c} According to reference 15, compounds 20 and 23 have been interchanged.
^{d} Sign changed as described in reference 15.
^{e} Relative to C-3.

The value of $^1\Delta$ for aromatic hydrocarbons falls between +0.24 and +0.31 ppm. A slight variation between α- and β-substitution in naphthalene is observed. The lower limit is given by 2-D-azulene,[39] a non-alternant hydrocarbon (Table 8). Substituent effects were first studied by Bell et al.,[40] who found that both $^1\Delta$ and $^2\Delta$ data are influenced by substituents. $^1\Delta$ values for ortho-substituted compounds correlate well with Taft's σ_o, while meta- and para-substituted compounds correlate with σ_m and σ_p, respectively. It was thus suggested that $^1\Delta$ is caused by two factors, the vibrational effect, which accounts for the major part, and a smaller contribution (ranging from -0.046 to $+0.032$ ppm) caused by the electronic demand of the substituent. An increase in the electron density at a given carbon gives an increased isotope effect.

Analysis of $^1\Delta$ of ortho compounds was suggested as a means of obtaining σ_o, as anisotropic and steric effects are expected to cancel out.[40] No further work along this line has, however, appeared.

The dramatic increase observed in going from benzene to tropylium ion[42] (Tables 7 and 8) is explained by a shortening of the C—H bond in the latter. An increase in $^1J(C-H)$ is also found;[42] formation of a chromium carbonyl compound leads to a corresponding drastic decrease.[43]

TABLE 7

Deuterium isotope effects on ^{13}C nuclear shieldings of benzenoid compounds; Csp^2DCsp^2 type (in ppm).

	$^1\Delta$	$^2\Delta$	$^3\Delta$	Ref.
Benzene-1-D	+0.289	+0.110	+0.001	40
	+0.292[c]	+0.108[c]	+0.008[c]	106
	+0.184[b]			106
	+0.297[c]	+0.110[c]	+0.014[c]	42
Benzene-1,3,5-D_3	+0.088[b]			106
Benzene-D_6	+0.57			94
	+0.53			93
	+0.5			21
	+0.66			12
$CH_3OC_6H_4$-2-D	+0.338	+0.092 (C-3)		40
		+0.037 (C-1)		
$CH_3C_6H_4$-2-D	+0.321	+0.114 (C-3)		40
		+0.086 (C-4)		
ClC_6H_4-2-D	+0.295	+0.106 (C-3)		40
		+0.106 (C-1)		
CNC_6H_4-2-D	+0.243	+0.103 (C-1)		40
$CH_3C_6H_4$-3-D	+0.29[a]	+0.19[a]		100
	+0.295	+0.110 (C-2)		40
		+0.111 (C-4)		40
$C(CH_3)_3C_6H_4$-3-D	+0.31[a]	+0.12[a]		100
$CH_3OC_6H_4$-3-D	+0.284	+0.102 (C-2)		40
		+0.111 (C-4)		
ClC_6H_4-3-D	+0.276	+0.106 (C-2)		40
		+0.101 (C-4)		
CNC_6H_4-3-D	+0.30	+0.15 (C-2)		45
		+0.10 (C-4)		
CNC_6H_4-4-D	+0.50	+0.10		45

TABLE 7 (*cont.*)

	$^1\Delta$	$^2\Delta$	$^3\Delta$	Ref.
$CH_3C_6H_4$-4-D	+0.294	+0.123		40
$CH_3OC_6H_4$-4-D	+0.292	0.118		40
ClC_6H_4-4-D	+0.284	0.101		40
$CF_3C_6H_4$-4-D	+0.279	0.113		40
1-Hydroxy-2-benzoic				91
acid methyl ester				
-3-D		+0.04 (C-2)		
-5-D		+0.1 (C-5)		91
-3,5-D_2		+0.2 (C-5)		91
C_6D_5Br	+0.32[a] (C-1)			12
	+0.33[a] (C-2)			
	+0.65[a] (C-3)			
	+0.54[a] (C-4)			
$C_6D_5CD_3$	+0.32[a] (C-1)			12
	+0.33[a] (C-2)			
	+0.65[a] (C-3)			
	+0.65[a] (C-5)			
$C_6D_5NO_2$	+0.11 (C-1)			12
	+0.65 (C-2)			
	+0.54 (C-3)			
	+0.54 (C-4)			
Scylatone-5-D	+0.2 (C-5)			85
Ferrocene-1-D_{10}	+0.40[a]			85
Benzenechromium				43
tricarbonyl-D_7[d]	+0.198[e]	+0.118[e]		
Tropylium ion-D_6[f]	+0.408[e]	+0.133[e]		42
Tropyliumchromium				
tricarbonyl-D_6[f]	+0.225	+0.097	+0.002	43

[a] Sign reversed compared to original paper.
[b] Relative to C-2.
[c] Relative to C-4.
[d] $^4\Delta = +0.032$ ppm relative to $^3\Delta$.
[e] Relative to $^3\Delta$.
[f] $^4\Delta = +0.035$ ppm relative to $^3\Delta$.

$^1\Delta$ values of heteroaromatics are clearly controlled by the nature of the heteroatom and possibly also by the degree of aromaticity (Table 9).

2. $^2\Delta$

The effects observed in olefins and aromatics are of the same magnitude ~0.1 ppm (Tables 6–8). In polycyclic aromatics Martin *et al.*[38] observed that $^2\Delta$ varies from +0.05 to 0.12 ppm and that in a compound with two

TABLE 8

Deuterium isotope effects on ^{13}C nuclear shieldings of polycyclic aromatics (in ppm).

	$^1\Delta$	$^2\Delta$	$^3\Delta$	Ref.
Naphthalene-1-D	+0.31	+0.12		38
Naphthalene-2-D	+0.28	+0.12 (C-1)		38
		+0.08 (C-3)		
Naphthalene-2,3-D$_2$	+0.38^a	+0.11^a		39
Phenanthrene-9-D	+0.31	+0.12 (C-10)	+0.04 (C-8)	38
		+0.07 (C-8a)		
Benzo[e]phenanthrene-2-D	+0.27	+0.11 (C-1)		38
		+0.10 (C-3)		
Hexahelicene-7-D		+0.10 (C-8)	+0.05 (C-6)	38
		+0.05 (C-6a)		
Anthracene-9,10-D$_2$	+0.38	+0.09 (C-4a)	+0.04 (C-1)	38
Phenanthrene-9,10-D$_2$	+0.2	+0.05 (C-8a)		107
Phenanthrene-4,5-D$_2$	+0.3	+0.07 (C-4a)		107
		~+0.1 (C-3)		
Pyrene-1,3,6,8-D$_4$	+0.32	+0.23 (C-2)	+0.06 (C-10)	108
		+0.08 (C-10a)		
Azulene-2-D^c	+0.24^a	+0.10^a	−0.01	39
Fluorene-1-D^d				109
Fluorenone-1-D^e		+0.09^a (C-2)		109
1-Phenylazo-2-naphthol-8-D		+0.1 (C-7)		110

^{a} Sign reversed compared to original paper.
^{b} $^5\Delta$(C-7) = +0.012^a ppm.
^{c} $^5\Delta$(C-5) = +0.01^a ppm; $^6\Delta$ = +0.009^a ppm.
^{d} $^4\Delta$(C-8a) = +0.14 ppm. Should probably be $^2\Delta$(C-9a).
^{e} $^4\Delta$(C-8a) = +0.09 ppm.

ortho carbons the larger Δ value is associated with the larger π-bond order or length of the C—C bond. As seen in Table 8, this feature is also observed in other polyaromatic hydrocarbons.

Bell *et al.*[40] have observed, in substituted benzene derivatives, that the value of $^2\Delta$ for a carbon with a substituent in the *para* position shows some correlation with σ_p, but carbons with substituents in *meta* or *ortho* positions do not. The value of $^2\Delta$ is considered to be dominated by the effects that the substituents have on the vibrational modes of the ring carbons.

Kitching *et al.*[44] report no splitting due to a $^2\Delta$ effect in 4-deuterio-1-cyanonaphthalene, whereas Lebel *et al.*[45] claim such an effect in 4-deuteriobenzonitrile. From [5] it is apparent that chlorine, methyl and cyano groups have little effect, whereas an oxygen diminishes the size of $^2\Delta$, a trend quite similar to that observed for the Csp^3DCsp^3 situation. It is also clear that more data are needed in this area.

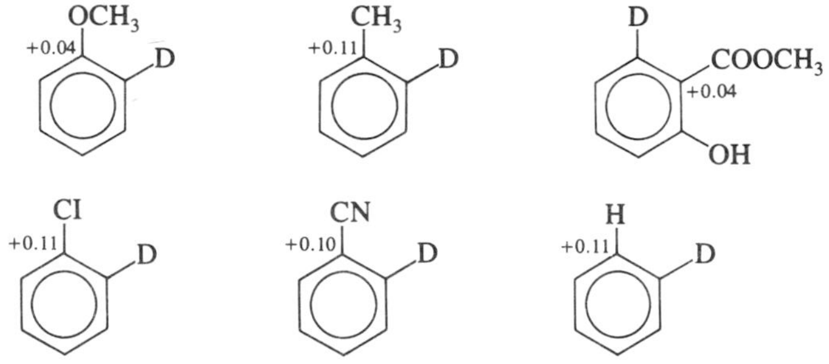

P. E. HANSEN

OCH$_3$ +0.04 D

CH$_3$ +0.11 D

D COOCH$_3$ +0.04 OH

Cl +0.11 D

CN +0.10 D

H +0.11 D

[5]

TABLE 9

Deuterium isotope effects on ^{13}C nuclear shieldings of heteroaromatics with deuterium at ring carbons.

	$^1\Delta$	$^2\Delta$	$^3\Delta$	Ref.
Pyridine-D$_5$	+0.54^a (C-2)			12
	+0.55			93
	+0.65^a (C-3)			12
	+0.29			93
	+0.54^a (C-4)			12
	+0.49			93
Indazole		+0.1^b		111
2,4-Dimethyl-3-ethyl-pyrrole-1,5-D$_2$	+0.34 (C-5)			112
2-Thiomethylthiophene-3-D	+0.2	−0.1 (C-2)		113
2-Bromothiophene-5-D	+0.2	−0.1 (C-3)^c	−0.1 (C-2)	113
3-Bromothiophene-2-D	−0.1	−0.1 (C-3)	−0.3 (C-4)	113
			−0.3 (C-5)	
3-Bromothiophene-4-D	−0.2	0 (C-3)	−0.2 (C-2)	113
		−0.2 (C-5)		
Dibenzothiophene-1-D^d		+0.1^a (C-2)		109
Dibenzothiophene-3-D		+0.09^a (C-2)		109
		+0.09^a (C-2)		
Dibenzothiophene-5-oxide-3-D		+0.07^a (C-2)		
		+0.09^a (C-4)		109
Dibenzothiophene sulphoxide-3-D	+0.345	+0.09^a (C-2)		109
		+0.13^a (C-4)		
Dibenzofuran-1-D^e	+0.23^a	+0.09 (C-2)		109
Dibenzofuran-1-D^f	+0.23^a	+0.14 (C-2)		109

^{a} Sign reversed compared to original paper.
^{b} Observed in general in indazoles.
^{c} The nuclear shielding of C-3 in the deuterated compound is assumed to be 130.8 ppm.
^{d} $^4\Delta$(C-8a) = +0.1^a ppm.
^{e} Solvent, CDCl$_3$.
^{f} Solvent, DMSO-d_6.

TABLE 10

Deuterium isotope effects on ^{13}C nuclear shieldings of heterocycles (in ppm).

Heterocycles	$^2\Delta$	$^3\Delta$	Ref.
1,2,3,4,6,7,12,12b-octa-hydroindolo[2,3-a]quino-			82
lizine	$+0.21^a$ (C-2)	$+0.15^a$ (C-12a)	
-1-D$_2$^{b}	$+0.08^a$ (C-12b)		
-3-D$_2$^{b}	$+0.16^a$ (C-2)		82
	$+0.09^a$ (C-4)		
-4-D$_2$^{b}	$+0.20^a$ (C-3)		82
-6-D$_2$^{b}	$+0.27^a$ (C-7)		82
-7-D$_2$^{b}	$+0.11$ (C-6)		82
-12b-D$_1$^{b}	$+0.21$ (C-1)		82
	-0.09 (C-12a)		
-9-D$_1$^{b}	$+0.09$ (C-8)		82
	$+0.05$ (C-10)		
8,13-Diazaestrone steroid			81
-9-D^c	$+0.09$ (C-10)		
	$+0.22$ (C-11)	$+0.13$ (C-7)	
-14-D^{c,d}	$+0.38$ (C-15)	$+0.09$ (C-16)	81
		$+0.32$ (C-7)	
		$+0.24$ (C-12)	
-17-D$_2$	$+0.23$ (C-16)		81

^{a} Sign reversed compared to original paper.
^{b} Many long-range effects are reported.
^{c} See [19].
^{d} $^4\Delta$(C-10) = +0.30 ppm; $^4\Delta$(C-6) = +0.12 ppm; $^5\Delta$(C-5) = +0.18 ppm; $^5\Delta$(C-9) = +0.27 ppm; $^5\Delta$(C-17) = +0.25 ppm.

TABLE 11

Deuterium isotope effects on ^{13}C nuclear shieldings of compounds with Csp^2 and Csp carbons (in ppm).

	$^1\Delta$	$^2\Delta$	Ref.
Csp^{2}DCsp			30
Phenylacetylene-β-D$_1$ ^{b}			
CspDCsp^3			
CH$_3$CN-D$_3$		-0.01^a	12
CspDCsp			
CH$_3$(CH$_2$)$_4$C≡CD	$+0.22$	$+0.50$	22

^{a} Signs changed compared to original paper.
^{b} $^6\Delta$ = +0.0076 ppm.

3. $^3\Delta$, $^4\Delta$ and $^5\Delta$

$^3\Delta$ values are quite small in benzene derivatives, but are larger and hence easier to observe in polycyclic aromatics in a *peri* position [6] (Table 8).

In 2-deuterioazulene a negative $^3\Delta$C-3a(D) value is found.[39] Whether or not the cause is the existence of a five-membered ring, as in the alkanes, is only speculation.

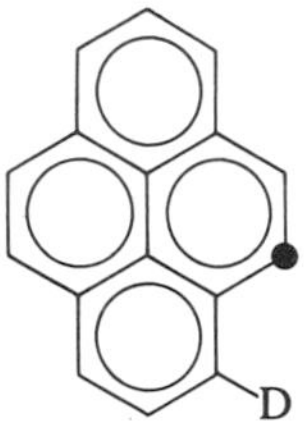

+0.06 ppm

[6]

Günther *et al.*[42] have observed in the tropylium ion,[42] and in π-benzene-chromium tricarbonyl,[43] that $^4\Delta > ^3\Delta$, whereas the π-tropyliumchromium tricarbonylcation shows the expected attenuation (Table 7). In 2,3-dideuterionaphthalene a $^5\Delta + ^6\Delta$ isotope effect is observed and 2-deuterioazulene shows isotope effects over five and six bonds.[39]

E. Csp^{2}DCsp^3 and CspDCsp effects

An interesting compound, phenylacetylene-β-D$_1$, shows a $^6\Delta$ effect.[30] In 1-deuteriohept-1-yne $^2\Delta > ^1\Delta$,[22] which has been confirmed.[48]

F. Csp^{3}DX effects

When X is oxygen the situation is observed in alcohols, carbohydrates and acids. As these hydrogens are labile, the deuteriated compounds are easy to prepare (Section XVIII).

1. $^2\Delta$

Alcohols show fairly constant $^2\Delta$ values[47] (Table 12). Some typical $^2\Delta$ values of carbohydrates have been reported,[49] as shown in [7]. $^2\Delta$ values of carbohydrates are similar except for C-6. This carbon has a relatively large shift, 0.15 ppm, considering that it has only a single $^2\Delta$ contribution.

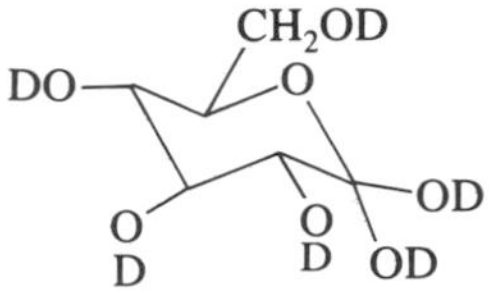

$^2\Delta$C-1(OD)$= +0.11$ ppm
$^2\Delta$C-4(OD)$= +0.14$ ppm
$^2\Delta$C-6(OD)$= +0.15$ ppm
$^3\Delta$C-3(OD-4)$= +0.03$ ppm
$^3\Delta$C-2(OD-1 *trans*)$= +0.06$ ppm
$^3\Delta$C-2(OD-1 *cis*)$= +0.03$ ppm

[7]

TABLE 12

Deuterium isotope effects on ^{13}C nuclear shieldings; deuterium on oxygen heteroatom; carbons of sp^3 type (in ppm).

Csp^3DX	Heteroatom	$^2\Delta$	$^3\Delta$	Ref.
Butanol	O	$+0.12^a$	$+0.05^a$	47
2-Methylpropanol	O	$+0.12^a$	$+0.04^a$	47
2-Ethoxyethanol	O	$+0.12^a$	$+0.03^a$	47
2-Propyn-1-ol	O	$+0.10^a$	$+0.04^a$	47
Benzyl alcohol	O	$+0.12^a$	$+0.04^a$	47
Cinnamyl alcohol	O	$+0.12^a$	$+0.04^a$	47
exo-Norborneol	O	$+0.12^a$	$+0.10^a$ (C-6)	47
			$+0.04^a$ (C-2)	47
Hexafluoroacetone hydrate-D$_2$	O	$+0.21^a$	$+0.16^a$	12
Trifluoroacetic acid-D$_1$	O	$+0.33^a$	$+0.11^a$	12
Acetic acid-D$_1$	O		$+0.02$	63
Pentane-2,4-dione-D$_1$	O		$+0.04$	61
N-Nitrosodiethanolamine-D$_1$	O	$+0.13$		54
Salol-D$_1$ ^{c}	O			74

^{a} Signs changed compared to original paper.
^{b} In trifluoroacetic acid.
^{c} $\Delta(CH_2) = -0.10$ ppm. The observed isotope effect is probably of equilibrium type.

Concievably this larger isotope effect could be accounted for by preferential hydration by D_2O between the ring oxygen and the C-6 OH group.[49]

2. $^3\Delta$

This type of isotope effect may be observed at C-5 in O-deuteriated carbohydrates and falls between $+0.03$ and $+0.06$ ppm (Table 13). In glucosylated sugars, such as 3-O-methyl-D-glucopyranose [8], $^3\Delta$C-3(OD) may be obtained directly. In this, as well as in α- and β-nigerose, and in α- and β-2,6-di-O-(3-nitropropanoyl)-D-glucopyranose, the value of $^3\Delta$ is twice as large as expected[49] when compared to $^3\Delta$C-5(OD) of aldohexoses (Table 13).

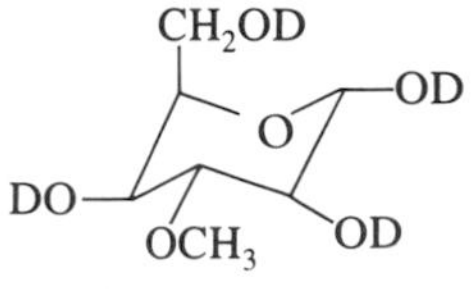

[8]

TABLE 13

Deuterium isotope effects on ^{13}C nuclear shieldings of carbohydrates deuterated at oxygens (in ppm).

Csp^{3}DX	Heteroatom	$^2\Delta^b$	$^3\Delta$	Ref.
β-D-Glucopyranose	O	+0.10^c (C-1)	+0.05^f (C-5)	50
		+0.15^d (C-1)		49
		+0.14^d (C-1)	+0.06^d (C-5)	48
		+0.19^c (C-2)	+0.02^e (C-5)	50
		+0.21^d (C-2)		49
		+0.15^c (C-3)		48
		+0.20^d (C-3)		48, 49
		+0.13^c (C-4)		50
		+0.15^d (C-4)		48
		+0.12^c (C-6)		50
		+0.15^d (C-6)		48
α-D-Glucopyranose	O	+0.10 (C-1)	+0.05^f (C-5)	50
		+0.15^d (C-1)		49
		+0.13^d (C-1)	+0.06^d (C-5)	48
		+0.17^c (C-2)	+0.01^e (C-5)	50
		+0.20^d (C-2)		49
		+0.20^d (C-2)		48
		+0.17^c (C-3)		50
		+0.20^d (C-3)		49
		+0.20^d (C-3)		48
		+0.13^c (C-4)		50
		0.15^d (C-4)		49
		+0.15^d (C-4)		48
		+0.12^c (C-6)		50
		+0.15^d (C-6)		48
α-D-Mannopyranose	O	+0.10^c (C-1)	+0.05^c (C-6)	50
		+0.15^c (C-2)		
		+0.12^c (C-3)		
		+0.14^c (C-4)		
		+0.12^c (C-6)		
β-D-Mannopyranose	O	+0.10^{b,c} (C-1)	+0.05^c (C-5)	50
		+0.13^{b,d} (C-2)		
		+0.17^{b,c} (C-2)		
		+0.16^{b,c} (C-3)		
		+0.13^{b,c} (C-4)		
		+0.12^{b,c} (C-6)		
α-D-Galactose	O	+0.14^d (C-1)	+0.07^d (C-5)	48
		+0.20^d (C-2)		
		+0.19^d (C-3)		
		+0.16^d (C-4)		
		+0.16^d (C-6)		
β-D-Galactose	O	+0.13^d (C-1)	+0.07^d (C-5)	48
		+0.24^d (C-2)		
		+0.19^d (C-3)		

TABLE 13 (*cont.*)

Csp^{3}DX	Heteroatom	$^2\Delta^b$	$^3\Delta$	Ref.
		$+0.17^d$ (C-4)		
		$+0.16^d$ (C-6)		
Methyl-α-D-glucopyra-noside^g	O	$+0.14^c$ (C-2)	$+0.02^c$ (C-1)	50
		$+0.15^d$ (C-2)	$+0.01$ (C-1)	48, 49
		$+0.19^c$ (C-3)	$+0.05^f$ (C-5)	50
		$+0.20^d$ (C-3)	$+0.07$ (C-5)	48, 49
		$+0.14^c$ (C-4)	$+0.02^e$ (C-5)	50
		$+0.15^d$ (C-4)		48, 49
		$+0.12^c$ (C-6)		50
		$+0.15^d$ (C-6)		48, 49
Methyl-β-D-glycopyra-noside^h	O	$+0.13^c$ (C-2)	$+0.05$ (C-1)	50
		$+0.15^d$ (C-2)	$+0.01^d$ (C-1)	48
		$+0.19^c$ (C-3)	$+0.05$ (C-5)	50
		$+0.21^d$ (C-3)	$+0.07$ (C-5)	48
	O	$+0.13^e$ (C-4)	0.00^e (C-1)	50
		$+0.16^d$ (C-4)		48
		$+0.12^c$ (C-6)		50
		$+0.15^d$ (C-6)		48
Methyl-α-D-galactopyra-noside	O	$+0.15^d$ (C-2)	$+0.05^d$ (C-5)	48
		$+0.18^d$ (C-2)	0.00^d (C-1)	
		$+0.15^d$ (C-4)		
		$+0.14^a$ (C-6)		
Methyl-β-D-galactopyra-noside^j	O	$+0.11^c$ (C-2)	$+0.01^c$ (C-1)	50
		$+0.17^d$ (C-2)	0.00^d (C-1)	48
		$+0.14^c$ (C-3)	$+0.03^c$ (C-5)	50
		$+0.18^d$ (C-3)	$+0.05^d$ (C-5)	48
		$+0.10^c$ (C-4)		50
		$+0.15^d$ (C-4)		48
		$+0.10^c$ (C-6)		50
		$+0.15^d$ (C-6)		48
Methyl-α-D-mannopyra-noside^i	O	$+0.14^{b,c}$ (C-2)	$+0.03^c$ (C-1)	50
		$+0.17^{b,c}$ (C-3)	$+0.05^c$ (C-5)	
		$+0.14^{b,c}$ (C-4)		
		$+0.12^{b,c}$ (C-6)		
3-*O*-Methyl-α-D-gluco-pyranose	O	$+0.18^d$ (C-2)		49
		$+0.12^d$ (C-3)		
		$+0.13^d$ (C-4)		
3-*O*-Methyl-β-D-gluco-pyranose	O			
		$+0.18^d$ (C-2)		49
		$+0.12^d$ (C-3)		
		$+0.12^d$ (C-4)		

TABLE 13 (*cont.*)

Csp^{3}DX	Heteroatom	$^2\Delta^b$	$^3\Delta$	Ref.
α-D-Cellobiose	O	+0.15^d (C-2)	+0.0^d (C-1)	48
		+0.20^d (C-3)	+0.05^d (C-5)	
		+0.15^d (C-4)	+0.05^d (C-4')	
		+0.15^d (C-6)	+0.04^d (C-5')	
		+0.11^d (C-1')		
		+0.21^d (C-2')		
		+0.15^d (C-3')		
		+0.14^d (C-6')		
β-D-Cellobiose	O	+0.13^d (C-1')	0.0^d (C-4')	48
		+0.22^d (C-2')	0.0^d (C-5')	
		+0.16^d (C-3')		
		+0.14^d (C-6')		
α-D-Lactose	O	+0.16^d (C-2)	0.0^d (C-1)	48
		+0.19^d (C-3)	0.07^d (C-5)	
		+0.16^d (C-4)	+0.03^d (C-4')	
		+0.15^d (C-6)	+0.03^d (C-5')	
		+0.15^d (C-1')		
		+0.23^d (C-2')		
		+0.15^d (C-3')		
		+0.15^d (C-6')		
β-D-Lactose	O	+0.13^d (C-1')	+0.03^d (C-4')	48
		+0.22^d (C-2')	+0.02^d (C-5')	
		+0.15^d (C-3')		
		+0.12^d (C-6')		
Nigerose^k	O	+0.13^d (C-2)	+0.13^d (C-3)	49
		+0.13^d (C-4)		
Nigerose	O	+0.15^d (C-2)	+0.13^d (C-3)	49
		+0.13^d (C-4)		
Methyl-β-D-lacto-pyranoside	O	+0.15^d (C-2)	+0.0^d (C-1)	48
		+0.19^d (C-3)	+0.05^d (C-5)	
		+0.16^d (C-4)	0.0^d (C-1')	
		+0.17^d (C-6)	0.0^d (C-4')	
		+0.16^d (C-2')	0.0^d (C-5')	
		+0.14^d (C-3')		
		+0.15^d (C-6')		
Sucrose		+0.12^{b,c} (C-2)	+0.05^f (C-1)	50
		+0.16^d (C-2)	+0.02^d (C-1)	48
		+0.15^{b,c} (C-3)	+0.01^e (C-1)	50
		+0.20^d (C-3)	+0.06^d (C-5)	48
		+0.12^{b,c} (C-4)	+0.05^c (C-5)	50
		+0.17^d (C-4)		48
		+0.15^{b,c} (C-6)	+0.05^c (C-5)	50
		+0.15^d (C-6)		48
		+0.10^{b,c} (C-1)^m	+0.05^f (C-2)	50
			+0.00^c (C-2)	50
		+0.15^d (C-1)^m	+0.03^d (C-2)	48

TABLE 13 (*cont.*)

Csp³DX	Heteroatom	$^2\Delta^b$	$^3\Delta$	Ref.
		$+0.17^d$ (C-3)^m	$+0.05$ (C-5)	50
		$+0.12^e$ (C-3)^m		50
		$+0.16^d$ (C-3)^m	$+0.06$ (C-5)	48
		$+0.15^{b,c}$ (C-4)^m		50
		$+0.16^d$ (C-4)^m		48
		$+0.10^{b,c}$ (C-6)^m		50
		$+0.14^d$ (C-6)^m		48
α,α-Trehalose	O	$+0.10^c$ (C-2)	0.00^c (C-1)	50
		$+0.15^d$ (C-2)	0.00^d (C-1)	48
		$+0.16^c$ (C-3)	$+0.01^f$ (C-5)	50
		$+0.20^d$ (C-3)	$+0.07^d$ (C-5)	48
		$+0.12^c$ (C-4)	$+0.04^e$ (C-5)	50
		$+0.15^d$ (C-4)		48
		$+0.10^c$ (C-6)		50
		$+0.13^d$ (C-6)		48
α,β-Trehalose^d	O	$+0.13$ (C-2α)	0.00 (C-1α)	48
		$+0.19$ (C-3α)	$+0.07$ (C-5α)	
		$+0.15$ (C-4α)	$+0.00$ (C-1β)	
		$+0.14$ (C-6α)	$+0.00$ (C-5β)	
		$+0.14$ (C-2β)		
		$+0.21$ (C-3β)		
		$+0.15$ (C-4β)		
		$+0.14$ (C-6β)		
α-Isomaltose	O	$+0.23^b$ (C-2′)	$+0.07$ (C-5′)	49
		$+0.24^b$ (C-3′)	$+0.00$ (C-1′)	
		$+0.18^b$ (C-4′)	$+0.00$ (C-6)	
		$+0.16^b$ (C-6′)	$+0.07$ (C-5)	
		$+0.17^b$ (C-1)		
		$+0.21^b$ (C-2)		
		$+0.23^b$ (C-3)		
		$+0.18^b$ (C-4)		
β-Isomaltose	O	$+0.23^b$ (C-2′)	0.00 (C-1′)	49
		$+0.24^b$ (C-3′)	$+0.07$ (C-5′)	
		$+0.18^b$ (C-4′)	$+0.07$ (C-5)	
		$+0.16^b$ (C-6′)	0.00 (C-1)	
		$+0.17^b$ (C-1)		
		$+0.23^b$ (C-2)		
		$+0.23^b$ (C-3)		
		$+0.18^b$ (C-4)		
α-D-Xylose	O	$+0.15^d$ (C-1)	$+0.07$ (C-5)	48
		$+0.20^d$ (C-2)		
		$+0.21^d$ (C-3)		
		$+0.17^d$ (C-4)		
β-D-Xylose	O	$+0.15^d$ (C-1)	$+0.07$ (C-5)	48
		$+0.23^d$ (C-2)		
		$+0.21^d$ (C-3)		
		$+0.17^d$ (C-4)		

P. E. HANSEN

TABLE 13 (*cont.*)

Csp³DX	Heteroatom	$^2\Delta^b$	$^3\Delta$	Ref.
α-D-Arabinose	O	$+0.15^d$ (C-1)	$+0.07$ (C-5)	48
		$+0.23^d$ (C-2)		
		$+0.19^d$ (C-3)		
		$+0.17^d$ (C-14)		
β-D-Arabinose	O	$+0.14^d$ (C-1)		48
Methyl-α-D-arabino-pyranoside	O	$+0.17^d$ (C-2)	$+0.04^d$ (C-1)	48
		$+0.19^d$ (C-3)	$+0.06^d$ (C-5)	
		$+0.17^d$ (C-4)		
Methyl-β-D-arabino-pyranoside	O	$+0.17^d$ (C-2)	$+0.03^d$ (C-1)	48
		$+0.20^d$ (C-3)	$+0.03^d$ (C-5)	
		$+0.17^d$ (C-4)		
Fructose^{n,r}	O	$+0.18^d$ (C-1)	$+0.06$ (C-6)	48
		$+0.12^d$ (C-2)		
		$+0.18^d$ (C-3)		
		$+0.19^d$ (C-4)		
		$+0.16^d$ (C-5)		
Talose^{o,r}	O	$+0.14^d$ (C-1)	$+0.06$ (C-5)	48
		$+0.20^d$ (C-2)		
		$+0.16^d$ (C-3)		
		$+0.16^d$ (C-4)		
		$+0.15^d$ (C-6)		
Talose^{n,r}	O	$+0.13^d$ (C-1)		48
		$+0.20^d$ (C-2)		
		$+0.16^d$ (C-3)		
		$+0.17$ (C-4)		
		$+0.16$ (C-6)		
Fructose^{p,d,r}	O	$+0.20$ (C-3)	$+0.10$ (C-2)	48
		$+0.18$ (C-4)	$+0.07$ (C-5)	
		$+0.13$ (C-6)		
Fructose^{q,d,r}	O	$+0.18$ (C-1)	$+0.11$ (C-2)	48
		$+0.17$ (C-3)	$+0.05$ (C-5)	
		$+0.15$ (C-4)		
		$+0.14$ (C-6)		
Talose^{p,d,r}		$+0.16$ (C-1)	$+0.05$ (C-4)	48
		$+0.18$ (C-2)		
		$+0.16$ (C-3)		
		$+0.18$ (C-5)		
		$+0.19$ (C-6)		
Talose^{q,d,r}		$+0.13$ (C-1)	$+0.06$ (C-4)	48
		$+0.18$ (C-2)		
		$+0.15$ (C-3)		
		$+0.18$ (C-6)		
3-Deoxy-D-*ribo*-hexo-pyranose		$+0.15$ (C-1)	$+0.16$ (C-3)	49
		$+0.17$ (C-2)	$+0.08$ (C-5)	

TABLE 13 (*cont.*)

Csp^{3}DX	Heteroatom	$^2\Delta^b$	$^3\Delta$	Ref.
		+0.17 (C-4)		
		+0.17 (C-6)		
3-Deoxy-β-D-*ribo*-hexo-pyranose		+0.15 (C-1)	+0.17 (C-3)	49
		+0.19 (C-2)	+0.07 (C-5)	
		+0.14 (C-4)		
		+0.16 (C-6)		
3-Deoxy-α-D-*ribo*-hexo-furanose		+0.15 (C-1)	+0.07 (C-3)	49
		+0.15 (C-2)	0.00 (C-4)	
		+0.14 (C-5)		
		+0.19 (C-6)		
3-Deoxy-β-*ribo*-hexo-furanose		+0.18 (C-1)	+0.07 (C-3)	49
		+0.17 (C-2)	+0.05 (C-4)	
		+0.14 (C-5)		
		+0.20 (C-6)		
6-Deoxy-α-D-galactose (fucose)^s		+0.13^d (C-1)	+0.03^d (C-5)	48
		+0.20^d (C-2)		
		+0.18^d (C-3)		
		+0.16^d (C-4)		
6-Deoxy-β-D-galactose		+0.14^d (C-1)	+0.04^d (C-5)	48
		+0.22^d (C-2)		
		+0.19^d (C-3)		
		+0.17^d (C-4)		
1,6-Anhydro-D-glucose		+0.17^d (C-2)	+0.05^d (C-1)	48
		+0.21^d (C-3)	+0.03^d (C-5)	
		+0.17^d (C-4)		
α-D-Glucofuranic acid δ-lactone		+0.13^d (C-1)	0.00^d (C-3)	48
		+0.12^d (C-2)	0.00^d (C-4)	
		+0.11^d (C-5)	0.00^d (C-6)	
β-D-Glucofuranoic acid δ-lactone		+0.16^d (C-1)	+0.04^d (C-3)	48
		+0.14^d (C-2)	0.00^d (C-4)	
		+0.11^d (C-5)	0.00^d (C-6)	
2-Amino-2-deoxy-α-D-glucopyranose	O	+0.22^b (C-1)	+0.10 (C-5)	49
		+0.27^b (C-3)		
		+0.15^b (C-4)		
		+0.17^b (C-6)		
2-Amino-2-deoxy–β-D-glucopyranose	O	+0.22 (C-1)	+0.08 (C-5)	49
		+0.26 (C-3)		
		+0.15 (C-4)		
		+0.14 (C-6)		

TABLE 13 (*cont.*)

Csp³DX	Heteroatom	²Δᵇ	³Δ	Ref.
2-Amino-2-deoxy-α-D-gluco-pyranose hydrochloride		+0.22ᵇ (C-1)		49
		+0.22ᵈ (C-1)		
		+0.34ᵈ (C-2)		
		+0.27ᵇ (C-3)		
		+0.27ᵈ (C-3)		
β-Amino-2-deoxy-α-D-glucopyranose hydro-chloride		+0.22ᵇ (C-1)		49
		+0.22ᵈ (C-1)		
		+0.26ᵇ (C-3)		
		+0.26ᵈ (C-3)		
α-Ascorbic acidᵗ	O	+0.20 (C-5)	+0.10ᵛ (C-4)	49
		+0.20 (C-6)		
α-Ascorbic acidᵘ	O	+0.16 (C-5)	+0.03 (C-4)	49
		+0.20 (C-6)		
α-D-Glucopyranuronic acid	O	+0.13 (C-1)	+0.14 (C-5)	48
		+0.19 (C-2)		
		+0.20 (C-3)		
		+0.19 (C-4)		
α-D-Glucopyranurate		+0.13 (C-1)	+0.00 (C-5)	48
		+0.18 (C-2)		
		+0.18 (C-3)		
		+0.14 (C-4)		
β-D-Glucopyranuronic acid		+0.13 (C-1)	+0.14 (C-5)	48
		+0.20 (C-2)		
		+0.21 (C-3)		
		+0.19 (C-4)		
β-D-Glucopyranurate		+0.13 (C-1)	0.00 (C-5)	48
		+0.20 (C-2)		
		+0.19 (C-3)		
		+0.14 (C-4)		
1,2,3,4-Tetra-O-acetyl-β-D-glucopyranoseʷ	O	+0.151 (C-6)		53
1,2-O-Isopropylidene-α-D-glucofuranoseˣ		+0.11 (C-3)	0.00 (C-1)	50
		+0.11ᵇ,ᶜ (C-5)	+0.03ᶜ (C-2)	
		+0.14ᵇ,ᶜ (C-6)	+0.03ᶜ (C-4)	
1,2,5,6-Di-O-isopropylidene-α-D-glucofuranoseʸ	O	+0.194 (C-3)	−0.01 (C-4)	53
			−0.01 (C-2)	
1,3,4-Tri-O-acetyl-α-D-xylo-pyranoseᶻ	O	+0.175 (C-2)	+0.063 (C-3)	53
			−0.02 (C-1)	
1,4-Di-O-acetyl-α-D-xylo-pyranose		+0.135 (C-3)	+0.02 (C-3)	53
1,2,3,4-Tetra-O-acetyl-6-O-(1,3,4,6-β-D-glucopyra-				

TABLE 13 (*cont.*)

Csp^3DX	Heteroatom	$^2\Delta^b$	$^3\Delta$	Ref.
nosyl)-β-D-glucopyranose*	O	+0.110 (C-2)	+0.06 (C-1) +0.006 (C-3)	53
Methyl-2,3,4-tri-O-acetyl-6-O-(1,2,3,4-tetra-O-acetyl-β-D-glucopyranosyl)-β-D-glucopyranoside	O	+0.170 (C-6)		53
2-Acetamido-1,3,4,6-tetra-O-acetyl-2-desoxy-β-D-glucopyranose	O		+0.02 (C-1)	53
2-Acetamido-1,3,4-tri-O-acetyl-2-desoxy-β-D-glucopyranose	O	+0.124 (C-5) +0.170 (C-6)	+0.07 (C-2) +0.122 (C-4)	53
Methyl-6-O-benzoyl-α-D-glucopyranoside	O	$+0.20^d$ (C-3) $+0.14^d$ (C-4)	0.00^d (C-5) 0.00^d (C-6)	48
Methyl 6-bromo-4-O-benzoyl-α-D-glucopyranoside		$+0.15^d$ (C-2) $+0.12^d$ (C-3)	0.00^d (C-1) 0.00 (C-4)	48

^{a} Signs changed compared to original paper.
^{b} Strictly speaking a combination of $^2\Delta$ and $^3\Delta$.
^{c} Solvent, DMSO-d_6. Similar values are measured in D_2O.
^{d} DIS measurements.
^{e} Solvent, D_2O.
^{f} Solvent, DMSO-d_6.
^{g} $\Delta(CH_3) = 0.00^{d,c}$ ppm; $\Delta(CH_3) = +0.06$ ppm (DIS).
^{h} $\Delta(CH_3) = +0.01$ ppm;^{d} $\Delta(CH_3) = 0.00$ ppm.^{e}
^{i} $\Delta(CH_3) = 0.0^{d,e}$ ppm.
^{j} $\Delta(CH_3) = +0.02^{d,e}$ ppm.
^{k} α-Anomer.
^{l} β-Anomer.
^{m} Fructofuranosyl.
^{n} β-Pyranose.
^{o} α-Pyranose.
^{p} α-Furanose.
^{q} β-Furanose.
^{r} Equilibrating sugar.
^{s} $^4\Delta$(C-6) = 0.07 ppm.
^{t} pH = 2.5.
^{u} pH = 6.5.
^{v} Combination of two different $^3\Delta$ values.
^{w} Effects of ~ -0.03 ppm are reported in the case of C-1, C-3 and C-4.
^{x} Isopropylidene part Δ(C-1) = +0.02 ppm; Δ(C-2) = 0.0 ppm.
^{y} $^4\Delta$(C-5) = +0.05 ppm.
^{z} $^4\Delta$(C-4) = −0.05 ppm.
* Effects of $\sim \pm 0.05$ ppm are reported for several carbons.

The pyranose form of α- and β-deoxy-D-*ribo*-hexopyranose [9] shows unusually large $^3\Delta$C-3(OD) isotope effects. The corresponding furanoses give $^3\Delta$C-3(OD) values of between +0.07 and +0.10 ppm. A large $^3\Delta$ effect is observed in the simple diol, 1,3-propanediol.[49]

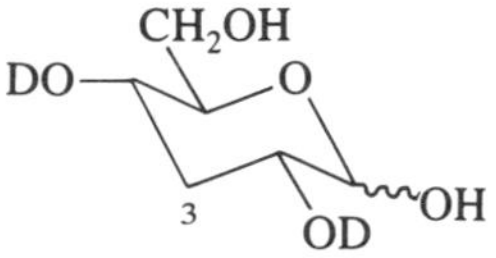

[9]

The magnitudes of $^3\Delta$C-2(OD) data are found to depend on the position of the OD group at C-1. No such effect is observed for other carbons.[48]

Bock, Gagnaire and Vignon[53] have observed a large number of negative isotope effects in acetylated carbohydrates.

In glycosides a $^4\Delta$CH$_3$(OD) effect is observed. Preferential solvation is suggested to be the reason for this observation.[48]

3. *Uses*

Isotope effects have been used extensively for assignment purposes in carbohydrates[48-53] and have led to some reassignments.[50] They may also be used to assign glucosylated sugars, amino-deoxysugars and sugar phosphates and sulphates and may be used to elucidate substitution patterns in, for example, acetylated carbohydrates.[49] Deuterium isotope effects serve also to elucidate and confirm the structure of *N*-nitrosoethanolamine.[54]

G. Csp²DO effects

This combination is primarily encountered in phenols and in acid derivatives.

1. *Phenols*

These show many unusual isotope effects concerning magnitudes, signs and long-range nature, as seen in Tables 14 and 15. The large $^2\Delta$ data can usually be ascribed to hydrogen bonding (Section III.K.1). Unusual signs are encountered in many cases. Newark and Hill[47] observed a few negative $^5\Delta$ values in substituted phenols, all with a methyl group at C-4 and furthermore with a *t*-butyl group at the *ortho*-position. They remarked that $^2\Delta$ and $^3\Delta$ are positive and that $^5\Delta$ is negative, and that they are similar in direction to the difference in nuclear shielding between phenol and the phenoxide ion; they also suggested that the isotope effect in phenols may in part be related to the greater ionization (lower pK_a) of the protiated compound compared with the deuteriated phenol. Some of the other compounds may be characterized by the structural elements shown in [10].

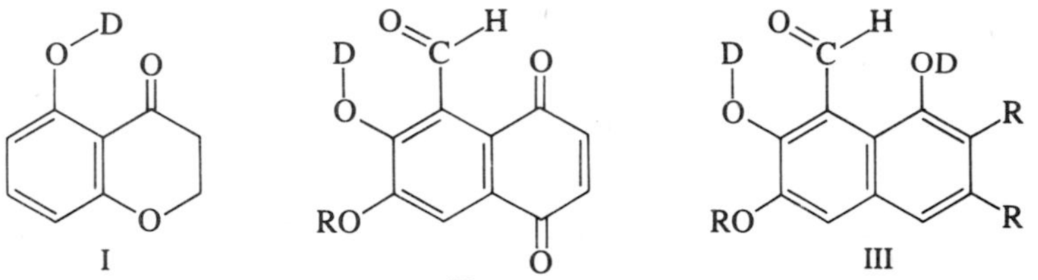

I

II

III

[10]

TABLE 14

Deuterium isotope effects on ^{13}C nuclear shieldings of non-hydrogen-bonded phenols (in ppm).

Csp^2DX	Hetero-atom	$^2\Delta$	$^3\Delta$	Ref.
2-Methylphenol	O	$+0.14^a$	$+0.07^a$ (C-2) $+0.10^a$ (C-6)	47
3-Methylphenol	O	$+0.12^a$		47
4-Methylphenol	O	$+0.13^a$	$+0.08^a$ (C-2) $+0.08^a$ (C-6)	47
2-Ethylphenol	O	$+0.12^a$	$+0.05^a$ (C-2) $+0.11^a$ (C-6)	47
2-t-Butylphenol	O	$+0.14^a$	$+0.07^a$ (C-2) $+0.10^a$ (C-6)	47
2-Ethoxyphenol	O	$+0.14^a$	$+0.09^a$ (C-2) $+0.10^a$ (C-6)	47
2,6-Di-t-butylphenol	O	$+0.16^a$	$+0.11^a$ (C-2) $+0.11^a$ (C-6)	47
2-t-Butyl-4-methylphenol^b	O	$+0.14^a$	$+0.08^a$ (C-2) $+0.10^a$ (C-6)	47
2-Methyl-4-t-butylphenol	O	$+0.13^a$	$+0.07^a$ (C-2) $+0.10^a$ (C-6)	47
2-t-Butyl-4,6-dimethyl phenol^c	O	$+0.14^a$	$+0.05^a$ (C-2) $+0.08^a$ (C-6)	47
2,6-Di-t-butyl-4-methyl phenol^d	O	$+0.16^a$	$+0.08^a$ (C-2) $+0.08^a$ (C-6)	47
2,3,5-Trimethylphenol	O	$+0.14^a$	$+0.07^a$ (C-2) $+0.10^a$ (C-6)	47
1-Pyrenol	O	$+0.12$		114
Tyrosines^e	O	$\sim +0.10$		62

^{a} Signs changed compared to original paper.
^{b} $^4\Delta = -0.02^a$ ppm.
^{c} $^4\Delta = -0.04^a$ ppm.
^{d} $^4\Delta = -0.03^a$ ppm.
^{e} In proteins.

TABLE 15

Deuterium isotope effects on ^{13}C nuclear shieldings of hydrogen-bonded phenols and enols (in ppm).

Csp^{2}DX	Hetero-atom	$^2\Delta$	$^3\Delta$	Ref.
Pentane-2,4-dione[b]	O	+0.66[a]		60
		+0.64		61
		+0.72		79
3-Methylpentane-3,4-dione[b]	O	+0.75[a]		60
1-Phenyl-1,3-butane dione[b]	O	+0.51 (C-2)		61
		+0.64 (C-4)		
2,2,6,6-Tetramethyl-3,5-dione[b]	O	+0.70		60
Dibenzoylmethane[b]	O	+0.59[a]	−0.06	60
		+0.59		61
1,3-Diphenyl-1,3-propane-dione	O	+0.72		79
2,2,6,6-Tetramethyl-heptane-3,5-dione	O	+0.72		79
3-Ethoxycarbonylpen-tane-2,4-dione	O	+1.00		79
Acetyl benzoyl methane	O	+0.73		79
Salicylaldehyde		+0.22[a] (C-1)	+0.06 (C-2)	60
		+0.27[a] (C-1)	+0.15[a] (C-6)	58
			+0.11[a] (C-2)	58
Deoxyherqueinone diace-tate	O	+0.6 (C-5)		75
3,6-O,O-Dimethylversi-conal acetate		+0.22 (C-1)		76
		+0.25 (C-8)		
3-Hydroxy-6-methylpico-linamide[c]		+0.29[a] (C-3)	+0.07[a] (C-4)	59
3-Hydroxypicolinamide[d]		+0.31[a] (C-3)	+0.08[a] (C-4)	59
Secalonic acid G[e,f,g]	O	+0.16[a] (C-1)	+0.06[a] (C-2)	57
		+0.17[a] (C-1)	−0.06[a] (C-9a)	
			+0.07 (C-2')	
			−0.04 (C-9a')	
Hemigossypolone	O	+0.33[a] (C-6)	−0.17[a] (C-5)	58
		−0.17[a,h] (C-7)		
7-Methoxyhemigossypolone	O	+0.33[a] (C-6)	+0.07[a] (C-7)	58
Hemigossypol	O	+0.71[a] (C-6)		58
		+0.22[a] (C-7)		
		−0.16[h] (C-4)		
7-Methoxyhemigossypol[i]	O	+0.78[a] (C-6)	+0.23[a] (C-7)	58
		−0.09[a,h] (C-4)		

TABLE 15 (*cont.*)

Csp^{2}DX	Hetero-atom	$^2\Delta$	$^3\Delta$	Ref.
Gossypol	O	+0.71^a (C-6)	+0.16^a (C-5)	58
		+0.21^a (C-7)	+0.15^a (C-8)	
		+0.30 (C-4)		
2,3-Dihydromenthol-2-carb-aldehyde^n	O	+1.46		79
Heliocide H$_1$	O	+0.50^a (C-6)	+0.09^a (C-8)	58
		+0.14^a (C-7)		
Heliocide B$_1{}^j$	O	+0.35^a (C-6)	+0.07^a (C-7)	58
Heliocide H$_2$	O	+0.42^a (C-6)	+0.16 (C-8)	58
		+0.21^a (C-7)		
Norlichexanthone	O	+0.55^a (C-1)	−0.05^a (C-2)	56
			−0.15^a (C-9a)	
		+0.04^a (C-6)^k		
Methylsalicylate^l	O	+0.18 (C-2)	+0.08 (C-3)	61
			+0.02 (C-2)	
5-Hydroxyflavonone	O	+0.39^a (C-4)	+0.06^a (C-4)	55
			+0.07^a (C-6)	
Averufin		+0.25 (C-1)		77
		+0.21 (C-8)		
Salol^m	O	+0.38	~0 (C-1)	74
			+0.10 (C-3)	

^{a} Signs changed compared to original paper.
^{b} Enol form.
^{c} $^5\Delta$(C-6) = −0.06^a ppm.
^{d} $^5\Delta$(C-6) = −0.05^a ppm.
^{e} As a diformate derivative.
^{f} Δ(C-3) = +0.01^a ppm; Δ(C-4) = −0.04^a ppm; Δ(C-4a) = −0.05^a ppm; Δ(C-3′) = 0.0^a ppm; Δ(C-4′) = −0.04^a ppm; Δ(C-4a′) = −0.03^a ppm.
^{g} Similar values are reported for secalonic acid A and E.
^{h} The deuterium causing the effect is not necessarily that which is two bonds away.
^{i} Δ(C-1) = +0.1^a ppm.
^{j} Δ(C-8) = −0.075 ppm.
^{k} Within the limits of uncertainty.
^{l} $^4\Delta$(CO) = +0.08 ppm.
^{m} Possibly in a tautomeric equilibrium.
^{n} $^4\Delta$(CO) = −0.38 ppm.

In all of these compounds hydrogen bonding between an OH group and a neighbouring C=O group may occur. However, this feature alone is not sufficient to account for the observed results. Salicylaldehyde (Table 15) shows normal positive isotope effects, although a very large two-bond isotope effect is observed due to hydrogen bonding. In a 5-hydroxyflavonone (naringenin, of structure pattern I) Wehrli[55] observed positive isotope effects for the carbons *ortho* to the hydrogen-bonded OH group, whereas in a very similar compound, norlichexanthone (also of structure pattern I), Sundholm[56] observed negative isotope effects for both *ortho* carbons. In secalonic acid G, again a compound of structure type I, Kurobane *et al.*[57] observed negative isotope effects, but in this case for the *ortho*, the *meta* and the *para* carbons. It is interesting that no effect is observed in ring C [11], which is expected to show a large $^2\Delta$ effect, because of hydrogen bonding (Section III.K.1).

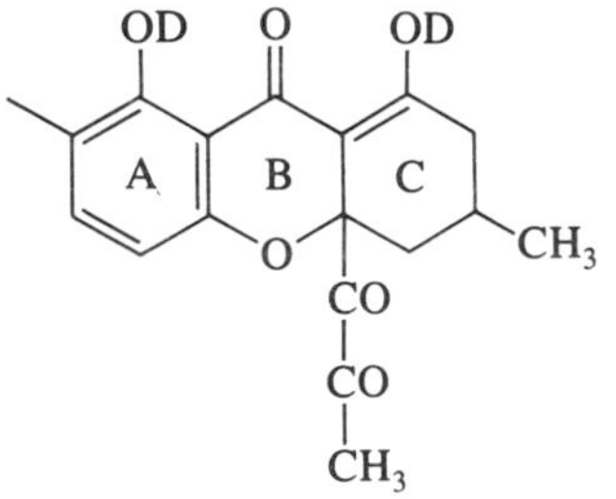

[11]

In the naphthoquinone structure II unusual effects are observed.[58] If R is H, C-5 and C-7 are shifted to high frequency, whereas if R = CH₃, C-7 is shifted to low frequency.

For compounds with structures like [10, III], C-4 and C-8 show negative isotope effects and C-2 and C-3 show unusually large positive two- and three-bond isotope effects. However, in a dimer (R, being identical to [10, III]), the shifts of C-4 and C-8 are positive. A common denominator for these observations is not obvious, except that tautomerism may possibly play a role (Section XVI).

Both 3-hydroxypicolinamide and the 6-methyl derivative show a small negative isotope effect at C-6.[59] The two-bond isotope effect of C-3 indicates intermediate hydrogen-bonding strength (Section III.K.1).

The negative isotope effects observed in types [10, I]–[10, III] have been ascribed to hyperconjugative effects.[31] It is, however, on such a basis, not possible to explain why salicylaldehyde and other similar compounds do not show negative isotope effects. At least one extra structural element seems to be needed to account for these effects. From the existing example this can be an oxygen *ortho* or *para* to the C=O group, as in [12, A] or [12, B].

Oxygen-substituents at the *ortho* or *para* positions increase the electron density at the C=O group, thereby increasing the chance of hydrogen bonding and possibly also proton transfer. That tautomerism is involved is also indicated by the finding that a negative effect is found in norlichexanthone [13, A] but not in naringenin [13, B]. In the latter, tautomerism is not possible.

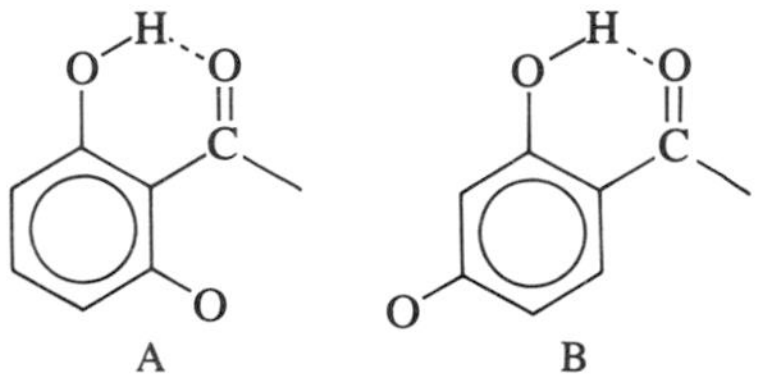

[12]

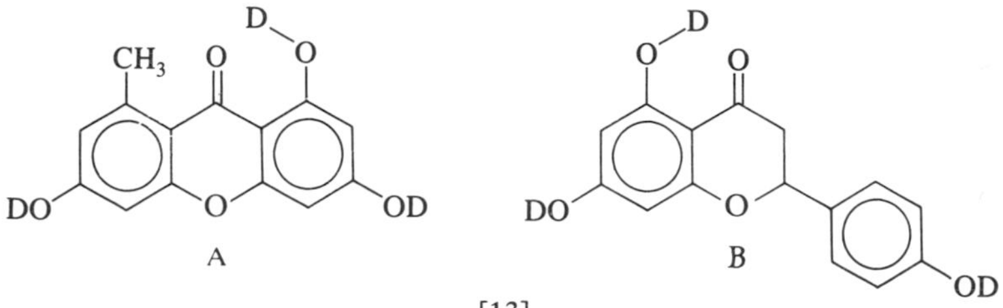

[13]

From the above comments it is rather difficult to give precise ranges for the isotope effects found in phenols. For ordinary phenols values of $^2\Delta$ are between 0.12 and 0.15 ppm[47] (Table 14) and $^3\Delta$ values are between 0.05 and 0.11 ppm.[47] Although the isotope effects may be difficult to predict, they have a large potential for assignment purposes both of the ^{13}C spectra of simple phenols incorporated in large structures, such as enzymes and proteins,[62] and of complex structures of biosynthetic origin containing phenolic groups, as their assignment potential is not related so much to their magnitude as to the fact that the effect may be made to disappear both by water addition and by heating.[56]

The unusual features of these isotope effects may, when understood better, be used for the description of phenolic systems.

2. Acids

This heading is far from precise as phenols also have acidic properties. However, in this case we are dealing primarily with carboxylic acids and amino acids. As the carbon must be sp^2 hybridized, only $^2\Delta$ effects are in question.

Ladner et al.[63] have analysed the deuterium isotope effects of amino acids dissolved in trifluoroacetic acid-d_1. Both deuterons attached to acid, amino and amide groups may be observed and consequently be used for assignment purposes. A linear regression analysis gives $^2\Delta C(OOD) = +0.226$ ppm. A value slightly smaller than observed in acetic acid and in trifluoroacetic acid (Table 16). The isotope effect on the carbonyl carbons of fully deuteriated acetic acid is given in Table 5.

TABLE 16

Deuterium isotope effects on ^{13}C nuclear shielding of acids and amino acids; Csp^2-type carbons.

Csp^{2}DX	Hetero-atom	$^2\Delta$	$^3\Delta$	Ref.
Acetic acid	O	+0.27		63
Trifluoroacetic acid	O	+0.33^a		12
Glycine	O	+0.30^b		63
Glycine^c	O	+0.28		64
Alanine	O	+0.40^b		63
Alanine^c	O	+0.33		64
Valine	O	+0.30^b		63
Lysine	O	+0.30^b		63
Lysine^c	O	+0.30		64
Proline	O	+0.29^b		63
Glycinylgycine	O	+0.25^b		63
Glycinylalanine	O	+0.23^b		63
Glycinylphenylalanine	O	+0.24^b		63
Glycinylvaline	O	+0.27^b		63
Ascorbic acid^e	O	+0.17 (C-2) +0.32 (C-3)	+0.12 (CO)	49
Ascorbic acid^f	O	0.14 (C-2) 0.13 (C-3)^g	+0.00 (CO)	49
α-D-Glucopyranuroic acid	O	+0.245 (C-6) +0.25 (C-6)^h +0.03 (C-6)^i		48 49 49
β-D-Glucopyranuroic acid	O	+0.23 (C-6) +0.23 (C-6)^h +0.03 (C-6)^i		48 49 49

^{a} Signs changed compared to original paper.
^{b} Includes a small contribution of $^3\Delta \sim 0.03$ ppm due to D-substitution at the amino group.
^{c} In D_2O. AH$_2$ species.
^{e} pH = 2.5.
^{f} pH = 6.5.
^{g} Given as 0.13 ppm in one table and as 0.03 in another.
^{h} pH = 1.8.
^{i} pH = 7.8.

A study of ascorbic acid [14] at different pH values gives a large effect at C-3 at pH 2.50 which vanishes at pH 6.50 and simultaneously the effect at C-4 decreases from 0.1 to 0.03 ppm, which is consistent with ionization of the 3-OD group at pH 6.5. A similar feature is observed for α- and β-glucopyranuroic acid at pH 1.80 and pH 7.80.[49] The large values for C-3 are most probably due to hydrogen bonding to the C=O group (Section III.K.1).

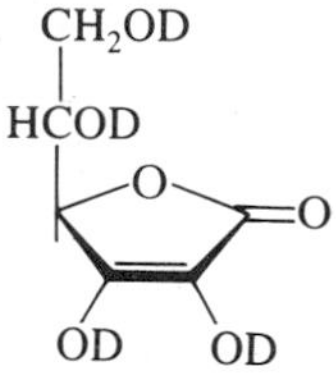

CH$_2$OD
|
HCOD

[structure with O, O, OD, OD]

[14]

H. Csp^{3}DN effects

Such effects are primarily investigated in amino acids. Data are given in Table 18. These are combined effects due to $^2\Delta C(ND)$ and $^3\Delta C(OOD)$. A regressional analysis of data obtained from amino acids dissolved in trifluoroacetic acid gives $^2\Delta C(ND) = +0.093$ ppm and $^3\Delta C(ND)$.[63] Ladner et al.[63] make no distinction between sp^2 and sp^3 hybridized carbons. For amino acids dissolved in D$_2$O, equilibrium isotope effects may play a role at pH values close to the pK_a of the amine protons[62,64] (Section XVI.D).

I. Csp^{2}DN effects

This combination is encountered in heteroaromatics, in amides and hence in peptides.

1. Amides

Deuterium isotope effects in amides were originally studied in order to provide an assignment method, in particular to be used in the often complex task of assigning the carbonyl carbons of peptides.[65]

$^2\Delta$ values may be found in Table 17. A slightly smaller isotope effect is observed for carbonyl groups in primary amides than for that in peptides. In aromatic secondary amides, such as acetanilide, the $^2\Delta$ value of the aromatic C-1 carbon is of the same magnitude as that of the carbonyl carbon.[66]

The observation of an isotope effect in a mixture of deuteriated and protiated compounds depends on the rate of exchange, which is usually sufficiently slow in amides for observation to be possible. Doublets are not observed, however, for the carbonyl carbons in cyclic amides, such as pyridone, succinimide and phthalimide; this is probably because of exchange.[66]

An interesting observation has been made in isoleucinomycin; both the carbonyl carbons of L-Lac and D-Hui show negative isotope effects (high frequency shifts) upon deuteriation of the amide nitrogen.[67]

Hawkes et al.[68] not only observed a value of $^2\Delta$ in viomycin, but, in favourable cases, a value for $^3\Delta CO(ND)$. The former are positive, whereas the sign of the latter can usually not be determined because of the small

TABLE 17

Deuterium isotope effects on ^{13}C nuclear shieldings of amino acids and amides; carbon sp^2 hybridized; deuterium on nitrogen (in ppm).

Csp^{2}DX	Hetero-atom	$^2\Delta$	$^3\Delta$	Ref.
Glycine[b]	N		+0.07	64
Glycine[c]			−0.15	64
Alanine[b]	N		+0.12	64
Alanine[c]			−0.19	64
Lysine[b]	N		+0.06	64
Lysine[c]			−0.23	64
Glycinylglycine[d]	N	+0.15[e]		63
Glycinylalanine[d]	N	+0.25[e]		63
Glycinylphenylalanine[d]	N	+0.17[e]		63
Glycinylvaline[d]	N	+0.24[e]		63
Glutathione	N	+0.10		66
N-Acetylglycine	N	+0.08 (CO) +0.09 (CH$_2$)	+0.02	66
N-Acetylvaline methyl ester	N	+0.10 (CH)		66
N-Acetylphenylalanine ethyl ester	N	+0.08 (CO) +0.08 (CH)	+0.02 (COOR)	66
N-Benzoylglycine	N	+0.07 (CO) +0.10 (CH$_2$)		66
N-Benzoylalanine	N	+0.08 (CO) +0.10 (CH)		66
N-Acetylvalinamide	N	+0.08 (CO)	+0.06 (CONH$_2$)	66
Acetanilide	N	+0.09 (CO) +0.10 (C-1)	+0.09 (C-2)	66
o-Acetotoluidide	N	+0.08 (CO) +0.11 (C-1)	0.09 (C-2)	66
m-Acetotoluidide	N	+0.09 (CO) +0.09 (C-1)	+0.09 (C- +0.09 (C-6)	66
p-Acetotoluidide	N	+0.09 (CO) +0.09 (C-1)	+0.09 (C-2)	66
o-Acetoaniside	N	+0.08 (CO) +0.09 (C-1)	+0.08 (C-2) +0.10 (C-6)	66
m-Acetoaniside	N	+0.10 (CO) +0.10 (C-1)	+0.08 (C-2) +0.09 (C-6)	66
p-Acetoaniside	N	+0.10 (CO) +0.09 (C-1)	+0.10 (C-2) +0.10 (C-6)	66
o-Acetoacetaniside	N	+0.09 (CO) +0.09 (C-1)	+0.07 (C-2) +0.13 (C-6)	66
o-Acetophenetidide	N	+0.08 (CO) +0.10 (C-1)	+0.08 (C-2) +0.12 (C-6)	66
Benzoylacetanilide	N	+0.09 (CO) +0.09 (C-1)	+0.10 (C-2)	66

TABLE 17 (*cont.*)

Csp^{2}DX	Hetero-atom	$^2\Delta$	$^3\Delta$	Ref.
2-Acetamidonaphthalene	N	+0.09 (CO)	+0.09 (C-1)	66
		+0.09 (C-2)	+0.09 (C-3)	
3-Acetamidoquinoline		+0.09 (CO)	+0.08 (C-4)	66
Dicyanocobyrinic acid				69
hexamethyl ester		+0.06 (CO)^f		

^{a} Signs changed compared to original paper.
^{b} In D_2O. AH species.
^{c} In D_2O. A$^-$ species.
^{d} In trifluoroacetic acid.
^{e} $^3\Delta$ contribution from deuterium at the amino groups.
^{f} Per deuterium.

magnitude. The value of $^2\Delta$ depends on the exchange rate, which again depends on pH. By observing the collapse at varying pH the assignment of CO groups can be linked to exchange rates observed in the ^1H spectrum.

In trifluoroacetic acid-d_1 the carbonyl groups of amides, as well as the other carbons bound to the amide nitrogen, show a varying isotope effect which is ascribed to different degrees of protonation of the amide group which gives two contributions rather than one. A difference between Gly-Ala and Gly-Val on one hand and Gly-Gly and Gly-Phe on the other is observed[63] (Table 18).

TABLE 18

Deuterium isotope effects on ^{13}C nuclear shieldings of compounds deuterated at nitrogen; carbons sp^3 hybridized (in ppm).

Csp^{3}DX	Hetero-atom	$^2\Delta$	$^3\Delta$	Ref.
Ethylamine	N	+0.32	+0.13	63
Glycine^b	N	+0.29		63
Glycine^c	N	+0.30		64
Glycine^d		+0.25		64
Glycine^e		+0.08		64
Alanine^b	N	+0.31	+0.13	63
Alanine^c		+0.31	+0.20	64
Alanine^d		+0.20	+0.20	64
Alanine^e		+0.17	+0.04	64
Lysine^{b,f}	N	+0.29 (C-α)	+0.10 (C-β)	63
		+0.33 (C-ε)	+0.14 (C-δ)	
Lysine^{c,g}		+0.26 (C-α)	+0.13 (C-β)	64
		+0.29 (C-ε)	+0.16 (C-δ)	

TABLE 18 (*cont.*)

Csp^{3}DX	Hetero-atom	$^2\Delta$	$^3\Delta$	Ref.
Lysine[d,b]		+0.28 (C-α)	+0.21 (C-β)	64
		+0.28 (C-ε)	+0.14 (C-δ)	
Lysine[e,i]		−0.23 (C-α)	+0.20 (C-β)	64
		+0.19 (C-ε)	−0.19 (C-δ)	
Proline[b]	N	+0.20 (C-α)	+0.01 (C-β)	64
		+0.19 (C-δ)	+0.01 (C-γ)	
Glycinylglycine[b]	N	+0.27 (C-α NH$_2$)		63
		+0.18 (C-α NH)		
Glycinylalanine[b]	N	+0.24 (C-α gly)		63
		+0.12 (C-α ala)		
Glycinylphenylalanine[b]	N	+0.25 (C-α gly)		63
		+0.14 (C-α phe)	+0.02 (C-β)	
Glycinylvaline[b]	N	+0.24 (C-α gly)		63
		+0.17 (C-α val)		
2-Amino-2-deoxy-α-D-glucopyranose	N	+0.34 (C-2)		49
2-Amino-2-deoxy-β-D-glucopyranose	N	+0.37 (C-2)		49
2-Amino-2-deoxy-α-D-glucopyranose	N	+0.34 (C-2)		49
2-Acetamido-1,3,4,6-tetra-*O*-acetyl-2-deoxy-β-D-glucopyranose	N	+0.100 (C-2)		53
2-Acetamido-1,3,4-tri-*O*-acetyl-2-deoxy-β-D-glucopyranose	N	+0.07 (C-2)		53
2-Methyl-2-phenyl-aziridine	N	+0.10 (C-3)		116
Viomycin[j]	N	+0.07[a] (C-1)[k]	(−)[e] 0.0019	68
		+0.06[a] (C-10)[k]		
		+0.14[a] (C-11)[l]		
		+0.12[a] (C-12)[l]	−0.08[a]	
		+0.09[a] (C-14)[k]		
		+0.07[a] (C-17)[k]	(−)[m] 0.02	
		+0.07[a] (C-21)[k]	(−)[m] 0.02	
		+0.07[a] (C-25)[k]	(−)[m] 0.02	
		+0.08[a] (C-28)[k]		
		+0.08[a] (C-2)		
		+0.10[a] (C-3)		
		+0.10[a] (C-5)		
		+0.10[a] (C-18)		
		+0.07[a] (C-22)		
		+0.08[a] (C-26)	(−)[m] 0.02(7)	
		0.00 (C-30)		
		0.09[a] (C-34)		
		0.10[a] (C-36)		

Isotope effects over three bonds are observed in viomycin,[68] but they are also observable in secondary amides, such as N-acetylvaline methyl ester and N-acetylphenylalanine ethyl ester, whereas no effects on the carbonyl carbons of the acid groups are observed in hippuric acid and N-benzoyl-alanine.[66] In these compounds a $^3\Delta C\text{-}1_{arom}(ND)$ value of ~ 0.03 ppm is observed (Table 17). In aromatic secondary amides a $^3\Delta C\text{-}2_{arom}(ND)$ value of ~ 0.09 ppm is reported.[66] The effect over three bonds is apparently transmitted more easily through double bonds.

Values of $^n\Delta$ of amides have, as already shown, been used for assignment purposes of peptides, but may also be used to distinguish acetamido carbonyl groups from other acetyl groups in acetylated aminosugars.[53] Furthermore, these data may be used to identify the amide carbonyl carbon of dicyanocobyrinhexamethyl-C-amide.[69]

2. Heteroaromatics

Various compounds containing pyrrole moieties have been investigated. Wray et al.[70] have shown that the isotope effect of carbons next to nitrogen range from 0.09 to 0.19 ppm in [15, A] and from 0.13 to 0.24 ppm in [15, B], except that the carbonyl group shows no isotope effect. This is rather surprising as bilirubin[81] [15, C], for which [15, B] is a building block, shows an isotope effect for C-1 (the carbonyl group) (Table 19). In bilirubin $^2\Delta$ effects are very similar for C-6 and C-9 both belonging to the pyrrole moiety; the same is true for the $^3\Delta$ values of C-7 and C-8. $^2\Delta$ and $^3\Delta$ data of this ring are, furthermore, very similar to those observed in pyrrole[72] and indole-3-butyric acid[73] (Table 19).

Oldfield et al.[62] have used deuterium isotope effects for the assignment of enzymes. Carbons close to labile protons, such as C^ξ of arginine, C^γ of histidine, C^ξ of tyrosine, C^{ϵ_2} and C^γ of tryptophan, are identified. Examples

TABLE 18 Footnotes

^{a} Signs changed compared to original paper.
^{b} In trifluoroacetic acid-d_1.
^{c} In D_2O. AH_2 species.
^{d} In D_2O. AH species.
^{e} In D_2O. A^- species.
^{f} $^4\Delta(C\text{-}\gamma) = +0.01$ ppm.
^{g} $^4\Delta(C\text{-}\gamma) = -0.01$ ppm.
^{h} $^4\Delta(C\text{-}\gamma) = 0.0$ ppm.
^{i} $^4\Delta(C\text{-}\gamma) = -0.04$ ppm.
^{j} In $H_2O\text{-}D_2O$.
^{k} Carbonyl carbon.
^{l} Olefinic carbons.
^{m} Sign uncertain.

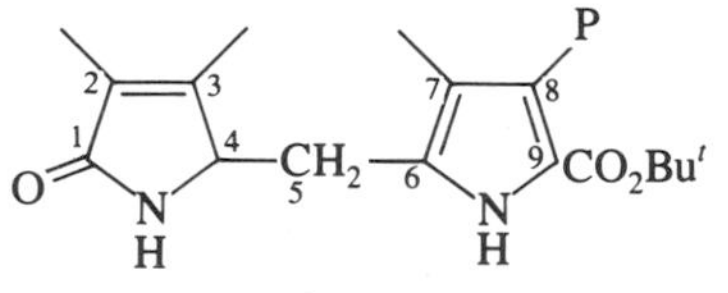

[15]

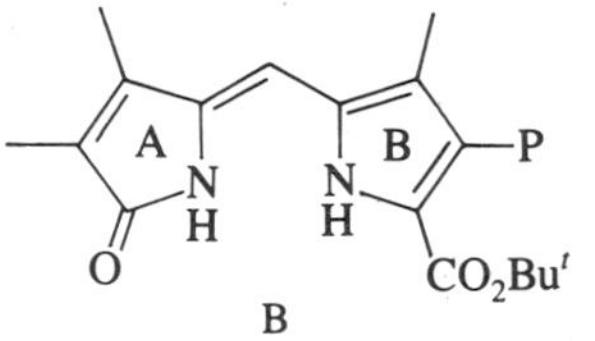

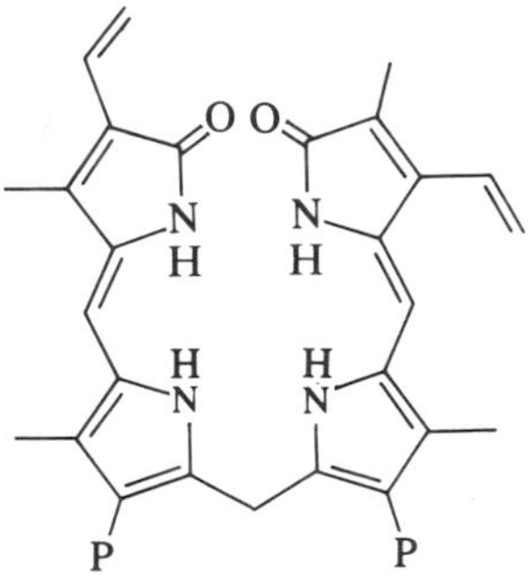

TABLE 19

Deuterium isotope effects on ^{13}C nuclear shieldings of heteroaromatics with deuterium on nitrogen (in ppm).

Csp^{2}DX	Hetero-atom	$^2\Delta$	$^3\Delta$	Ref.
Pyrrole	N	+0.177	+0.034	72
		+0.15		70
Indole	N	+0.1		111
		+0.15		70
Indole-3-butyric acid	N	+0.167 (C-2)	+0.030 (C-3)	73
		+0.167 (C-7a)	+0.030 (C-3a)	
			+0.046 (C-7)	
Bilirubin	N	+0.147 (C-4)	+0.048 (C-3)	71
		+0.167 (C-6)	+0.03 (C-7)	
		+0.164 (C-9)	+0.027 (C-8)	
		+0.104 (C-1)		
L-Tryptophan[b]	N	+0.13 (ε-2)	+0.06 (C-γ)	62
		~0 (δ-1)		
L-Arginine[b]	N	+0.19 (C-δ)		62
Gly-His-Gly	N	+0.09 (C-γ)[c]		62
		−0.03 (C-γ)[d]		

[a] Signs changed compared to original paper.

[b] Isotope shifts for Trp and Arg. His and Tyr are also reported in hen egg white lysozyme, horse heart ferrocytochrome c, horse heart ferricytochrome and horse heart cyanoferricyto-chrome c.

[c] pH = 3.0.

[d] pH = 9.6.

are taken from ferricytochrome *c*. In all cases the pH is sufficiently far from the pK_a to ensure that the residue is fully in one form, thus avoiding a possible change in the equilibrium in going from H_2O to D_2O.

J. Csp″DS effects

This type of isotope effect has been studied in β-thioxoketones and in β-thioxocarboxylic acids and esters.

Most of the β-thioxoketones show unusually large isotope shifts due to equilibrium effects, as shown in [16] and in Table 20. In the β-thioxacids tautomerism does not take place and an idea of the direct effect can be estimated from these compounds. Long-range effects are observed in the β-thioxyesters.[74]

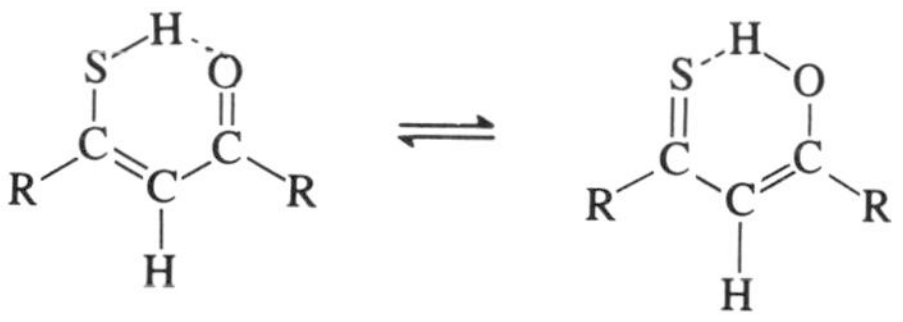

[16]

K. Special features

1. *Hydrogen bonding*

As seen from Tables 15 and 20, hydrogen bonding may drastically affect $^2\Delta$ values. The $^2\Delta$ data of hydrogen-bonded phenols and enols fall within a broad range. The β-diketones,[60,61] deoxyherqueinoneacetate[75] and hemigossypol[58] [10, II] show two-bond isotope effects larger than +0.6 ppm. Structures such as [10, I] and [10, III] show isotope effects of intermediate value, from 0.35 to 0.55 ppm, although not without exceptions: secalonic acid[57] [11], gives a $^2\Delta$ (C-1, O, D) value of only +0.16 ppm. Furthermore, secalonic acid contains a β-diketone structure, C-9, C-8a, C-8, so one would, based on the above, expect a rather large isotope effect for C-8. The observation of two peculiarities in this molecule possibly points towards a structure which is an average of two or more tautomeric structures. Small but distinctive effects are observed in 1,8-dihydroxy-9,10-anthraquinone structures[76] [17], in salicylaldehyde[60,61] and in methyl salicylate.[61]

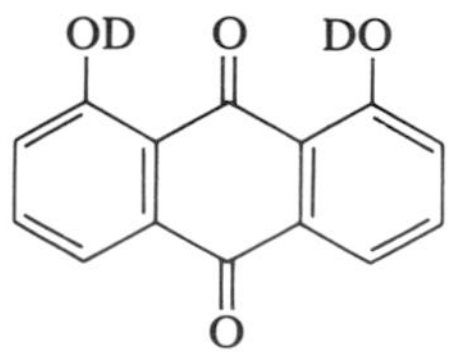

[17]

TABLE 20

Deuterium isotope effects on ^{13}C nuclear shieldings of β-thioketones and β-thioxyesters[a].

Compounds	C-1 (C=O)	C-2 (C—H₂)	C-3 (C=S)	C-1'
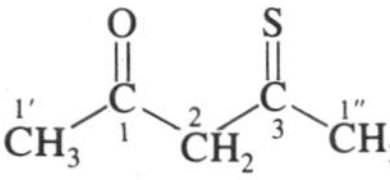	+2.33	+0.45	−7.50	+0.93
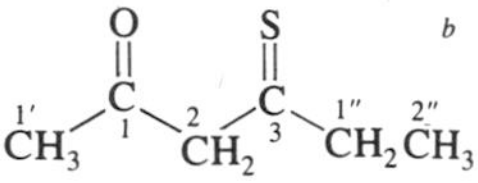	+2.60	+0.39	−7.97	+0.97
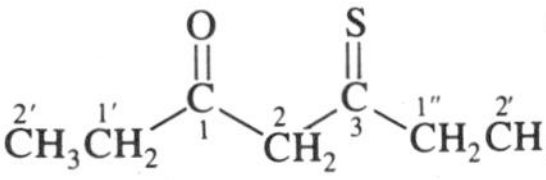	+2.37	+0.48	−7.88	+0.82
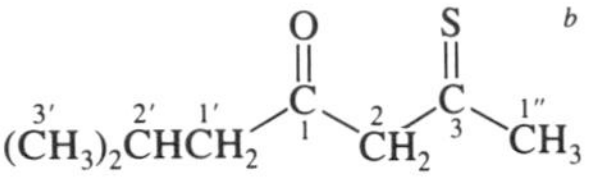	+2.43	+0.44	−7.61	+0.81
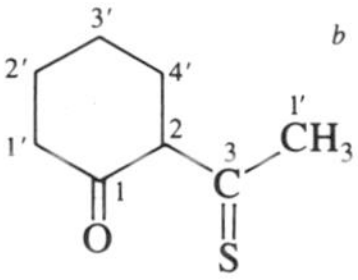	+2.28[e]	+0.42	−5.86[e]	+0.68[e]
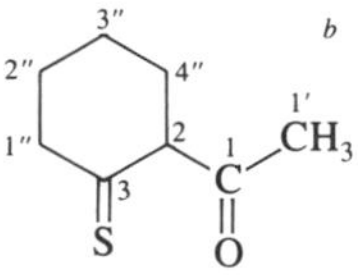	+0.23	0	−0.40	−0.08
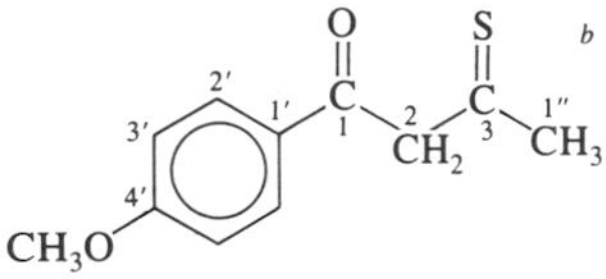	+1.76	+0.38	−6.01	+0.54
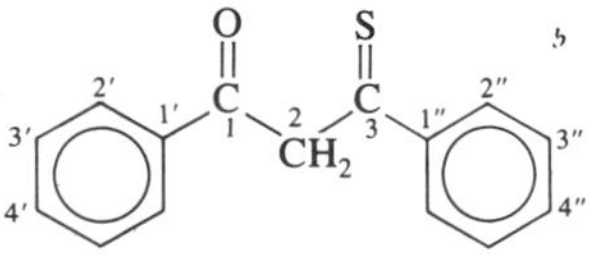	+1.74	+0.33	−4.52	+0.39
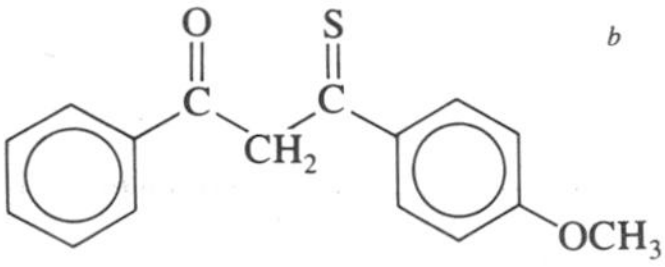	+1.38	+0.11	−3.17	+0.24

TABLE 20 (*cont.*)

C-2'	C-3'	C-4'	C-1"	C-2"	C-3"	C-4"	Other	Ref.
—	—	—	−1.33	—	—	—	—	74
—	—	—	−1.21	−0.03	—	—	—	74
−0.26	—	—	−1.18	0	—	—	—	61
−0.16	0	—	−1.33	—	—	—	—	61
+0.03^d	+0.11^d	+0.08	−0.95^e	—	—	—·	—	74
—	—	—	−0.05	0	0	−0.03	—	74
+0.10	−0.04	−0.01	−1.10	—	—	—	0^c	74
+0.08^f	0^f	−0.03	−0.53	+0.04^f	0^f	−0.05	—	74
+0.03^f	−0.05^f	0	−0.54	−0.05^f	0	−0.08	−0.06^c	74

TABLE 20 (*cont.*)

Compounds	C-1 (C=O)	C-2 (C—H₂)	C-3 (C=S)	C-1'

Structure images and data rows:

	C-1 (C=O)	C-2 (C—H₂)	C-3 (C=S)	C-1'
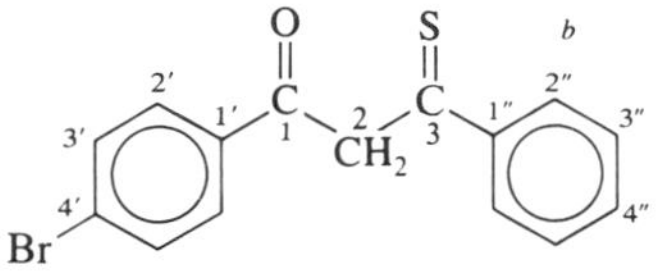	+1.69	+0.32	−4.69	+0.39
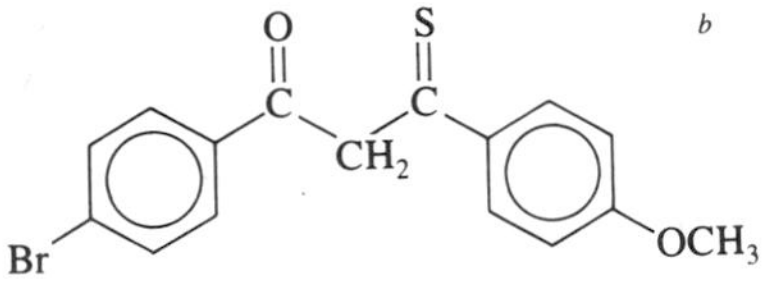	+1.58	+0.23	−3.46	+0.32
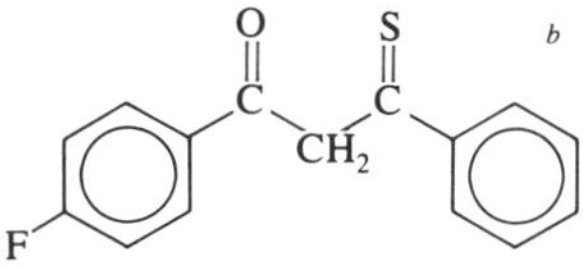	+1.79	+0.27	−4.85	+0.44
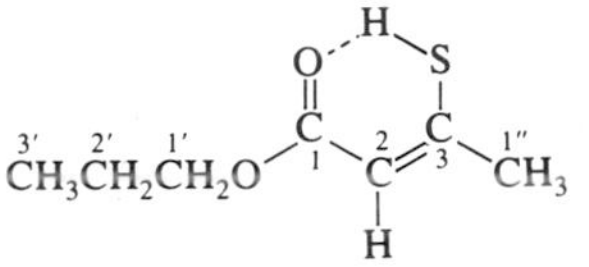	+0.04	−0.06	+0.04	−0.01

[a] The isotope effects refer to exchange of the enolic hydrogen.
[b] Structures are drawn without indication of the preferred tautomeric form.
[c] OCH_3 signal.
[d] May be exchanged.

β-Thioxoesters[61,74] and some β-thioxoketones[74] likewise show small effects (Table 20). In the former case the $^2\Delta$ effect is correlated to the percentage of hydrogen-bonded *cis* and non-hydrogen-bonded *trans* isomers[61] [18].

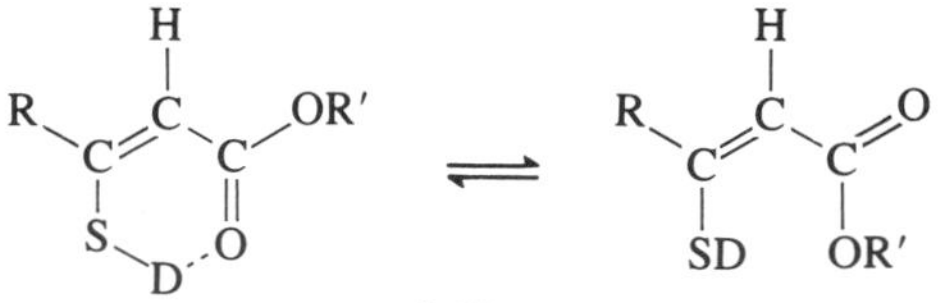

[18]

In some hydrogen-bonded compounds a small $^4\Delta CO(OD)$ effect or $^4\Delta CO(SD)$ value is observed.[61] Whether this feature is a general one is not

TABLE 20 (*cont.*)

C-2'	C-3'	C-4'	C-1"	C-2"	C-3"	C-4"	Other	Ref.
+0.03[f]	−0.04	+0.06	−0.57	+0.10[f]	0[f]	−0.06	—	74
+0.09[f]	0	+0.06	−0.51	0[f]	+0.05	−0.06	0[c]	74
+0.12	−0.03	0	−0.53	+0.05	0	−0.06	—	61
0	0	0	—	—	—	—	—	61

[e] The spectrum of the non-deuteriated compound shows a doublet. The shifts given refer to the low-field set.

[f] Assignment of closely lying aromatic signals is done to give the closest internal consistency between chemical shifts.

yet known, as it may have been overlooked elsewhere because of its smallness.

The observation of unusually large deuterium isotope effects, $^2\Delta C(OD)$, of some hydrogen-bonded systems (Table 15) often parallels the primary isotope effects $^P\Delta(H, D)$ and $^P\Delta(H, T)$ observed in the same systems (Table 45). The unusually large primary isotope effects are explained by a change in vibrational motion upon isotopic substitution leading, for medium strength hydrogen bonds, to a two-minima potential with different equilibrium bond lengths for H, O and T[78,79] (Section XVII). The difference in equilibrium bond length may possibly be the direct cause of the secondary two-, three- or *n*-bond isotope effects. In the case of carbon it is less obvious to predict how a lengthening of the O—H bond will perturb the ^{13}C nuclear shielding. A reasonable estimate should be obtainable from studies of phenols in solvents of different hydrogen-bonding properties.

Effects of hydrogen bonding have also been reported in the case of $^1\Delta C(^{18}O)$. Diakur et al.[80] have observed, in averufin, a $^1\Delta C\text{-}9(^{18}O)$ value of 0.029 ppm, whereas the effect at C-10 is 0.041 ppm. The low value observed in (1-p-aminophenyl)ethanol is ascribed to hydrogen bonding.[49]

In summary, $^2\Delta C(OD)$ data are a good indicator of hydrogen bonds. Whether they can also be developed in a more quantitative way to express hydrogen-bond strength is under investigation.[61]

2. Stereochemical dependence

In 2-D-adamantane [2] Aydin and Günther[18] have observed a dependence of $^3\Delta$ very similar to that of $^3J(C{-}H)$, which means that $^3\Delta$ (in which the carbon and the deuterium are antiperiplanar) is larger than $^3\Delta$, in which the carbon and the deuterium are gauche (Table 1). Hansen and Led[20] observed no such effect in 3-D-cyclobutane, admittedly a ring system with strained bonds. A recent report[34] refers to this phenomenon in 13-D-dicyanocobyrinic acid heptamethyl ester and finds an angle of 96° corresponding with an isotope effect of ~ 0 ppm, whereas the angle of 24° corresponds with an effect of $+0.069$ ppm, a trend contrary to that found by Aydin and Günther.[18]

Morris and Murray[33] have observed that $^3\Delta C\text{-}5(D\text{-}3) > {}^3\Delta C\text{-}7(D\text{-}3)$ (Table 2) in 4-substituted camphors, although both are antiperiplanar to a deuterium.

$^3\Delta C\text{-}2(OD)$ values of carbohydrates depend on the position of the OD group; such effects are not observed for other carbons.[48]

In summary, $^3\Delta C(D)$ data may depend on geometrical factors, but the trends are at present not obvious.

3. Long-range effects

Ernst[34] has observed a significant $^4\Delta$ value in dicyanocobyrinic acid heptamethyl ester between $CH_3\text{-}15'$ and D-13 (Table 2).

Some unusually large, long-range effects are observed in some deuteriated 8,13-diazasteroids.[81] The effects are given in Table 10 and shown in [19].

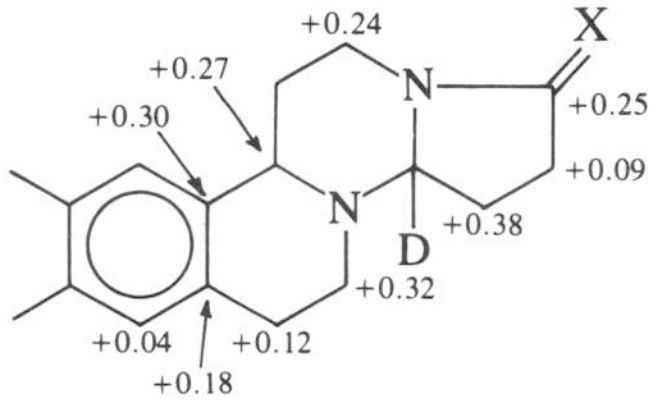

[19]

The C-14—H-14 bond is *trans*-diaxial relative to both the N-8 and N-13 lone pairs. The long-range effects and this fact are linked together.

Gribble *et al.*[82] (Table 10) have observed long-range effects in a collection of specifically deuteriated indolo[2,3-a]quinolizidines. However, the large isotope effect of the aromatic carbons in the $6\text{-}D_2$ derivative is conspicuous as these carbons are five to seven bonds away. In the $1\text{-}D_2$ derivative, in which the $^n\Delta$, $n > 4$, values are small and within the experimental uncertainty, an interesting $^3\Delta\text{C-12a}(\text{D-1})$ value of the peri type (Table 10) is observed.

4. *Unusual effects*

Negative isotope effects are observed in the paramagnetic metallocenes $(C_5D_5)_2CO$ (-7.4 ppm) and $(C_5D_5)_2V$ (-7.3 ppm) relative to the diamagnetic $(C_5D_5)Fe$. The intrinsic effect is thus eliminated. $(C_5D_5)_2Fe^+$ and $(C_5D_5)_2Cr$ show positive isotope effects. The percentage isotope effect is fairly constant for all four compounds. It is suggested that the isotope shifts are produced by different spin delocalization mechanisms.[83]

Hawkes *et al.*[68] have observed a negative $^3\Delta$ value in the fragment $H_2NCONH\underline{C}H=C-N-\underline{D}$ found in viomycin.

Negative isotope effects are found in acetone[32] and in other α-substituted ketones (Section III.C.1), in a peptide[67] (Section III.I.1), in phenols (Section III.C.1), in carbohydrates[53] (Section III.A.3), in α-deuterated alkylaromatics (Section XV.B), and they are often the usual ones when equilibrium isotope effects are treated (Section XVI).

5. *Uses: biosynthetic studies*

This use of deuterium isotope effects requires special mention. Both $^1\Delta$ and $^2\Delta$ isotope effects have been used to detect the position of incorporation of deuterium from the enriched precursor. One-bond isotope effects are difficult to observe because of the low intensity and signal splitting into a triplet. However, in an early study[84] ethanol-1-^{13}C-1,1'-D_2 was administered to bile fistula rats and the presence of deuterium in the isolated methyl cholate and methyl chenodeoxycholate was monitored by $^1\Delta$. This was made possible by simultaneous decoupling of both the ^1H and D nuclei. The method shows that it is possible to detect the presence of deuterium, but it has proved difficult to quantify the results. As the magnitude of the Δ values is well known, the number of deuterium nuclei can be determined. Decoupling has been used in the case of palmitic acid methyl ester.[85] Despite the intensity problem, incorporation of acetate-2-D_3-2-^{13}C has been used in biosynthetic studies. The observation of a triplet is, of course, in this case sufficient to establish incorporation, but one-bond isotope effects are also reported.

Acetate-2-D_3-2-^{13}C was used in the study of scylatone, ($+$)-rugolosin and 2-hexyl-5-propylresorcinal.[86] Incorporation of this precursor was also used in the study of terrein. The chain-starter methyl groups in polyketide biosynthesis may be identified in this way.[87] Use of acetate-2-D_3-2-^{13}C also reveals that C-7 retains two hydrogens in the biosynthesis of the sesquiterpenoid, dihydrobotrydial. C-7 is thus not part of an aromatic ring in an intermediate step.[88]

More complex precursors may also be used. Incorporation of mevalonic acid-4-D_2-4-^{13}C into dihydrobotrydial confirms that a 1,3-hydride shift takes place during biosynthesis.[89] This precursor was also used to follow 1,2 hydride shifts in ascochlorin.[90]

Two-bond isotope effects may be observed without deuterium decoupling. Incorporation of acetate-2-D_3-1-^{13}C was used to monitor the fate of deuterium in the biosynthesis of 6-methylsalicylic acid.[91]

IV. DEUTERIUM ISOTOPE EFFECTS ON ^1H NUCLEAR SHIELDING

Deuterium isotope effects are, except in molecular hydrogen, over at least two bonds. Furthermore, the range for ^1H nuclear shieldings is small and the observed effect is usually quite small and is thus given here in ppb (parts per billion).

The discussion is divided into two parts; direct substituent effects are treated here, whereas equilibrium isotope effects including lanthanide induced shifts are dealt with in Sections XVI.A, XVI.B, XVI.C and XVI.F.

A. Aliphatics

The deuteriated methanes show a $^2\Delta$ value of ~ 1.9 ppb, but the effect is far from additive.[118] A similar value is reported for deuteriated 2-butanone.[119] A study of deuteriated isobutane gives the isotope effects over three bonds, as shown in [20]. The effect per D is thus in one case 9 ppb and in the other 1.3 ppb, although it is over the same number of bonds and in the same molecule. Sauer et al.[121] have found a similar trend in the fragments $CDCH_3$ $^3\Delta = 7.2$ ppb and CD_3CH $^3\Delta = 4.2$ ppb. Acetoin with a hydroxy group at the carbon gives a similar trend.[122] Bengsch et al.[123] have reported isotope effects in deuteriated ethanols which confirm the trends mentioned above. In addition, they report $^3\Delta D(H)$ data. The values are very similar to those observed for $^3\Delta H(D)$.

$$DC(CH_3)_3 \qquad HC(CD_3)_3$$
$$9 \text{ ppb} \qquad\qquad 12 \text{ ppb}$$
$$[20]$$

Jensen and Schaumburg[124] have reported isotope effects in ethyl bromide-2-D and ethyl bromide-1,2-D_2. The value of $^3\Delta$ in the former, 7.7 ppb, is much larger than those mentioned above. If we analyse the data for the latter, it is seen that no additivity exists.

Haddon and Jackman[125] have derived isotope effects from deuteriated t-butylcyclohexanes. The magnitude of $^2\Delta$ is very similar to that reported in open chain compounds. As seen in [21], $^3\Delta$ depends, to some extent, on the geometry, and the effect appears to parallel that of couplings between analogously oriented protons. A similar conclusion was drawn by Batiz–Hernandez and Bernheim.[1]

A very small $^4\Delta$ value was also derived.

The difference in the magnitude of $^3\Delta$ in deuterated $CCl_3CH_2CH_2Cl$ (Table 21) is thought to show that the isotope effect depends on the nature of the environment of the ^1H nucleus.[126]

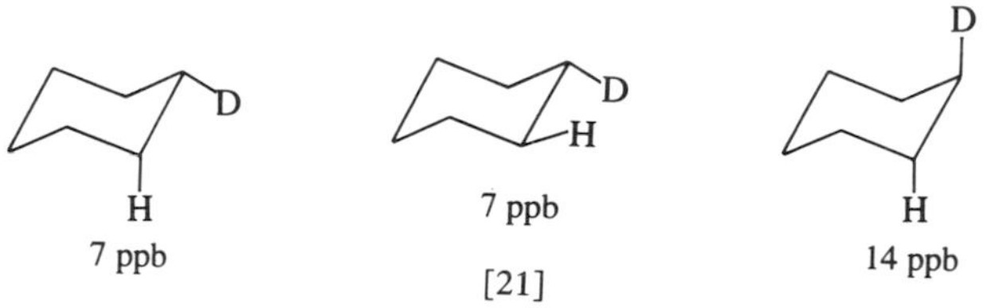

[21]

In erythro- and threo-ethyl-1,2-dibromopropionate [22] a difference in the shielding of the hydrogen in the two compounds is found,[127] showing that $^3\Delta$ is stereospecific. Isotope effects are also observed in threo-1,2-dibromo-3,3-dimethyl-butane-1-D.

[22]

B. Olefins

Specificially deuteriated styrenes reveal that $^3\Delta H(D)$ gem $> ^3\Delta H(D)$ trans $> ^3\Delta H(D)$ cis[129], as seen in Table 22. In vinyl acetate, $^3\Delta H(D)$ trans $> ^3\Delta H(D)$ cis.[121] This trend is similar to that observed for $^3\Delta F(D)$. Baird[129] suggested this trend to be quite general and that the effect is caused by different zero-point vibrations for the isotopically substituted species. An unusually large $^3\Delta$ value is observed in the fragment $CH_2DCH=$.[121]

A $^3\Delta H$-2(D-1) value of 5 ppb is reported in 1-D-cyclobutene[20] and of 7.2 ppb in the phycocyanobiline fragment $CH_3CD=$.[121] This trend is thus opposite to that found for aliphatic compounds.

TABLE 21

Deuterium isotope effects on ^1H nuclear shieldings; ^1H on sp^3 or sp^2 hybridized carbons, deuterium on sp^3 hybridized carbons (in ppb).

	$^2\Delta$	$^3\Delta$	$^4\Delta$	Ref.
$DCsp^3HCsp^3$				
CDH_3	19			118
CD_2H_2	27			118
CD_3H	45			118
CD_3CH_3		24		120
$(CH_3)_3CD$		9		118
$(CD_3)_3CH$		12^c		120
$(CD_3)_3C(CH_3)_3$			3	120
$(CD_3)_3CC(CH_3)_3$			0	120
Cyclohexanes^d	19±2	$7^e±2$	3.5±1	125
		$14^f±1$		
CD_3OD	37^b			21
CD_3CH_2OH	18^c	4^c		123
CH_3CD_2OH	7.8^c	18.5		123
CH_2DCH_2Br	16.4	7.7		124
CHD_2CHDBr	39.4	27.9		124
CH_3CHDF	19	7.8		140
CH_2DCH_2F	18.1	2.9		140
CH_2DCHDF	24.0^g			140
	23.9^h			
CHD_2CHDF	25.2^g			140
	42.0^h			
$CH_3CHDOCOCH_3$	16	7		121
$CH_3CD_2OCOCH_3$		8^c		121
$CH_3CDOHCOOH$		7.4		121
$CH_2DCDOHCOOH$	16.2^c	7.0^c		121
$CHD_2CDOHCOOH$	16.2^c	7.1^c		121
$CHD_2CHOHCOOH$		4.3^c		121
$CD_3CHOHCOOH$		4.2^c		121
$(CD_3)_2CHOCH_3$		4.5^c		121
$(CD_3)_2CDOCH_3$		6.5^c		121
$H_3CCDOHCOCH_3$		8		122
$CH_3OCHOHCOCH_3$		5		122
$ArCH_2D^i$	16.7			121
$ArCHD_2{}^i$	16.1^c			121
$CCl_3CHDCHCl_2$	10.7	3.7		126
$CCl_3CH_2CDCl_2$		2.4		126
$CCl_3CHDCDCl_2$	12.2			126
CD_3COCD_3	34^b			21
CD_2HCOCD_3	36			141
CD_3SOCD_3	$+38.9^b$			21
4-Methyl[2.2]-p-cyclophane-3′-$D_3{}^l$				31
4-Methyl[2_4](1, 2, 4, 5)cyclophane-3′-$D_3{}^m$				31

TABLE 21 (*cont.*)

	$^2\Delta$	$^3\Delta$	$^4\Delta$	Ref.
1, 2, 3, 4, 5, 6-Hexa-*O*-acetyl-D-galactiol-1-D	+17.6			142
2,6-Diisopropyl-4-methylpyrylium-α-D ion		+5.8		356
5-Oxabenzocycloheptene-3,4,6,7-D$_4$			5	136
Sn(CH$_3$)$_3$CH$_2$D	15			104
Sn(CH$_3$)$_3$CHD$_2$	30			104
Sn(CH$_3$)$_2$CD$_3$CH$_2$D	17^j			104
Sn(CH$_3$)$_2$CD$_3$CHD$_2$	34^k			104
Sn(CH$_3$)(CD$_3$)$_2$CH$_2$D	17^j			104
Sn(CH$_3$)(CD$_3$)$_2$CHD$_2$	34^k			104
Sn(CD$_3$)$_3$CH$_2$D	20^j			104
Sn(CD$_3$)$_3$CHD$_2$	38^k			104
DCsp^{3}HCsp^2				
CH$_2$DCHO	17	0		121
Phycocyanobilin (CH$_2$DCH=)	15			121
Cyclobutene-1-D		5^a		20

^{a} Signs reversed compared to original puper.

^{b} Compared to the protiated compound.

^{c} Per deuterium.

^{d} Regressional data. A footnote claims that the presence of deuterium causes deshielding. The data presented in the paper show the opposite trend, which has also been chosen here.

^{e} H and D eq, eq or eq,ax.

^{f} H and D ax,ax.

^{g} H at C-1. Combined effects.

^{h} H at C-2. Combined effects.

^{i} These measurements are the average of phycocyanobilin and the 1 and 3 methyl groups of methylphephorbide.^{a}

^{j} Hydrogen at CH$_2$D group.

^{k} Hydrogen at CHD$_2$ group.

^{l} ΔH-2*syn*(CD$_3$)=4.0^a ppb; ΔH-15(CD$_3$)=6.0^a ppb.

^{m} ΔH-2*syn*(CD$_3$) = 4.7^a ppb; ΔH-13(CD$_3$) = 5.8^a ppb.

C. Benzenes

Ford *et al.*[130] have investigated the origin of negative isotope effects previously observed in neat liquids of a mixture of deuteriated nitrobenzenes.[131] By extrapolating the measurements to infinite dilution they find no experimentally significant deuterium isotope effect in pentadeuteriobenzene and ascribe the negative isotope effects to differences in the degree of association between deuteriated and non-deuteriated molecules. In the

TABLE 22

Deuterium isotope effects on ^1H nuclear shieldings; ^1H on sp^3 or sp^2 hybridized carbons; deuterium on sp^2 hybridized carbons (in ppb).

	$^2\Delta$	$^3\Delta$	$^4\Delta$	Refs.
$DCsp^2HCsp^3$				
$\quad CH_3CDO$		+6		121
$\quad$ Phycocyanobilin				
$\quad\quad (CH_3CD=)^b$		+7.2		121
$DCsp^2HCsp^2$				
$\quad CH_2{=}CDOCOCH_3$		0^e		121
		$+5^f$		
$\quad CHD{=}CHOCOCH_3$	+12	0^e		121
		$+13^f$		
$\quad$ Benzene-1-D$_1$	+0.3	−1.3	−1.4	139
$\quad$ Benzene-D$_5{}^c$	−3.2			132
$\quad$ Benzene-D$_5$	-0.3^m			130
	-0.4^n			
	-0.2^o			
	$+0.6^p$			
	-2.83^q			
$\quad$ 1-Bromobenzene-D$_2{}^d$		+2.55	1	134
$\quad$ Styrene-β-D^g	+17	+13		129
$\quad$ Styrene-β-D^h	+18	+6		129
$\quad$ Styrene-α-D^i	$+5^j$	$+12^k$		129
$\quad$ Styrene-α,β-D$_2{}^h$	$+28^l$			129
$\quad$ Styrene-α,β-D$_2{}^g$	$+18^l$			129
$\quad$ Styrene-β,β′-D$_2$		$+15^l$		129
$DCspHCsp$				
$\quad$ Acetylene-1-D		+16		143

^{a} Signs reversed compared to original paper.
^{b} Methyl groups of phycocyanobilin and the 1 and 3 methyl groups of methylpheophorbide.^{a}
^{c} Solvent, C_6H_6. The value is independent of the ratio $C_6H_6 : C_6D_6$.
^{d} Obtained from dideuteriated bromobenzene.
^{e} H and D, *cis*.
^{f} H and D, *trans*.
^{g} Phenyl groups and deuterium are *cis*.
^{h} Phenyl groups and deuterium are *trans*.
^{i} Phenyl groups and deuterium are geminal.
^{j} Hydrogen and deuterium *cis*.
^{k} Hydrogen and deuterium *trans*.
^{l} Combined effect.
^{m} Infinite dilution, neopentane.
^{n} Infinite dilution, cyclohexane.
^{o} Infinite dilution, CCl$_4$.
^{p} Infinite dilution, acetone.
^{q} Neat liquid.

neat liquid an isotope effect of -2.5 ppb is found. Yonemitsu[132] has investigated pentadeuteriobenzene and finds a value of -3.2 ppb, which is independent of the ratio between benzene and pentadeuteriobenzene.

In benzene-D_1 Yonemitsu and Kubo[133] have observed negative $^4\Delta$ and $^5\Delta$ values, whereas $^3\Delta$ is positive. They present the following explanation. The high frequency shift observed is uncommon and is attributed to the predominant effects of increased electron density for the carbon skeleton related to the π-electron system. Positive $^3\Delta$ and $^4\Delta$ values are found in deuteriated bromobenzene.[134]

Isotope effects in benzene derivatives are very sensitive to solvent and concentration and should probably be treated as equilibrium isotope effects.

D. Steric effects

Anet and Dekmezian[102] have observed, in 2,2,5,5-tetramethyl-1,3-dioxane-2',2'-D_6 [23 A], an isotope effect at the methylene hydrogens of C-4 of 2.6 ppb which is an effect over five bonds. It cannot be ascribed to an equilibrium effect as both conformers are identical, but is explained as a steric effect as follows. Steric repulsion leads to a high frequency shift. Since the repulsion in the H–D system is less than in the H$\cdots$H system because of the lower vibrational amplitude of a C—D bond, a low frequency shift is predicted in agreement with observation. The internuclear distance is estimated to be ~ 0.2 nm. In the half-cage acetate [23, B] the observed

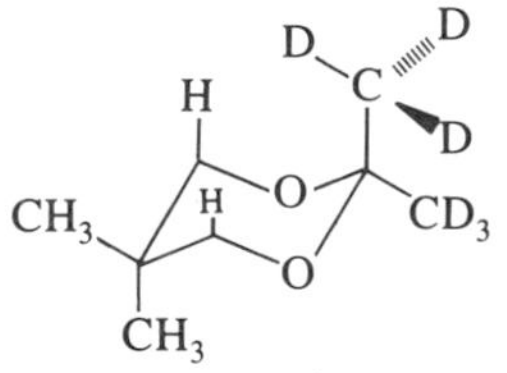

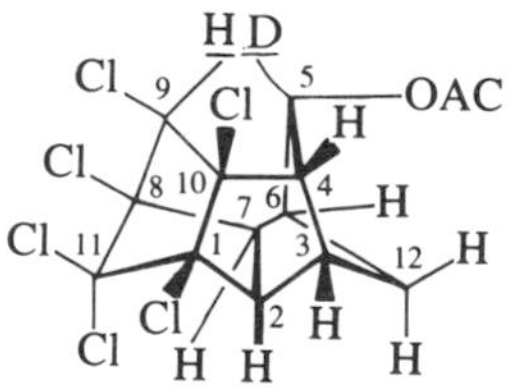

[23]

splitting is 13.7 ppb. The distance between the involved nuclei is, in this case, 0.160 nm. Based on these observations the authors suggest that the steric deuterium isotope effect occurs whenever the H$\cdots$D distance is distinctly less than the sum of the two van der Waals radii. The finding that C-2 in [23, B] shows no isotope effect is, however, difficult to explain as the carbon of a strained C—H bond shows a larger shift to low frequency than the hydrogen shows to high frequency. Canuel and St-Jacques[136] have obtained evidence for an isotope effect over four bonds in 5-oxobenzo-cycloheptane-3,4,6,7-D_4.

In 4-methyl[2.2]paracyclophane-3′-D_3, long-range isotope effects on H-2 *syn* and H-15 of 4.0 and 6.0 ppb are observed.[31] H-1 *syn* shows a broadening due to a small isotope effect of around 1 ppb. In the 4-methyl-1,2,4,5-cyclophane-3′-D_3, H_2-*syn* is shielded by 4.7 ppb and H-13 is likewise shielded by 5.8 ppb. H-13 in the latter is in the same position as H-15 in the former.[31] It is interesting to notice that C-15 and C-13 also show isotope effects (Section XV.B).

E. $D_n XH_{n-1}$ compounds

One of the few negative isotope effects reported by Batiz–Hernandez and Bernheim[1] concerns the deuteriated ammonium ions. Fraenkel et al.,[137] who first noted this finding, suggest that the unusual isotope effect is due to electrostatic effects arising from hydrogen bonding with the polar solvent.

TABLE 23

Deuterium isotope effects on nuclear shielding of ^1H nuclear shieldings of hydrogens bonded to other nuclei than carbons (in ppm).

	$^2\Delta$	$^3\Delta$	Refs.
HCsp^{3}DX			
Tertiary amines		−0.018 to −0.032	131
NH$_2$D^a	+0.029		146
NHD$_2$^{a}	+0.053		146
NH$_3$D$^{+\,a}$	−0.013		138
NH$_3$D$^{+\,b}$	−0.015		138
NH$_2$D$_2^{+\,b}$	−0.030		138
NHD$_3^{+\,b}$	−0.045		138
LiBH$_3$D	+0.011		144
	+0.013		357
LiBH$_2$D$_2$	+0.025		357
LiBHD$_3$	+0.038		357
NaBH$_3$D	+0.015		145
NaBH$_2$D$_2$	+0.029		145
NaBHD$_3$	+0.041		145
KBH$_3$D	+0.016		357
KBH$_2$D$_2$	+0.030		357
KBHD$_3$	+0.045		357
HXDCsp^3			
Aniso-actinomycin C-α-D		−0.02	358

^{a} Enriched with ^{15}N.
^{b} ^{14}N.
^{c} Sign not defined.

This proposition was refuted by Shporer and Loewenstein,[138] who investigated the effect on ^{11}B and ^{14}N nuclear shieldings and find in both cases low frequency (positive shifts). The $^2\Delta H(D)$ effect is measured in $^{15}NH_{4-n}D_n^+$, thus leaving out any possible effect of the ^{14}N quadrupolar nucleus. Shporer and Loewenstein[138] base their explanation on electric field effect arguments.

Leyden and McCall[139] have noticed a small difference for the methyl group hydrogens of ammonium ions. In the case of tertiary ammonium ions a negative isotope effect is found for the methyl proton.

F. ΔT(T) effects

In benzene-1,4-T_2 a high frequency isotope effect is observed, $^5\Delta T(T) = -4$ ppb. The introduction of an extra tritium may change H–H distances. The symmetry difference between benzene-T and benzene-1,4-T_2 possibly plays a role.[147]

G. ΔD(H) effects

These types of effect, observed in deuteriated alcohols, are discussed in Section IV.A above.

V. DEUTERIUM ISOTOPE EFFECTS ON ^{19}F NUCLEAR SHIELDING

The shift range of ^{19}F is very large and isotope effects are hence easy to observe. Some long-range interactions may possibly be observed with this combination.

A. $^1\Delta$ effects

A $^1\Delta F(D)$ effect has been observed in hydrogen fluoride. Schaumburg *et al.*[148,149] have investigated the ^{19}F nuclear shielding of HF in aqueous solutions containing KF. Use of H_2O and D_2O systems yields values for $^1\Delta$ of DF (4.7 ppm), DF_2^- (0.7 ppm), and $F(D_2O)_x$ (2.96 ppm). Hindermann and Cornwell[150] found a $^1\Delta$ of DF in the gaseous phase of 2.47 ppm; HF and DF are monomers in the gas phase. Hydrogen bonding thus leads to an appreciable increase in the isotope effect, similar to the finding for $^1\Delta P(D)$ and $^2\Delta C(OD)$ (Section III.K.1).

For fluorocomplexes, in which the first coordination sphere is not penetrated by water, Radley and Reeves[151] find that the isotope effect depends on the charge of the anion. If this is taken into account, almost the same value, 0.35 ppm, is arrived at for all the compounds considered.

A similar observation is reported by Scherbakov et al.,[152] who furthermore used the isotope effects to determine the hydration numbers of a series of fluorocomplexes. For $[BeF(H_2O)_3]^+$, $[BeF_2(H_2O)_2]$, and $[BeF_3(H_2O)]^-$, in which the first coordination sphere is penetrated, much larger isotope effects are found.[151]

B. $^2\Delta$ and $^3\Delta$ effects

$^2\Delta F(D)$ and $^3\Delta F(D)$ data in ethyl fluorides are given by Jensen and Schaumberg.[140] A study of deuteriated fluorocyclohexane at two temperatures discloses that $^3\Delta F(D)$ is dependent on geometry and that $^3\Delta F(D)$ trans $> ^3\Delta F(D)$ cis;[153,154] a trend similar to that observed for $^3\Delta H(D)$. As D and F are further away from each other in the trans configuration than in the cis, such results exclude the possibility that isotope effects may originate primarily in electric field or van der Waals effects. Instead the origin is suggested to be due to inductive effects. Deuterium is assumed to be more electronegative than hydrogen, as judged from the finding that DCOOH is more acidic than HCOOH. The isotope effect is thus considered to be a simple inductive substituent phenomenon and, as inductive effects are transmitted better through bonds in a trans geometry, the difference is explained.[154]

In fluoroolefins it is found that $^3\Delta F(D)$ trans $> ^3\Delta F(D)$ cis and that the value of $^3\Delta$ decreases with an increasing number of fluorines. Phillips and Wray[155] have proposed that a domination in long-range bond orders, produced by the shorter C—D bond compared to the C—H bond, is a possible cause for the observed difference.

In derivatives of 4-fluoro-4'-X-cis or trans-stilbene-α-D, negative isotope effects over six-bonds are observed (Table 26). In addition, in 4-fluoro-4'-fluoro-cis- or -trans-stilbene the isotope effect over seven bonds, experienced by the 4'-fluorine, is positive.[156] The negative effect is parallel to that observed in 4-F-methylbenzene,[157] which means that the origin could be hyperconjugative (vide infra).

By comparing the isotope effects in the cis- and trans-stilbenes the value of Δ remains constant in the cis series, but varies in the trans series. The magnitude in the latter can be correlated with a Hammett σ_p for the X-substituent. The positive slope indicates that the trans olefinic bond is more transmissive in the deuteriated than in the protiated compound and is hence more planar.[156]

Traficante and Maciel[157] have observed, in p-fluorotoluene-D₃, a long-range isotope effect on ^{19}F, but not in the m-fluoro derivative. Similarly, large $^6\Delta$ values have been observed in methyl-p-fluorophenylcarbonium ions[17,159] (Table 25). (For a discussion, see Section XV.)

TABLE 24

Deuterium isotope effects on ^{19}F nuclear shieldings of aliphatic compounds (in ppm).

Compounds	$^2\Delta$	$^3\Delta$	$^c\Delta$	Ref.
Ethyl fluoride-1-D	+0.645			140
Ethyl fluoride-2-D		+0.244		140
Ethyl fluoride-1,2-D$_2$	+0.902[b]			140
Cyclohexylfluoride-F$_{ax}$-2-D *trans*		+0.35		153
Cyclohexylfluoride-F$_{eq}$-2-D *trans*		+0.15		153
Cyclohexylfluoride-F$_{ax}$-2-D *cis*		+0.18		153
Cyclohexylfluoride-F$_{eq}$-2-D *cis*		+0.13		153
Cyclohexylfluoride-F$_{ax}$-1-D	+0.72			153
Cyclohexylfluoride-F$_{eq}$-1-D	+0.81			153
Cyclohexylfluoride-F$_{ax}$-2,2,6,6-D$_4$		+1.06		153
Cyclohexylfluoride-F$_{eq}$-2,2,6,6-D$_4$		+0.60		153
Bicyclo[2.2.1]heptanyl-1-fluoride-2-D *cis*		+0.15		153
Bicyclo[2.2.1]heptanyl-1-fluoride-2-D *trans*		+0.40		153
trans-1-Methyl-3-trifluoromethylcyclohexane-1'-				96
D$_3$			+0.0294	
D$_2$			+0.0181	
D$_1$			+0.0093	
trans-1-Hydroxymethyl-3-trifluoromethylcyclohexane-1'-D$_2$			+0.0282[d]	96
			+0.030[e,f]	
			+0.0289[e,g]	
			+0.0296[h,f]	
			+0.0289[h,g]	

[a] Signs reversed compared to original paper.
[b] Combined effect $^2\Delta + {}^3\Delta$.
[c] Equilibrium isotope effects.
[d] Solvent, acetone.
[e] Solvent, toluene.
[f] Temperature, 273 K.
[g] Temperature, 308 K.
[h] Solvent, methanol.

TABLE 25 (*cont.*)

Deuterium isotope effects, $^7\Delta$, on ^{19}F nuclear shieldings of F—⟨○⟩—C$\big\langle^{\text{CD}}$ derivatives (in ppm).

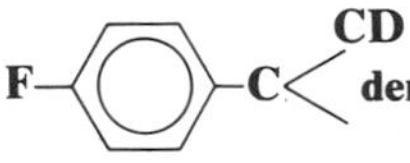

FCsp^{2}DCsp^3		$^7\Delta$	Ref.
F—⟨○⟩—C$^+$(CD$_3$)(CD$_3$)	-β-D$_6$	$-0.44^{a,b}$ -0.440^c -0.461^d	158
	-β-D$_3$	-0.226^e -0.230^d	158
	-β-D$_2$	-0.0142^e	158
	-β-D$_1$	-0.068^e	158
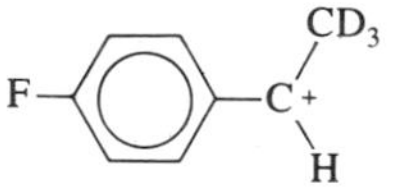 F—⟨○⟩—C$^+$(CD$_3$)(H)	-β-D$_3$	-0.315^f	158
	-β-D$_2$	-0.198^g	158
	-β-D$_1$	-0.090^g	158
	-β-D$_3$	-0.330^h	158
	-β-D$_2$	-0.204^h	158
	-β-D$_1$	-0.092^h	158
	-β-D$_3$	-0.335^i	158
	-β-D$_2$	-0.203^i	158
	-β-D$_1$	-0.091^i	158
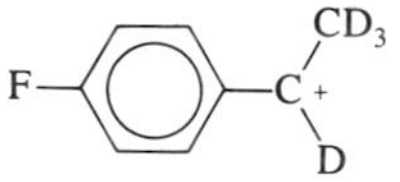 F—⟨○⟩—C$^+$(CD$_3$)(D)		$-0.33^{b,f}$	158
F—⟨○⟩—C$^+$(D)(CH$_3$)		0	158
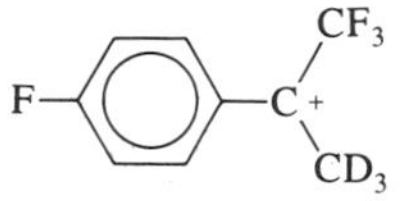 F—⟨○⟩—C$^+$(CF$_3$)(CD$_3$)		-0.40^f	158
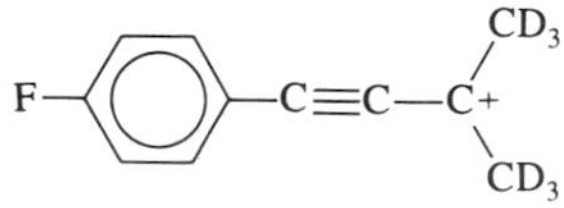 F—⟨○⟩—C≡C—C$^+$(CD$_3$)(CD$_3$)		-0.28	158
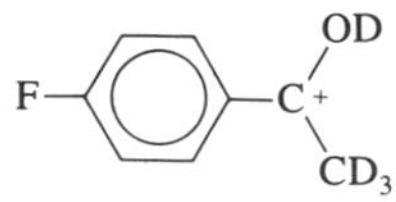 F—⟨○⟩—C$^+$(OD)(CD$_3$)		-0.11	158
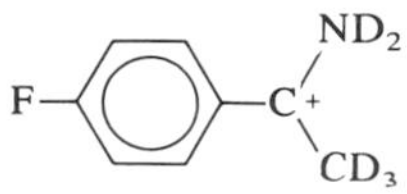 F—⟨○⟩—C$^+$(ND$_2$)(CD$_3$)		$+0.04$	158

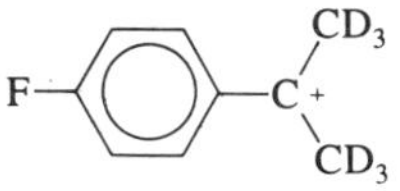

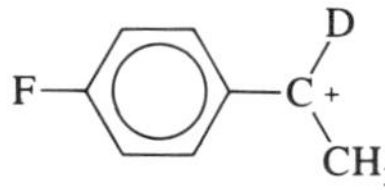

TABLE 25 (*cont.*)

FCsp^{2}DCsp^3	$^7\Delta$	Ref.
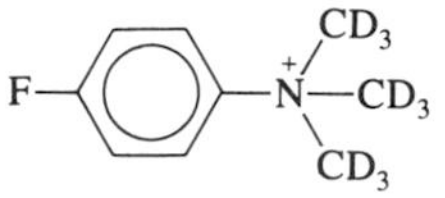 F—C$_6$H$_4$—N$^+$(CD$_3$)$_3$	-0.008^j	158
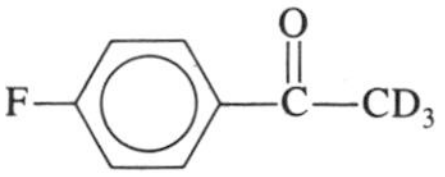 F—C$_6$H$_4$—C(=O)—CD$_3$	-0.008^k	158
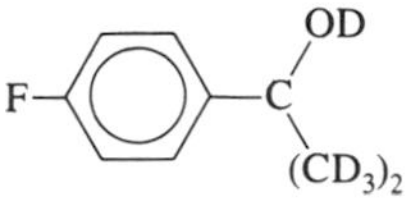 F—C$_6$H$_4$—C(OD)(CD$_3$)$_2$	-0.008^h	158
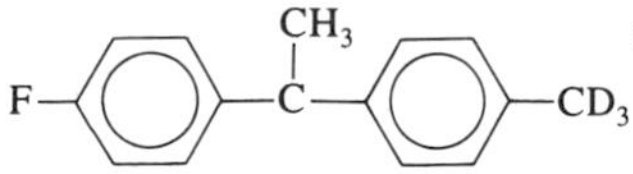 F—C$_6$H$_4$—C(CH$_3$)—C$_6$H$_4$—CD$_3$ ^{l}		158

---

^{a} Sign reversed compared to original paper.

^{b} Temperature $-40\,°C$, but independent of temperature in the interval from -20 to $-60\,°C$.

^{c} Temperature, $-18\,°C$.

^{d} Temperature, $-60\,°C$.

^{e} Temperature, $-15\,°C$.

^{f} Temperature, $-40\,°C$.

^{g} Temperature, $-19\,°C$.

^{h} Temperature, $-83\,°C$.

^{i} Temperature, $-121\,°C$.

^{j} Solvent, D$_2$O.

^{k} Solvent, CCl$_4$. $^7\Delta=0$ in CH$_3$OH.

^{l} $^{11}\Delta = -0.06^a$ ppm.

In *N*-deuterofluoroacetamide, Pendleburg and Phillips[159] have observed that $^4\Delta F(ND)$ depends strongly on the number of fluorines and probably on other substituents also, shown as X and Y in [24]; for data see Table 24.

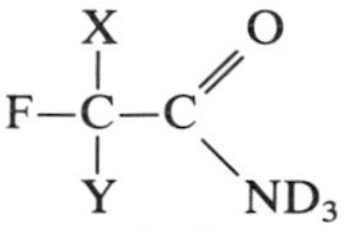

$$F-\underset{\underset{Y}{|}}{\overset{\overset{X}{|}}{C}}-C\overset{O}{\underset{ND_3}{\diagup}}$$

[24]

The value of $^4\Delta$ decreases from 0.136 ppm per D in the monofluoro derivative to 0.035 ppm in the trifluoro derivatives. Furthermore, the

P. E. HANSEN

TABLE 26

Deuterium isotope effects on ^{19}F nuclear shieldings of p-fluorine-substituted stilbenes (in ppm).

Compound	$^5\Delta$	Ref.
4-Fluoro-4'-dimethylamino- *trans*-stilbene-α-D	-0.005^a	156
4-Fluoro-4'-methoxy-*trans*- stilbene-α-D	-0.006^a	156
4-Fluoro-4'-methyl-*trans*- stilbene-α-D	-0.007^a	156
4-Fluoro-*trans*-stilbene-α-D	$-0.008\ (5)^a$	156
4,4'-Difluoro-*trans*-stil- bene-α-D	$-0.008\ (5)^a$	156
4-Fluoro-4'-chloro-*trans*- stilbene-α-D	-0.010^a	156
4-Fluoro-4'-bromo-*trans*- stilbene-α-D	$-0.008\ (5)^a$	156
4-Fluoro-4'-cyano-*trans*- stilbene-α-D	-0.012^a	156
4-Fluoro-4'-nitro-*trans*- stilbene-α-D	-0.013^a	156
4-Fluoro-4'-methoxy-*cis*- stilbene-α-D	-0.018^a	156
4-Fluoro-4'-methyl-*cis*- stilbene-α-D	-0.018^a	156
4-Fluoro-*cis*-stilbene-α-D	-0.019^a	156
4,4'-Difluoro-*cis*-stilbene- α-D	-0.018^a	156
4-Fluoro-4'-chloro-*cis*-stil- bene-α-D	-0.020^a	156
4-Fluoro-4'-bromo-*cis*-stil- bene-α-D	-0.020^a	156
4-Fluoro-4'-cyano-*cis*-stil- bene-α-D	-0.019^a	156
4-Fluoro-4'-nitro-*cis*-stil- bene-α-D	-0.019^a	156

^{a} Signs reversed compared to original paper.

monodeuterio derivatives show different values for the *cis* and *trans* isomers.[159,160] In the monofluoro derivative $^4\Delta$ *cis* > $^4\Delta$ *trans*, whereas $^4\Delta$ *trans* > $^4\Delta$ *cis*[159] in the di- and tri-fluoro derivative.

$^4\Delta$ values observed in deuteriated trifluoroacetyl derivatives of amino acids fall around 0.25 ppm in mixtures of H_2O and D_2O. Slightly larger values are observed in acetone[161] (Table 27).

Hull and Sykes[163] have used the deuterium effect as a probe for the degree of solvent exposure[164] of fluorine-containing tyrosines in

TABLE 27

Deuterium isotope effects on ^{19}F nuclear shieldings of fluorinated amides.

Compound	$^4\Delta$	Ref.
N-Trifluoroacetylglycine-D	$+0.026^b$	161
	$+0.033^c$	
	$+0.026^{d,e}$	
N-Trifluoroacetylalanine-D	$+0.028^{b,f}$	161
	$+0.029^c$	
	$+0.026^g$	
	$+0.031^d$	
N-Trifluoroacetylvaline-D	$+0.028$	161
Fluoroacetamide-D-cis^h	$+0.135^i$	159
	$+0.150^j$	
	$+0.146^k$	
Fluoroacetamide-D-$trans^h$	$+0.125^i$	159
	$+0.133^j$	
	$+0.122^h$	
Fluoroacetamide-D$_2$	$+0.263^i$	159
	$+0.287^j$	
	$+0.266^h$	
Difluoroacetamide-D-cis^h	$+0.060^i$	159
	$+0.070^j$	
	$+0.056^k$	
Difluoroacetamide-D-$trans^h$	$+0.068^i$	159
	$+0.064^k$	
Difluoroacetamide-D$_2$	$+0.125^i$	159
	$+0.134^j$	
	$+0.119^k$	
Trifluoroacetamide-D-cis^h	$+0.032^j$	159
	$+0.028^k$	
Trifluoroacetamide-D-$trans^h$	$+0.041^j$	159
	$+0.036^k$	
Trifluoroacetamide-D$_2$	$+0.073^j$	159
	$+0.065^k$	

[a] Signs reversed compared to original paper.
[b] Solvent, 1:1 mixture of H_2O and D_2O.
[c] Solvent, acetone.
[d] Solvent, dimethylsulphoxide.
[e] Solvent, ether.
[f] Solvent, acetonitrile.
[g] Solvent, dioxane.

[h] cis and trans refer to

[i] Solvent, methanol.
[j] Solvent, dichloromethane.
[k] Solvent, dimethylformamide.

fluorotyrosine alkaline phosphatase. The largest values observed are identical with that observed for m-fluorotyrosine itself (0.25 ppm from pH 6.3 to 10). In fluorotyrosine M13 coat protein the same technique was used to show that the fluorotyrosines of M13 in phospholipid vesicles are not accessible to solvent, whereas two m-fluorotyrosines show partial exposure to solvent when the protein is incorporated into deoxycholate micelles at pH 9 and three residues are exposed to solvent in sodium dodecyl sulphate micelles at pH 10.[165] A change in solvent from H_2O to D_2O was previously shown to generate an effect of 0.15 to 0.22 ppm on fluorocarbons.[166] Sykes et al.[165] have suggested different degrees of hydrogen bonding to the fluorines in H_2O and D_2O as the cause of the effect.

4-Trifluoromethylbenzenesulphonate-3,5-D_2 has been used as an enzyme label.[162] In the anion of the former a $^4\Delta CF_3(D)$ value of 0.02 ppm is found, whereas in the modified enzymes a small negative isotope effect of the same magnitude is observed. Change of solvent from H_2O to D_2O causes a deshielding of 0.11 ppm.

Based on the large $^4\Delta$ values observed in trifluoroacetyl derivatives one would also expect a contribution from the deuterium at the phenolic group. However, an almost constant value of the isotope effect of m-fluorotyrosine at different pH seems to indicate that the effect of deuterium on the hydroxyl group is small.

The $^4\Delta$ data observed in the trifluoroacetyl derivatives of amino acids in H_2O and D_2O are ascribed to D exchange of the amide proton, but alternatively they may be referred to as a general solvent effect.[162]

The negative solvent effect found[162] by dissolving 4-trifluoromethylbenzenesulphonate-labelled enzyme in D_2O could possibly be caused by an equilbrium effect on a nearby amino acid residue, as described for deuterium effects on ^{13}C nuclear shieldings of amino acids (Section XVI.D).

Scherbakov et al.[152] and Radley and Reeves[151] report on acid derivatives, but again the effect may be ascribed to exchange of the acidic proton.

In general, the effects of exchangeable protons and perturbation of amino acid equilibria should probably be taken into account when discussing solvent isotope effects on fluorine nuclear shielding.

VI. DEUTERIUM ISOTOPE EFFECTS ON ^{14}N, ^{15}N, ^{31}P, ^{11}B, ^{32}S, ^{45}Sc, ^{59}Co, ^{67}Zn, ^{77}Se, ^{89}Y, ^{119}Sn AND ^{207}Pb NUCLEAR SHIELDING

A. $^1\Delta N(D)$ effects

Litchman et al.[146] have reported a $^1\Delta^{15}N(D)$ value of 0.65 ppm per D in ammonia. A much smaller $^1\Delta^{14}N(D)$ value is found in the ammonium ion[138] (Table 28).

London et al.[167] have observed, in the fast exchange region (high pH), a low frequency shift of 1.2 ppm for the guanidino nitrogens of L-arginine when dissolved in D_2O rather than in H_2O.

B. $\Delta P(D)$ effects

Stec et al.[168] have remarked that, although the synthesis of compounds with deuterium bonded directly to phosphorus is straightforward, very few compounds have been investigated. This was so in 1971 and is still the case in 1982.

In phosphine-D_1 a $^1\Delta$ value of 2.54 ppm is obtained, together with a deviation from additivity[168] (Table 27). In some substituted phosphines, Borisenko et al.[170] have observed a $^1\Delta$ value of ~1.2 ppm per D (Table 27) a value which is much smaller than in phosphine itself. Borisenko et al.[170] also find, in general, a much larger deuterium isotope effect on three-coordinate than on four-coordinate phosphorus (Table 28) and relate this to a decrease in the isotope effect with an increase in the valence angle. They are able to correlate $^1\Delta P(D)$ with the values of $^1J(P—H)$ couplings (Section XV.D). Furthermore, for $(C_6H_5)PHD$ the isotope shift is calculated as a function of the variation in the valence angle caused by deuterium substitution. The decrease in the angle is assumed to be the same as in PH_2D going to PHD_2. The calculated isotope effect is in very good agreement with the experimental result. For three-coordinate phosphorus the variation in the valence angle is presumed to govern the isotope effect.

Stec et al.[168] have investigated the deuterium effect on dialkylphosphonates. The isotope effects in these, which are tetra-coordinate, are smaller than in the corresponding three-coordinated compounds. The choice of solvent influences the value of $^1\Delta$. In general, ^{31}P nuclear shieldings are highly sensitive to bond angle changes and insensitive to small variations in internuclear distances. A 1° increase of the $\widehat{OPO}$ angle is calculated to give an effect of 4.9 ppm, so small angle perturbations caused by deuterium substitution may cause a large isotope effect. These results are in agreement with those of Borisenkov et al.,[170] although the estimated consequences are not numerically alike.

McFarlane and Rycroft[171] have investigated dimethylphosphonate (dimethylphosphite) and observed likewise that the value of $^1\Delta P(D)$ depends on the solvent. The value of $^1\Delta$ is largest in dilute cyclohexane and smallest in neat liquid and methanol. Hydrogen bonding is assumed to diminish the isotope effect.

A high frequency shift of $^2\Delta = -0.12$ ppm is reported in glucose 6-phosphate at pH = 3.[49] The pK_a value of glucose 6-phosphate is around 2 and the effect is probably an equilibrium effect caused by a difference in pK_a of the deuteriated and non-deuteriated forms, as observed for some amino acids in the high pH range (Section XVI.D).

TABLE 28

Deuterium isotope effects on ^{15}N and ^{31}P nuclear shieldings (in ppm).

	$^1\Delta$	$^2\Delta$	Ref.
^{15}N			
Arginine[b]	+1.2		167
Ammonia, D	+0.68		146
Ammonia, D_2	+1.29		146
Ammonia, D_3	+1.96		146
^{14}N			
NH_3D^+	+0.30		138
^{31}P			
$(C_6H_5)_2POD$	+0.59		170
$C_6H_{13}OHPOD$	+0.54		170
$((C_2H_5)_2N)_2POD$	+0.47		170
HDPOOH	+0.39		170
D_2POOH	+0.72		170
$(C_4H_9O)_2PSD$	+0.44		170
$(OH)_2POD$	+0.27		170
$(CH_3O)_2POD$	+0.33		170
	+0.4		168
	+0.49[a,c]		
	+0.31[a,d]		
	+0.32[a,c]		
	+0.42[a,f]		
$(C_6H_5O)_2POD$	+0.31		168
$(C_2H_5O)_2POD$	+0.3		168
$(C_3H_7O)_2POD$	+0.4		168
$(iso\text{-}C_3H_7O)_2POD$	+0.4		170
C_6H_5PHD	+1.21		170
$C_6H_5PD_2$	+2.39		170
$(iso\text{-}C_4H_9O)_2PD$	+1.02		170
$(iso\text{-}C_3H_7)_2PD$	+1.30		170
$(C_6H_5)_2PD$	+1.29		170
PH_2D	+0.804[a]		169
PHD_2	+0.845[a,g]		169
PD_3	+0.888[a,g]		169
	+0.93 (+0.67)[i,q]		360
Diphosphane	+2.8 (+2.0)[i]		360
2-Methyl-4-amino-6-dimethylamino-1,3,5,6-triazaphosphorin	+0.5		361
Glucose-6-phosphate[h]		−0.12	49

C. $^1\Delta^{11}B(D)$ effects

Fairly constant values are found in lithium and sodium boranhydrides (Table 29).

D. $\Delta Se(D)$ effects

Jakobsen et al.[172] find a large $^1\Delta$ value in hydrogen selenide. The effect is additive (Table 29). In selenophene-2-D a $^2\Delta$ value is reported.[173]

E. $\Delta Me(D)$ effects

Deuterium effects on the nuclear shielding of various metals are given in Table 29.

VII. ^{13}C ISOTOPE EFFECTS ON ^1H NUCLEAR SHIELDING

Substitution of ^{13}C instead of ^{12}C invariably leads to an increase in nuclear shielding. Most, if not all, compounds investigated only contain ^{13}C in natural abundance. The only way to obtain $^n\Delta(H, C)$ data is from analysis of either satellite spectra or indirectly from analysis of fully coupled ^{13}C spectra (Section XVIII). From such an analysis only differences are usually obtained.

The value of $^1\Delta H(C)$ is very small (<4 ppb), and $^1\Delta > {}^2\Delta > {}^3\Delta$. The large relative errors as a consequence of the analysis do not permit a distinction between $^1\Delta$ at olefinic and methylenic carbon centres.[180]

Small variations within similar compounds are observed. 1,4-Dichloro-, dibromo- and diiodobenzenes show variations in $^1\Delta$.[181] Variations in $^2\Delta(H\text{-}3, C\text{-}2) - {}^4\Delta(H\text{-}5, C\text{-}2)$ and in $^2\Delta(H\text{-}2, C\text{-}3) - {}^4\Delta(H\text{-}6, C\text{-}3)$ are also observed.[182]

However, the greatest significance is in the analysis as the inclusion of $\Delta(H, C)$ leads to lower root mean square errors and thus to more precise determinations of couplings.

TABLE 28 Footnotes

[a] Sign reversed compared to original paper.
[b] N-2 of guanidino part. Deuterium effect observed in D_2O.
[c] Solvent, C_6H_{12}.
[d] Neat liquid.
[e] Solvent, CH_3OD.
[f] Solvent, CCl_4.
[g] Per deuterium.
[h] pD = 3.0.
[i] Temperature −80 °C. It is not quite clear from the original paper which values belong to phosphine and which to diphosphane.

TABLE 29

Deuterium isotope effects on ^{17}O, ^{77}Se, ^{119}Sn, ^{59}Co, ^{11}B, ^{14}N, ^{207}Pb, ^{66}Zn and ^{89}Y nuclear shielding (in ppm).

	Nucleus	$^1\Delta$	$^2\Delta$	$^3\Delta$	Ref.
D_2O	^{17}O	+3.08			174
Selenophene-2-D	^{77}Se		+1.682		173
Hydrogen selenide-D$_1$	^{77}Se	+7.02^a			172
Hydrogen selenide-D$_2$	^{77}Se	+14.04^a			172
$Sn(CH_3)_3CD_3$	^{119}Sn		+0.80		104
$Sn(CH_3)_2(CD_3)_2$	^{119}Sn		+1.91		104
$Sn(CH_3)(CD_3)_3$	^{119}Sn		+2.63		104
$Sn(CD_3)_4$	^{119}Sn		+2.86		104
$[Co(NH_3)_6]Cl_3$	^{59}Co		+5.2^b		175
$[Co(en)_3]Cl_3$	^{59}Co		+4.7^b		175
$LiBH_3D$	^{11}B	+0.13			144
		+0.139			145
$LiBH_2D$	^{11}B	+0.27			144
		+0.275			145
$LiBHD_3$	^{11}B	+0.423			145
$LiBD_4$	^{11}B	+0.558			145
$NaBH_3D$	^{11}B	+0.137			145
$NaBH_2D_2$	^{11}B	+0.271			145
$NaBHD_3$	^{11}B	+0.411			145
$NaBD_4$	^{11}B	+0.543			145
BH_3D^-	^{11}B	+0.17			138
BD_4^-	^{11}B	+0.58			138
$Zn^{2+\ c,d}$	^{67}Zn	−13.1			177
$Sc^{3+\ c}$	^{45}Sc	−6.2			179
$Pb^{2+\ c}$	^{207}Pb	−3.1			176
$Y^{3+\ c}$	^{89}Y	−4.3			178

^{a} Sign reversed compared to original paper.
^{b} Per deuterium. In D_2O.
^{c} Ions dissolved in H_2O and D_2O.
^{d} For $ZnBr_2$. Perchlorate, nitrate and sulphate salts show no isotope effects.

The inclusion of $\Delta(H, {}^{13}C)$ in the analysis of fully coupled ^{13}C spectra led to an improvement for C-2, but not for C-3.[183] For furan,[184] improvements are obtained for both carbons. In toluene the long-range isotope effects are calculated assuming that the size of $^2\Delta$ is the same in all cases.[185]

For a large number of monosubstituted benzenes[188] inclusion of $\Delta H(^{13}C)$ improves the calculations for C-2 and C-3, but not for C-1 and C-4.

The fluorobenzenes have been analysed.[186-189] For 1,3,5-trifluorobenzene, absolute values were originally given,[186] but these have since been changed to relative values.[188]

TABLE 30

^{13}C isotope effects on ^{1}H nuclear shieldings[b] (in ppb).

	$^1\Delta$	$^2\Delta$	$^3\Delta$	Ref.
Benzene-1-^{13}C	$1.6^{a,c}$			190
		$0.9^{a,d}$	0.3	185
Toluene-1-^{13}C		$0.8^{a,d}$	0.2	185
Fluorobenzene-2-^{13}C		$0.9^{a,e}$		182
		$1.4^{a,e}$		189
Fluorobenzene-3-^{13}C		$+0.07^{a,f}$		189
Aminobenzene-2-^{13}C		$+0.98^{g,e}$		182
Nitrobenzene-2-^{13}C		$+0.8^{a,e}$		182
Cyanobenzene-2-^{13}C		$+0.9^{a,e}$		182
Benzaldehyde-2-^{13}C		$+0.6^{a,e}$		182
Benzaldehyde-3-^{13}C		$+1.0^{a,f}$		182
Methoxybenzene-2-^{13}C		$+0.8^{a,e}$		182
Methoxybenzene-3-^{13}C		$-0.1^{a,f}$		182
Trimethylsilylbenzene-2-^{13}C		$+1.32^{a,e}$		182
Trimethylsilylbenzene-3-^{13}C		$+0.96^{a,f}$		182
1,2-Difluorobenzene-3-^{13}C	$+2.1^g$			189
1,2-Difluorobenzene-4-^{13}C	$+1.6^h$			189
1,3-Difluorobenzene-1-^{13}C		$+1.4^i$		189
1,4-Difluorobenzene-2-^{13}C	$+2.5^g$	$+0.9^e$	0.2^j	189
1,3,5-Trifluorobenzene-1-^{13}C^m		$+0.8^k$		189
1,3,5-Trifluorobenzene-2-^{13}C	$+1.6^l$			189
1,2,3,4-Tetrafluorobenz-ene-1-^{13}C		$+0.5^d$		189
1,2,4,5-Tetrafluorobenz-ene-3-^{13}C	$+1.7^g$			189
1,4-Dichlorobenzene-2-^{13}C	$+2.2$	$+1.8^k$	$+0.3^j$	181
1,4-Dibromobenzene-2-^{13}C	$+3.1$	$+1.5^k$	$+0.9^j$	181
1,4-Diiodobenzene-2-^{13}C	$+1.4\,(5)$	$+1.73^h$	$+0.4^j$	181
1,4-Dihydroxybenzene-2-^{13}C	$+2.1\,(7)$			191
p-Dimethoxybenzene-2-^{13}C	$+2.1\,(7)$			191
p-Benzoquinone	$+2.1\,(3)^n$			191
	$+2.6\,(7)^o$			
Pyrene-1-^{13}C	0.8^h			192
Pyrene-2-^{13}C	3.2^h			192
1,3,6,8-Tetradeutero pyrene-4-^{13}C	6.2^h			192
Biphenylene-1-^{13}C	$+1.7^{a,g}$			193
Biphenylene-2-^{13}C	$+2.5^{a,h}$			193
Pyridine-2-^{13}C		$+1.6^{a,h}$		183
2-Bromopyridine	$+2.9^l$			183
Pyridine-N-oxide-2-^{13}C	$+2.1^p$	$+1.1^q$	0.2	194
Pyridine-N-oxide-3-^{13}C	$+2.6^l$	$+1.1^k$		194
		$+0.4$		
Furan-2-^{13}C		$+1.6^d$		183
Furan-3-^{13}C		$+0.8^d$		183

TABLE 30 (*cont.*)

	$^1\Delta$	$^2\Delta$	$^3\Delta$	Ref.
Pyrrole-^{15}N-2-^{13}C		$+0.9^d$		172, 195
Pyrrole-^{15}N-3-^{13}C	$+1.4$	$+0.6^d$		172
Thiophene-2-^{13}C	$+1.3\,(7)^a$	$+0.9^{a,d}$		196
Thiophene-3-^{13}C	$+1.9^a$	$+0.8^{a,d}$		196
Allenes-3-^{13}C	$+3–4$			197
1,3-Butadiene-1-^{13}C	$+2.2^g$	$+1.0^d$		198
1,3-Dithiole-2-thione	$+3.5$	$+2.5$	$+1.8$	199
1,3-Dithiole-2-one	$+3.3$	$+2.3$	$+1.3$	199
1,3-Dioxole-1-^{13}C	$+4.0^r$	$+1.6^r$	$+0.8^r$	180
	$+3.6^s$	$+1.0^s$	$+0.1^s$	
1,3-Dioxole-4-^{13}C	$+3.3^r$		$+1.1^r$	180
	$+3.0^s$		$+1.8^s$	
1,3-Dioxolyl-1-^{13}C	$+3.5$	$+0.8$		180
1,3-Dioxolyl-4-^{13}C	$+3.1$	$+0.4$		180
Cyclopentadiene-1-^{13}C		$+1.1^{a,d}$		200
Cyclopentadiene-2-^{13}C		$+0.8^{a,d}$		200
Acetone-1-^{13}C	$+5$			141
Acetone-D$_5$-1-^{13}C	$+3$			141
Dimethylsulphoxide-1-^{13}C	$+4$			201
Dimethylsulphide-1-^{13}C	$+5$			201
Dimethyl ether	$+7.5$			201
Chloroform	$+4$			99
Methylene chloride	$+4$			201
cis-1,2-Dichloroethy-lene-1-^{13}C	$+3.8$			201
trans-1,2-Dichloroethy-lene-1-^{13}C	$+3.5$			201
Cyclohexane-D$_{11}$	$+1.8^u$	$+0.5^u$	$+0.2^u$	17
	$+2.6^v$	$+1^v$	$+0.4^v$	
	$+1.1^w$	$+0.9^w$		
Cyclohexane-1,1′,2,2′,3,3′,4,4′-D$_8$	$+1.4$			203
Butadieneiron tricarbonyl-1-^{13}C	$+6(\text{H-a})^x$ $-5(\text{H-b})$			204
Butadieneiron tricarbonyl-2-^{13}C	-5^y			204
Acetylene	$+0.3$			143

^{a} Sign reversed compared to original paper.
^{b} In ppb.
^{c} Relative to other hydrogens.
^{d} $^2\Delta-^3\Delta$.
^{e} $^2\Delta-^4\Delta$; $\nu(3)-\nu(5)$.
^{f} $^1\Delta-^4\Delta$; $\nu(2)-\nu(6)$.
^{g} $^1\Delta-^4\Delta$.
^{h} $^1\Delta-^2\Delta$.

VIII. ^{13}C ISOTOPE EFFECTS ON ^{13}C NUCLEAR SHIELDING

The $\Delta^{13}C(^{13}C)$ isotope effects are, as expected, small, and have only been reported in a limited number of compounds, as shown in Table 31.

The $^1\Delta$ values fall between 0 and +0.027 ppm. In benzene derivatives Wray et al.[206] have reported values between 0.008 and 0.025 ppm, a range also covering the aromatic carbons of ferrocene and 1,1-dimethyl-ferrocene.[208]

A range of 0.012–0.025 ppm is quoted for heteroaromatics.[203] The values of $^1\Delta C$-1(C-α) and $^1\Delta C$-α(C-1) are ~0.005 ppm is ferrocene, a value somewhat smaller than those of ΔC-1(C-2) or ΔC-2(C-3). Amino acids[209] and alcohols[210] show $^1\Delta$ values around 0.01 ppm, which is also the case for 1,4-dimethylcyclohexane-1'-^{13}C.[211]

A general trend is visible for non-substituted compounds. $^1\Delta$ increases in magnitude when the hybridization goes from sp^3 to sp^2 and possibly sp (no exact examples of the latter case are known). A few isotope effects over two bonds are reported. In ferrocenes[208] and in cyclobutene[20] they fall around 0.002–0.003 ppm and in benzene they are less than 0.004 ppm.[206]

In cyclobutanone a very interesting $^2\Delta$(C-3, C-1) value of -0.012 ppm is reported. No similar effects are observed in bromo- and chlorocyclobutane (Table 31). In strongly coupled systems $^1J(C-C)$ can only be calculated if $^1\Delta C(^{13}C)$ is known or can be reliably estimated.[217]

$^2\Delta C(^{13}C)$ data can be used to distinguish between $^2J(C-C)$ and $^4J(C-C)$ in double quantum coherence spectra recorded at high magnetic fields.[61]

TABLE 30 Footnotes

---

^{i} $^2\Delta H$-6(C-1)$-^4\Delta H$-4(C-1).

^{j} $^3\Delta-^4\Delta$.

^{k} $^2\Delta-^4\Delta$.

^{l} $^1\Delta-^3\Delta$.

^{m} In reference 186, absolute values are given.

^{n} Solvent, DMSO.

^{o} Solvent, CDCl$_3$.

^{p} $^1\Delta H$-2(C-2)$-^3\Delta H$-6(C-2).

^{q} $^2\Delta H$-3(C-2)$-^4\Delta H$-5(C-2).

^{r} Solvent, acetone-d_6.

^{s} Solvent, benzene-d_6.

^{t} Al-Rawi et al.[99] have observed a similar effect on the tritium resonance.

^{u} Average value.

^{v} Equatorial hydrogen.

^{w} Axial hydrogen.

^{x} Difference between H(1) and H(4).

^{y} Difference between H(2) and H(3).

TABLE 31

^{13}C isotope effects on ^{13}C nuclear shieldings (in ppm).

	$^1\Delta$	$^2\Delta$	Ref.
Thiophene-2-^{13}C	+0.03 (C-3)		210
Thiophene-3-^{13}C	+0.03 (C-2)		
3-Pentanol-1-^{13}C	+0.01 (C-2)		210
3-Pentanol-2-^{13}C	+0.01 (C-1)		
Acetylene	+0.01±0.05		214
Amino acids	+0.01[b]		209
Cyclobutanone-1-^{13}C	+0.027 (C-2)		212
Cyclobutanone-2-^{13}C	+0.002 (C-1)		
Cyclobutanone-2-^{13}C	+0.010 (C-3)		
Cyclobutanone-3-^{13}C	+0.013 (C-2)		
Cyclobutanone-1-^{13}C		+0.010 (C-3)	
Cyclobutanone-3-^{13}C		−0.012 (C-1)	
Bromocyclobutane-1-^{13}C	+0.018 (C-2)		212
Bromocyclobutane-2-^{13}C	+0.010 (C-1)		
Bromocyclobutane-2-^{13}C	+0.015 (C-3)		
Bromocyclobutane-3-^{13}C	+0.024 (C-2)		
Chlorocyclobutane-1-^{13}C	+0.015 (C-2)		212
Chlorocyclobutane-2-^{13}C	+0.013 (C-1)		
Chlorocyclobutane-2-^{13}C	+0.010 (C-3)		
Chlorocyclobutane-3-^{13}C	+0.010 (C-2)		
Oxetane-2-^{13}C	+0.011 (C-3)		212
Oxetane-3-^{13}C	+0.009 (C-2)		
Thiethane-2-^{13}C	+0.011 (C-3)		212
Thiethane-3-^{13}C	+0.005 (C-2)		
Ferrocene	+0.018		208
1,1-Dimethylferrocene-1-^{13}C	+0.018 (C-2) +0.005 (5) (C-α)	+0.002 (4) (C-3)	208
1,1-Dimethylferrocene-2-^{13}C	+0.020 (C-1) +0.009 (9) (C-3)	+0.002 (7) (C-α)	
1,1-Dimethylferrocene-3-^{13}C	+0.019 (C-2)	+0.003 (6) (C-1)	
1,1-Dimethylferrocene-α-^{13}C	+0.004 (8) (C-1)	+0.002 (4) (C-2)	
trans-Stilbene-α-^{13}C	+0.020 (C-β)		215, 216
Tetraphenylethylene-α-^{13}C	+0.020 (C-β)		215, 216
Tetra(*p*-bromophenyl)-ethylene-α-^{13}C	+0.020 (C-β)		216
cis-1,4-Dimethylcyclo-hexane-1'-$^{13}C^e$	+0.011 (5)[a] (C-1)[c] +0.011 (9)[a] (C-1)[d]		211
trans-1,4-Dimethyl-cyclohexane-1'-$^{13}C^f$	+0.016 (0)[a] (C-1)		211
Cyclobutene-1-^{13}C	+0.008 (C-4)[a]	+0.002 (C-3)[a]	20
Glucose-4-^{13}C	+0.017 (C-4)		213

[a] Sign reversed compared to original paper.
[b] With the exception of C=O, where the effect is zero.
[c] The enriched methyl group is equatorial.
[d] The enriched methyl group is axial.
[e] $^5\Delta(CH_3) = +0.007\ 58^a$ ppm, ^{c} $^5\Delta(CH_3) = +0.0021^a$ ppm.[d]
[f] $^5\Delta(CH_3) = +0.007\ 96^a$ ppm.

IX. ^{13}C ISOTOPE EFFECTS ON ^{19}F NUCLEAR SHIELDING

These are typically obtained from ^{13}C–^{19}F satellite spectra. $^1\Delta$ values fall in the range +0.079–0.163 ppm. Hinton and Jaques[218] have suggested that $^1\Delta$ is larger for fluorines attached to sp^3 carbons. This is not evident from Table 32. Substituent effects seem to determine the magnitude. A comparison of ethyl fluoride (0.082 ppm), heptafluorobutionyl chloride (+0.124 ppm),[220] hexafluoroacetone (+0.128 ppm)[219] and fluoropentachloroacetone (+0.163 ppm)[219] shows that the value of $^1\Delta$ increases with an increasing number of halides and that a chlorine gives a larger contribution than a fluorine atom.

Long-range effects are also observed in some cases. In fluorochloroacetones it is observed[219] that a variation in the composition of substituents at the other end of the molecule perturbs $^1\Delta F(^{13}C)$. In the fluorobenzenes[186,188,221,222] only very small variations, 0.079–0.093 ppm, are observed (Table 32).

In general, $^1\Delta > {^2\Delta} > {^3\Delta} \sim {^4\Delta}$. $^4\Delta$ values are only reported for fluorobenzenes.

The value of $^2\Delta$ depends on the number of electronegative substituents in the case of fluorines on sp^3 hybridized carbons. The value of $^2\Delta$ in ethylene is larger than those of sp^3 hybridized carbons and of the same magnitude as in fluorobenzenes. In the latter case the variation is small, +0.021–0.026 ppm.

X. ^{13}C ISOTOPE EFFECTS ON THE SHIELDING OF VARIOUS NUCLEI

The relevant data are given in Table 33. The most interesting are the $^1\Delta^{111}Cd(^{13}C)$ and $^1\Delta^{113}Cd(^{13}C)$ data, which are of the same magnitude and negative.[225] The ratio between $^1\Delta Cd(^{13}C)$ and $^1\Delta Hg(^{13}C)$ is close to 2.8, which is close to the ratio of the chemical shift ranges of mercury and cadmium.[226] This indicates that the unusual isotope effect is caused by a feature common to the two compounds considered.

XI. ^{15}N ISOTOPE EFFECTS ON ^{1}H AND ^{13}C NUCLEAR SHIELDING

A. $\Delta H(^{15}N)$ effects

These effects are usually small and of a magnitude comparable to that of $\Delta H(C)$, as seen in Table 34.

TABLE 32

^{13}C isotope effects on ^{19}F nuclear shieldings (in ppm).

	$^1\Delta$	$^2\Delta$	Ref.
Ethyl fluoride	+0.082		140
		+0.007 (8)	
3-Fluorothiophene	+0.089		223
2-Fluorothiophene	+0.103		223
2-Chloro-1,4-dibromo-1,2,2-trifluorobutane	+0.133[b]	+0.017[b]	218
Trifluoromethylbenzene	+0.129[e]	+0.007[e]	220
	+0.132[f]	+0.008[f]	
1,2-Dichloro-1,2-tetra-fluoroethane	+0.139[e]	+0.015[e]	220
	+0.141[f]	+0.017[f]	
Heptafluorobutionyl chloride	+0.124[c,e]	+0.019[c,e]	220
	+0.127[c,f]	+0.020[c,f]	
	+0.127[b,e]	+0.017[b,e]	
	+0.130[b,f]	+0.017[b,f]	
1,1,1-Trifluoro-2,2,2-trichloroethane	+0.124[e]	+0.013[e]	220
	+0.129[f]	+0.015[f]	
	+0.131	+0.014	224
Fluoropentachloro-acetone	+0.163[a]		219
1,3-Difluorotetrachloro-acetone	+0.159[a]		219
1,1',3'-Trifluorotri-chloroacetone	+0.152[a] (CF)		219
	+0.131[a] (CF$_2$)		
Pentafluorochloro-acetone	+0.144[a] (CF$_2$)		219
	+0.128 (CF$_3$)		
Hexafluoroacetone	+0.128 (CF$_3$)		219
1,1-Difluoro-2,2-di-chloroethylene	+0.098[e]	+0.027[e]	220
	+0.101[f]	+0.031[f]	
cis-1,2-Difluoro-1,2-dichloroethylene	+0.107[e]	+0.041[e]	220
	+0.110[f]	+0.047[f]	
1-Fluorotrichloro-ethylene	+0.109[e]	+0.028[e]	220
	+0.113[f]	+0.037[f]	
1-Fluorobenzene	+0.086		188
1,2-Difluorobenzene	+0.079		188
1,2-Difluorobenzene[q]	+0.079[a]	+0.024[a] (F-2)	187
		+0.021[a] (F-3)	
13-Difluorobenzene	+0.087		188
1,3-Difluorobenzene[h]	+0.088[a]	+0.023[a] (F-1)	222
		+0.026[a] (F-3)[i]	

TABLE 32 (*cont.*)

	$^1\Delta$	$^2\Delta$	Ref.
1,4-Difluorobenzene	+0.085		188
	+0.083[e]		220
	+0.086[f]		220
	+0.087[a,j]	+0.025[a,j]	222
1,2,3-Trifluorobenzene		+0.021[a,d]	221
1,2,4-Trifluorobenzene	+0.081 (C-1)		188
	+0.086 (C-2)		
	+0.087 (C-4)		
1,3,5-Trifluorobenzene[k]	+0.093		188
	+0.093[a]	+0.027[a]	186
1,2,3,4-Tetrafluoro-benzene[l]	+0.083 (C-1)		188
	+0.082 (C-2)		
1,2,3,5-Tetrafluoro-benzene[m]	+0.085 (C-1)		188
	+0.081 (C-2)		
	+0.085 (C-3)		
	+0.089 (C-5)		
1,2,4,5-Tetrafluoro-benzene[n]	+0.083		188
1,2,3,4,5-Pentafluoro-benzene	+0.085 (C-1)		188
	+0.083 (C-2)		
	+0.084 (C-3)		
Hexafluorobenzene[o]	+0.083		188

[a] Signs reversed compared to original paper.

[b] Fluorine at C-4.

[c] Fluorine at C-3.

[d] $^2\Delta F$-3(C-4) $- ^3\Delta F$-2(C-4).

[e] Temperature, 303 K.

[f] Temperature, 203 K.

[g] $^3\Delta F$-1(C-3) = 0.0014[a] ppm; $(^4\Delta F$-1(C-4) $+ ^3\Delta C$-2(C-3))/2 = +0.0008[a] ppm; reference 188 reports $^2\Delta F$-2(C-3) $- ^3\Delta F$-1(C-3) = +0.019 ppm.

[h] $^3\Delta F$-1(C-5) = 0.005[a] ppm; $^3\Delta F$-3(C-1) = +0.004[a] ppm; $^4\Delta F$-1(C-4) = +0.004[a] ppm.

[i] $^3\Delta F$-3(C-4).

[j] $^3\Delta$ = +0.006[a] ppm; $^4\Delta$ = +0.007[a] ppm; $^2\Delta - ^3\Delta$ = 0.019 ppm according to reference 188.

[k] $^1\Delta F$-1(C-1) $- ^3\Delta F$-3(C-1) = 0.086 ppm; $^2\Delta F$-1(C-2) $- ^4\Delta F$-5(C-2) = +0.019 ppm. In reference 188 $^3\Delta$ = +0.007[a] ppm and $^4\Delta$ = +0.008[a] ppm.

[l] $^2\Delta F$-2(C-1) $- ^3\Delta F$-3(C-1) = +0.024 ppm; $^2\Delta F$-1(C-2) $- ^3\Delta F$-4(C-2) = +0.023 ppm; $^2\Delta F$-4(C-5) $- ^4\Delta F$-1(C-5) = +0.019 ppm.

[m] $^2\Delta F$-3(C-4) $- ^4\Delta F$-1(C-4) = +0.018 ppm.

[n] $^1\Delta F$-1(C-1) $- ^3\Delta F$-5(C-1) = +0.083 ppm; $^2\Delta F$-2(C-1) $- ^3\Delta F$-5(C-1) = +0.024 ppm; $^3\Delta F$-5(C-1) $- ^4\Delta F$-4(C-1) = +0.003 ppm; $^2\Delta F$-2(C-3) $- ^3\Delta F$-1(C-3) = +0.017 ppm.

[o] $^1\Delta F$-1(C-1) $- ^3\Delta F$-3(C-1) = +0.083 ppm; $^2\Delta F$-2(C-1) $- ^3\Delta F$-3(C-1) = +0.023 ppm; $^3\Delta F$-3(C-1) $- ^4\Delta F$-4(C-1) = +0.004 ppm.

TABLE 33

^{13}C isotope effects on nuclear shieldings of various nuclei (in ppm).

	Nuclei	$^1\Delta$	$^2\Delta$	References
Methylpentaborane	^{11}B	+0.006		227
Tetramethyldiphosphine	^{31}P	+0.025^b		228
Bis(diphenylphos-phino)acetylene	^{31}P	+0.062	+0.004	229
Dimethylcadmium	^{111}Cd	−0.14		225
Dimethylcadmium	^{113}Cd	−0.14		225
Dimethylmercury	^{199}Hg	−0.4^c		226
K$_3$Co(CN)$_6$-^{13}C$_1$	^{59}Co	+0.85		230
K$_3$Co(CN)$_6$-^{13}C$_2$	^{59}Co	+1.70 (9)		230
K$_3$Co(CN)$_6$-^{13}C$_3$	^{59}Co	+2.56		230
Selenophene	^{77}Se	+0.257		173

^{a} Signs reversed compared to original paper.
^{b} Average isotope effect: $(^1\Delta + {}^2\Delta)/2$.
^{c} Slightly dependent on solvent and concentration.

TABLE 34

^{15}N isotope effects on ^{1}H nuclear shieldings (in ppb).

	$^1\Delta$	$^2\Delta$	Ref.
NH$_3$	1.83		201
Diazomethane		0.7^b	233
Salicylideneaniline^c		1.11 (CH)	232
1-Phenylazo-2-naphthylamine-2-^{15}N	−0.1		231
1-Phenylazo-2-naphthylamine-1-^{15}N^d			231

^{b} Or 1.4 ppb.
^{c} ΔOH(^{15}N) = −4.66 ppb.
^{d} ΔNH(^{15}N) = −80 ppb.

The large value observed in 1-(phenylazo)-2-naphthylamine-1-^{15}N, but not in the 2-^{15}N derivative,[231] is explained by assuming a diminished donor ability of ^{14}N compared to that of ^{15}N due to interaction of the quadrupolar charge of ^{14}N with the lone-pair electrons. In salicylidene a much smaller negative effect is observed at the OH proton.[232]

B. ΔC(^{15}N) effects

Wasylishen et al.[234] have reported $^1\Delta$ values in various cyanides. The other main source of data is from biological molecules which have ^{15}N enrichment. Examples are anthramycin-10-^{15}N,[236] streptothricin,[237] nicotine[238] and amino acids.[209] Only the value of $^1\Delta$N(C) has been reported in these molecules. In the derivatives shown in [25] both a $^1\Delta$ and a $^2\Delta$ value are observed. Very little difference is seen between the oxime and its anion (Table 35).

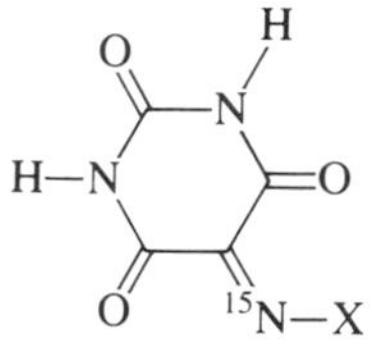

X = OH or O$^-$

[25]

TABLE 35

^{15}N isotope effects on ^{13}C nuclear shieldings (in ppm).

	$^1\Delta$	$^2\Delta$	Ref.
CN$^-$	+0.03		234
Cu(CN)$_4^{3-}$	+0.04		234
Ni(CN)$_4^{2-}$	+0.03 (7)		234
Pt(CN)$_4^{2-}$	+0.02		234
Cd(CN)$_4^{2-}$	+0.04		234
Mg(CN)$_4^{2-}$	+0.025		234
Pyrimidine-2(1H̱), 4(3H̱), 5(6H̱), 6-tetrone-5- oxime-5'-^{15}N(H$_3$vi)	+0.025^a	+0.006^a (C-6)^b	238
Anion	+0.028^a		238
fac-[Fe(H$_2$vi)$_3$]$^-$	+0.029^a		238
fac-[Ru(H$_2$vi)$_3$]$^-$	+0.030^a	−0.006^a (C-6)^b	238
Didehydroanhydroanthra mycin-10-^{15}N	+0.023		235
Streptothricin F	+0.015^c		236
Nicotine-1'-^{15}N	+0.012		237
Salicylidene-aniline-^{15}N	+0.023^a (C-7) +0.013^a (C-1')	+0.009^a (C-1)	232
Aniline	+0.014^a		232
Anilium ion	+0.008^a		232
Amino acids	+0.020		209

^{a} Sign reversed compared to original paper.
^{b} Assignment uncertain.
^{c} Guanidino carbon.

A crude trend may be seen in the data of Table 35. In compounds with C—N single bonds the value of $^1\Delta$ falls between 0.012 and 0.020 ppm; in compounds with C=N bonds the range is 0.023–0.029 ppm, whereas in the cyanide $^1\Delta > 0.02$ and most are larger than 0.03 ppm.

The isotope effects observed in salicylidene-aniline-^{15}N[232] fall nicely into these ranges: $^1\Delta N(C\text{-}7) = 0.023$ ppm (double bond) and $^1\Delta N(C\text{-}1') = +0.013$ ppm (single bond). A two-bond isotope effect is also observable. The usual isotope effect observed at the OH proton may hence be ascribed to hydrogen bonding.

XII. ^{18}O ISOTOPE EFFECTS ON ^{1}H AND ^{13}C NUCLEAR SHIELDING

A. $^1\Delta H(^{18}O)$ effects

Pinchas et al.[239] have investigated the solvent effect of $H_2{}^{18}O$ on the methyl protons of acetonitrile. A considerable negative isotope effect of 0.12 ppm is observed in 10–15% solutions of $H_2{}^{18}O$ in acetonitrile. The effect levels off at a content of 40% $H_2{}^{18}O$. The maximal effect, observed at 10%, is ascribed to the formation of a $CH_3CN(H_2{}^{18}O)_4$ complex. The negative isotope effect is explained by the poorer proton acceptor properties of ^{18}O than of ^{16}O. The authors conclude that the electronic structure of ^{18}O-containing molecules is not the same as that for ^{16}O-containing molecules. A study of the water protons in dilute solutions of organic solvents again gives a large negative isotope effect. In pure $H_2{}^{18}O$ and $H_2{}^{16}O$ the difference is much less[241] (Table 36).

B. $\Delta C(^{18}O)$ effects

The introduction of ^{18}O is usually performed either by catalysed exchange with $H_2{}^{18}O$ or by biosynthetic incorporation. It is obvious that oxygen exchange processes may be studied in this way.

Usually only one-bond isotope effects are observed (Tables 36–40), the basic properties of which have been studied.[80,242–245]

The magnitudes of Δ of typical compounds are given in [26]. They illustrate the following findings: in unhindered alcohols and ethers the value of $^1\Delta$ is ~0.025 ppm; in aldehydes and ketones $^1\Delta$ is ~0.050 ppm; in carboxylic acid derivatives the double-bonded oxygen gives a higher $^1\Delta$

$$
\begin{array}{ccc}
\mathrm{C}{=}^{18}\mathrm{O} & \mathrm{C}{=}\mathrm{O} & \mathrm{C}{-}^{18}\mathrm{O} \\
\diagup & \diagup & \diagup \quad \diagdown \\
\mathrm{O} & {}^{18}\mathrm{O} & \mathrm{O} \qquad \mathrm{C} \\
+0.038\ \mathrm{ppm} & +0.015\ \mathrm{ppm} & +0.029
\end{array}
$$

[26]

TABLE 36

^{18}O isotope effects on ^{1}H nuclear shieldings of H_2O and ^{13}C nuclear shieldings of metal carbonyls (in ppm).

	$^{1}\Delta$	Ref.
$^{1}\Delta H(^{18}O)$		
$\quad H_2{}^{18}O$	$+0.012^{c}$	239
$^{1}\Delta C(^{18}O)$		
$\quad W(C^{18}O)_5P(OMe)_3$	$+0.040^{a}$ (CO)	252
$\quad W(CO)_6$	$+0.040^{b}$ (CO)	251
$\quad W(CO)_5PPh_3$	$+0.040^{b}$ (CO)	251
$\quad Cr(CO)_5P(n\text{-}Bu)_3$	$+0.040^{a}$ (CO)	253
$\quad Mo(CO)_6$	$+0.045^{b}$ (CO)	253
$\quad Mo(CO)_5PPh_3$	$+0.040^{b}$ (CO)	253
$\quad Mo(CO)_5P(OMe)_3$	$+0.040^{b}$ (CO)	253

[a] Same value for axial and equatorial CO groups.

[b] Equatorial group.

[c] Higher values are found if recorded in dilute solutions of organic solvents.

value than that for a single-bonded ^{18}O and finally the effects observed in phenols are smaller than in aliphatic alcohols.[244,245] A closer study of each functional group shows that for alcohols $^{1}\Delta$ is larger in t-butyl alcohol than in primary and secondary alcohols.[242,244,245] Risley and Van Etten[244] have studied sterically hindered alcohols further to determine if the large effect is related to steric interaction. Tri-t-butyl carbinol gives a slightly smaller Δ value than t-butyl alcohol and so does triphenyl carbinol (Table 39). The authors conclude that steric interactions have little effect, whereas electronic substituent effects may be important. The scheme given by Risley and Van Etten[243] is shown in [27].

Vederas[245] has observed, for carbonyl compounds, the following trend in $^{1}\Delta$ values: ketones > aldehydes > esters > amide. Conjugation reduces the effect by 0.002–0.006 ppm. A study of p-substituted acetophenones shows that increasing electron donation by the *para* group generally produces smaller isotope effects.[80] The general trend of a decrease in the magnitude of the isotope effect, due to conjugation, is supported from the findings that saturated ketones display larger effects than unsaturated ketones and from the fact that the planar fluoronone and nearly planar anthrone have smaller effects than, for example, the benzophenones.

Ring size plays a small role in cyclic ketones but is important for cyclic alcohols.[80] No general correlation between the isotope effect and the carbon–oxygen bond-length exists. However, within more limited functional

TABLE 37

^{18}O isotope effects on ^{13}C nuclear shieldings of acids and acid derivatives (in ppm).

	$^1\Delta$	$^2\Delta$	Ref.
$C={}^{18}O^b$			
Ethyl acetate	+0.038		244
Propylcyclohexane carboxylate	+0.039		245
Cyclohexylcyclohexane carboxylate	+0.037		245
Propyl benzoate	+0.033		245
p-Bromophenacyl formate	+0.054[c]		243
p-Bromophenyl acetate	+0.037 (5)		254
Cyclohexane carboxamide	+0.036		245
N-Propylcyclohexane carboxamide	+0.033		245
N,N'-Dipropylcyclohexane carboxamide	+0.033		245
Benzamide	+0.030		245
N-Propyl benzamide	+0.028		245
N,N'-Dipropyl benzamide	+0.030		245
2-Pyrrolidone	+0.028		245
$C-{}^{18}O^d$			
Propyl benzoate	+0.015 (C=O)	+0.005	245
	+0.032 (CH$_2$)		
Propylcyclohexane carboxylate	+0.015 (CO)	+0.005	245
	+0.030 (CH$_2$)		
α-D-Glucosepentaacetate[e]	+0.010 (CO)		245
	+0.024 (CH)		
β-D-Glucosepentaacetate[e]	+0.010 (CO)		245
	+0.026 (CH)		
p-Bromophenacyl formate	+0.026 (CH$_2$)		243
p-Bromophenacyl acetate	+0.010[d] (CO)		254
	+0.048 (5)[c] (CO)		
n-Butyl acetate	+0.015 (CO)		244
	+0.029 (CH$_2$)		
Acetic acid	+0.025[f,g]		250
	+0.027[f,h]		
Glutamic acid	+0.02[f]		250
Benzoic acid	+0.031[i,f]		250
Sodium formate	+0.025[f]		250
syn-7-Carboxynorbornene	+0.025[f]		250
anti-7-Carboxynorbornene	+0.025[f]		250
Hibiscus acid dimethyl ester	+0.037[j]		257

[a] Sign reversed compared to original paper.
[b] Double-bonded oxygen enriched.
[c] Both oxygens enriched.
[d] Single-bonded oxygen enriched.

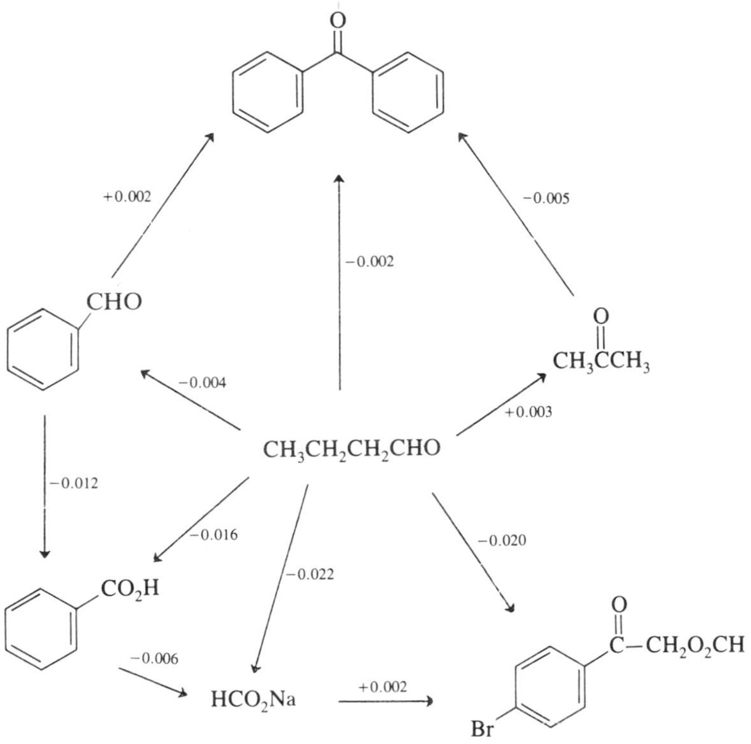

[27][243]

groups some correlation may be found. For single bonds a smaller isotope effect is related to a shorter bond. A similar relationship exists for conjugated and non-conjugated ketones.[243]

Risley and Van Etten[244] report that the ^{18}O isotope effect on the carbonyl carbon of a carboxylate group (0.038 ppm) compared with that of aldehydes and ketones (≈ 0.050 ppm), indicates that the CO group of a carboxylate does not behave as an isolated carbonyl group. This is very evident from the vibrational spectra and much of the information concerning substituent effects on $\Delta C(^{18}O)$ can probably be extrapolated from vibrational spectra.

Footnotes to Table 37 continued

[e] ^{18}O on C-1.
[f] Per oxygen.
[g] pH 2.0.
[h] pH 8.0.
[i] Solvent, D_2O.
[j] Solvent, $CDCl_3$.

P. E. HANSEN

TABLE 38

^{18}O isotope effects on ^{13}C nuclear shieldings of ketones and aldehydes (in ppm).

	$^1\Delta$	Ref.
Acetone[a]	+0.049	248
	+0.050	243
Di-t-butyl ketone	+0.054	244
Acetophenone	+0.049	245
	+0.047	80
Benzophenone	+0.045	80, 243
Cyclohexyl methyl ketone	+0.052	80
Pulegone	+0.045	80
p-Hydroxyacetophenone	+0.044	80
p-Aminoacetophenone	+0.044	80
p-Methoxyacetophenone	+0.045	80
p-Methylacetophenone	+0.046	80
p-Chloroacetophenone	+0.046 (5)	80
p-Nitroacetophenone	+0.047 (5)	80
m-Methylacetophenone	+0.046	80
o-Methylacetophenone	+0.047	80
Cyclopropanone	+0.049	80
Cyclobutanone	+0.051	248, 80
Cyclopentanone	+0.053	248
	+0.052	80
Cyclohexanone	+0.053	245, 248
	+0.051	80
Cycloheptanone	+0.052	80
Fluorenone	+0.041	80
Anthrone	+0.041	80
10,11-Dihydro-5H-dibenzo[a,d]cyclo-hepten-5-one	+0.048	80
6,7-Dihydro-5H-dibenzo[a,d]cyclo-octen-12-one	+0.044	80
Tropone	+0.037	80
Adamantanone	+0.051	80
5-H-Dibenzo[a,d]cyclohepten-5-one	+0.044 (5)	80
Dehydromethone	+0.045	80
Bicyclo[2.2.1]heptan-2-one	+0.050	80
Benzaldehyde	+0.042	245
	+0.043	243
Cyclohexane carbaldehyde	+0.045 (5)	80
Octanal	+0.047 (7)	245
n-Butyraldehyde	+0.047	243

[a] In reference 141 an isotope effect of -0.20 ppm (negative sign probably refers to deshielding) is reported for the $^1\Delta^{13}C(^{17}O)$ isotope effect.

TABLE 39

^{18}O isotope effect on ^{13}C nuclear shieldings of alcohols and ethers (in ppm).

	$^1\Delta$	Ref.
Primary alcohols		
Benzylalcohol	$+0.019^b$	245
	$+0.023$	243
Octanol	$+0.021^b$	245
	$+0.019^c$	80
Butanol	$+0.020$	244
Secondary alcohols		
Isopropyl alcohol	$+0.023^d$	243
Diphenylcarbinol	$+0.020$	245
1-Phenyl ethanol	$+0.023^b$	80, 245
	$+0.021^g$	
	$+0.020^h$	
1-Cyclohexyl ethanol	$+0.026^b$	245
	$+0.028^c$	
α-D-Glucose	$+0.015^f$	245
β-D-Glucose	$+0.013^f$	245
1-(p-Aminophenyl)ethanol	$+0.018^e$	80
1-(p-Hydroxyphenyl)ethanol	$+0.021\,(5)^e$	80
1-(p-Methoxyphenyl)ethanol	$+0.021\,(6)^b$	80
1-(p-Methylphenyl)ethanol	$+0.021^b$	80
1-(p-Chlorophenyl)ethanol	$+0.022^b$	80
1-(p-Nitrophenyl)ethanol	$+0.021\,(5)^b$	80
Cycobutanol	$+0.020$	80
Cyclopentanol	$+0.026\,(5)$	80
Cyclohexanol	$+0.020$	80
Cycloheptanol	$+0.026$	80
Cyclooctanol	$+0.029$	80
Tertiary alcohols		
t-Butanol	$+0.035^b$	242
	$+0.035^d$	243
2-Cyclohexylpropan-2-ol	$+0.032$	245
1-Methylcyclohexanol	$+0.031^b$	80
	$+0.033^c$	
	$+0.029^i$	
Tri-t-butylcarbinol	$+0.030^j$	244
Triphenylcarbinol	$+0.025$	244
Ethers		
Phenetole	$+0.016$ (CH)	243
	$+0.025$ (CH$_2$)	
p-Bromophenetole	$+0.016$ (CH)	243
	$+0.025$ (CH$_2$)	
Phenyl vinyl ether	$+0.018$ (CH arom.)	243
	$+0.018$ (CH vinyl)	
Tetrahydrofuran	$+0.024$	248

TABLE 39 (*cont.*)

	$^1\Delta$	Ref.
Trimethyl *n*-butylorthocarbonate[k]	+0.015 (C)	244
	+0.025 (CH$_2$)	
Dimethyl di-*n*-butylorthocarbonate[k]	+0.015 (C)[l]	244
Methyl tri-*n*-butylorthocarbonate[k]	+0.015 (C)[l]	244
Tetra-*n*-butylorthocarbonate	+0.015 (C)[l]	244

[a] Signs reversed compared to original paper.
[b] Solvent, CDCl$_3$.
[c] Solvent, methanol-d_4.
[d] Solvent, D$_2$O.
[e] Solvent, DMSO-d_6.
[f] ^{18}O at C-1.
[g] Solvent, benzene-d_6.
[h] Solvent, pyridine-d_5.
[i] Solvent, *t*-butanol-d_{10}.
[j] Solvent, acetone-d_6.
[k] ^{18}O-butyl oxygen.
[l] Propyl oxygen.

TABLE 40

^{18}O isotope effects on ^{13}C nuclear shieldings of miscellaneous compounds (in ppm).

	Type of bond	$^1\Delta$	Ref.
Griseofulvin	C=O	+0.042 (C-4′)	254
	C=O	+0.040 (C-3)	
	ϕ^a—O	+0.020 (C-2′)	
	ϕ—O	+0.015 (C-6)	
	ϕ—O	+0.017 (C-7a)	
	ϕ—O	+0.016 (C-4)	
Sterigmatocystin	C=O	+0.029 (C-1)	246
	ϕ—O	+0.010 (C-3)	
	ϕ—O	+0.021 (C-8)	
	ϕ—O	+0.015 (C-10)	
Averufin	ϕ—O	+0.011 (C-1)	247
	ϕ—O	+0.020 (C-3)	
	ϕ—O	+0.011 (C-6)	
	ϕ—O	+0.010 (C-8)	
	C=O	+0.029 (C-9)	
	C—O	+0.023 (C-1′)	
	C=O	+0.041 (C-10)	80
2,5,14-Trioxahexa-cyclo[5.5.2.1.2^{4,10}.0^{4,17}. 0^{10,17}]hepta-decane	C—O—C	+0.023	248

TABLE 40 (*cont.*)

	Type of bond	$^1\Delta$	Ref.
Trispiro[tricyclo-3.3.3. $0^{1,5}$]undecane-2,2′: 8,2″: 9,2‴-tris[oxirane]	C—O—C	+0.030	248
1-Oxaspiro[2,5] octane	C—O—C	+0.033	248
Erythromycin A 2′-benzoate	COO^{b}C	+0.04 (C-1)	255
	R—O	+0.02 (C-3)	
	C—O—C	+0.03 (C-5)	
	C=O	+0.05 (C-9)	
	C—OH	+0.02 (C-11)	
	C—O—C	+0.03 (C-13)	
Erythromycin B 2′-benzoate	COOR	+0.04 (C-1)	255
	R—O	+0.03 (C-3)	
	C—O—C	+0.03 (C-5)	
	C=O	+0.06 (C-9)	
	C—OH	+0.02 (C-11)	
	C—O—C	+0.04 (C-13)	
Monensin A	COO$^-$	+0.03 (C-1)	258
	C—O—C	+0.03 (C-3)	
	C—O—C	+0.03 (C-5)	
	C—OH	+0.02 (C-7)	
	C—O—C	+0.03 (C-9)	
	C—OH	+0.02 (C-25)	
Lasalocid A	C—O—C	+0.020 (C-15)	249
1,3-Bis-*o*-methyl lasalocid A	φ—O—C	+0.016 (C-3)	249
	C—O—C	+0.022 (C-15)	
	C=O	+0.055 (C-13)	
	COCH$_3$	+0.015 (COCH$_3$)	
	C=O	+0.037 (CO)	
11,22-Bis(*o*-trifluoroacetyl)lasalosid A	φ—O	+0.015 (C-3)	249
	C—O—C	+0.020 (C-15)	
	C—O—C	+0.042 (C-11)	
Phenol	φ—O	+0.016	244
Cytochalasin B	OC=O	0.036^c (C-23)	259
	NC=O	0.031 (C-1)	

^{a} φ represents an aromatic ring on which the observed carbon is the *ipso* carbon.
^{b} Carboxylic acid derivative.
^{c} Frequency of 100.6 MHz assumed in calculation of ppm value.

These findings and other interesting effects may be demonstrated in the multitude of compounds isolated from biosynthetic studies, as shown in Table 40.

In sterigmatocystin[246] and averufin[247] a $^1\Delta \sim 0.020$ ppm is observed for the bond φ—O—φ or φ—O—C. The small value for phenol is thus increased in anisoles and similar bond types.

In averufin the $^1\Delta C\text{-}9(^{18}O)$ value is smaller than $^1\Delta C\text{-}10(^{18}O)$ (Table 40). The oxygen of the former takes part in hydrogen bonding. The authors ascribe this to electron donation to C-9 through the π-electron system from the oxygens at C-1, C-3, C-6 and C-8. A more straightforward explanation is that hydrogen bonding reduces the size of $^1\Delta$ (Section III.K.1). $^1\Delta$ values of epoxides may be studied in 1-oxaspiro[2.5]octane and in more complex triepoxides.[248] The value of $^1\Delta$ is larger than in unhindered alcohols and ethers.

In lasalocid A and its 1,3-bis-O-methyl and 11,22-bis-(O-trifluoroacetyl) derivatives the resolution of the $^{13}C\text{--}^{16}O$ and $^{13}C\text{--}^{18}O$ signals improve in that order. This may possibly be related to intra-molecular hydrogen bonding.[249]

1. Exchange reactions

These may be studied and the kinetics determined. Risley and Van Etten have studied the acid-catalysed exchange of oxygen in t-butyl alcohol,[242,243] and the oxygen exchange of acetic acid in dilute acid.[250]

$^1\Delta$ data may also, with advantage, be used to study oxygen exchange in metal carbonyls.[251–253]

The treatment of $W(C^{16}O)_5P(OMe)_3$ with $^{18}OH^-$ in a phase transfer catalysis leads to ^{18}O enrichment preferentially for the equatorial carbonyl groups. The degree of enrichment is easily measured by means of ^{13}C NMR spectroscopy; ^{18}O produces an isotope shift, thus giving rise to two separate signals when using high magnetic fields (100.6 MHz).[251] Treatment of ^{18}O-enriched $W(C^{18}O)_5P(OMe)_3$ with $C^{16}O$ shows no incorporation of ^{16}O, but only a rearrangement. By this method it is demonstrated that $W(CO)_5P(OMe)_3$ rearranges in a non-dissociative process.[252]

^{18}O-enriched carbonyl compounds of chromium, molybdenum and tungsten are likewise produced by exchange with $^{18}OH^-$. Isotope effects are given in Table 36. Again a preferential enrichment of the equatorial carbonyl oxygen is observed.[253]

C. $^1\Delta C(^{17}O)$ effects

A negative isotope effect, -0.2 ppm, is observed in acetone.[141]

XIII. ^{18}O ISOTOPE EFFECTS ON NUCLEAR SHIELDING OF ^{31}P AND OTHER NUCLEI

A. $^1\Delta^{31}P(^{18}O)$ effects

Isotope effects of this type are almost exclusively of the one-bond type, as shown in Table 41. Their main use is in kinetic studies of oxygen

TABLE 41

^{18}O isotope effects on ^{31}P nuclear shieldings (in ppm).

	$^1\Delta$	Ref.
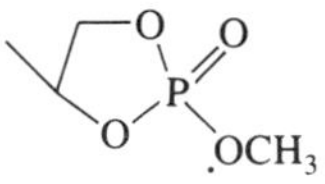	+0.018	273
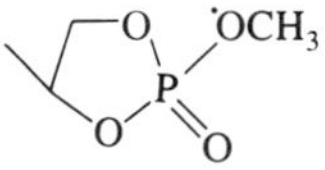	+0.018	273
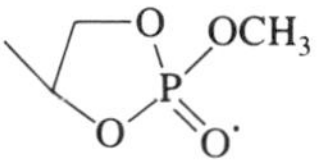	+0.043	273
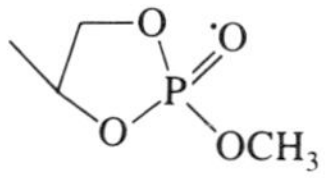	+0.043	273
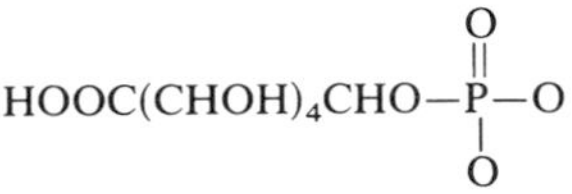 $HOOC(CHOH)_4CHO-P-O$	~+0.024[f]	268
α-D-Ribose 1-phosphate	+0.025[f]	263
Na_3PO_4	+0.019[f]	280
PO_4^{3-}	+0.0226[f]	263
PO_4^{3-}	+0.020[d]	279
	+0.0206[e]	260
$(MeO)_3PO$	+0.036	262
$(MeO)_2PO_2^{-\,i}$	+0.029	263
$(MeO)_2PO_3^{2-\,i}$	+0.023	262
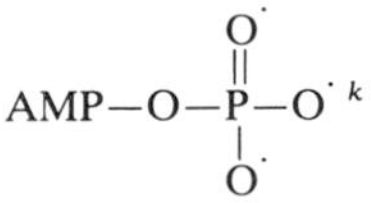 $AMP-O-P-O^{\,k}$	+0.0220	261
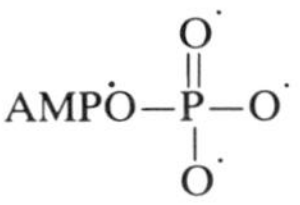 $AMP\dot{O}-P-O$	+0.0166 (α)[f,g] +0.0215 (β)[f,g]	261
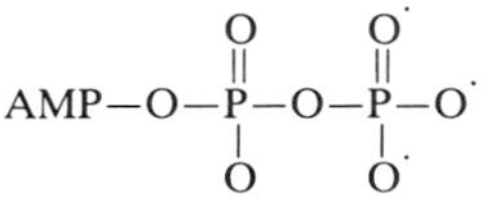 $AMP-O-P-O-P-O$	+0.0220 (γ)[f,g]	261
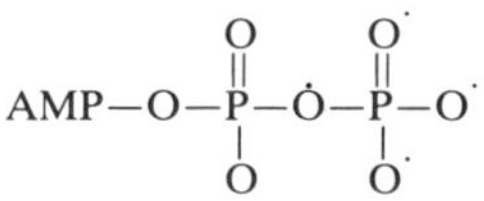 $AMP-O-P-\dot{O}-P-O$	+0.0165 (β)[f,g] +0.0220 (γ)[f,g]	261

TABLE 41 cont.)

	$^1\Delta$	Ref.
AMP—O—P—O—P—O (with Ȯ and O structure)	+0.0281 (β)[f,g] +0.0165 (β)[f,g,h] +0.0220 (γ)	261
AMP—O—P—Ȯ—P—O (structure)	+0.0285 (β)[f,g] +0.0163 (β)[f,g,h]	261
AMP—Ȯ—P—O—P—O (structure)	+0.0172 (α)[f,g] +0.0281 (β)[f,g] +0.0165 (β)[f,g,h]	261
AMP—Ȯ—P—Ȯ—P—O [j]	+0.0209 (γ) +0.0247 (α)	262 279
Ad—O—P—Ȯ—P—O—P—O [k] (with S)	+0.0165 (α) +0.0165 (β)	272
Ad—O—P—O—P—O—P—O (with S)	+0.0302 (α)	272
Ad—O—P—O (AMP)	+0.025 (α)	279
Ad—P—O—P—O (with S)	+0.037 (α)	275
Ad—P—Ȯ—P—O (with S)	+0.021 (α)	275
AMP—O—P—Ȯ—P—O (with S)	+0.021 (β)	275

TABLE 41 *cont.*)

	$^1\Delta$	Ref.
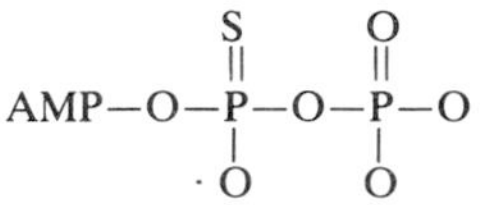 AMP—O—P—O—P—O (with S=, O= and O, O substituents)	+0.036 (β)	275
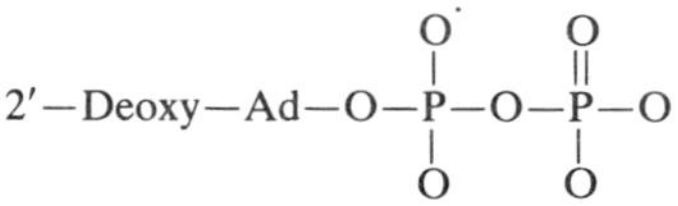 2′—Deoxy—Ad—O—P—O—P—O	+0.0272 (α)	276
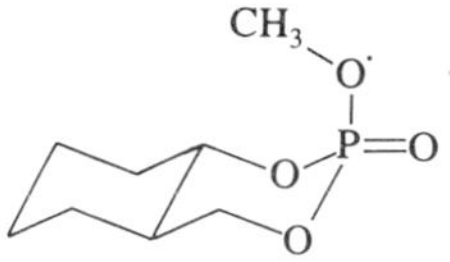 (cyclohexane CH₃—O—P=O ring structure)	+0.015	277
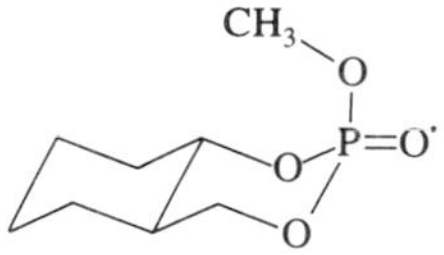 (cyclohexane CH₃—O—P=O' ring structure)	+0.040	277
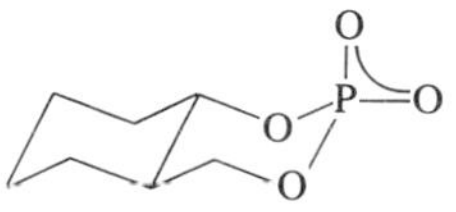 (cyclohexane P=O ring structure)	+0.026	277
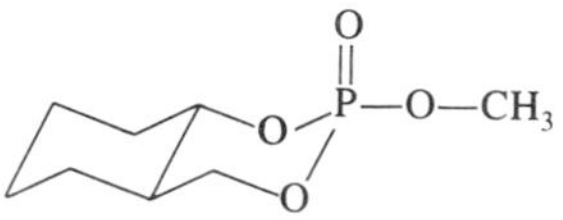 (cyclohexane P—O—CH₃ ring structure)		277
Cyclo-2′-deoxy-AMP	+0.038 (equatorial)[m] +0.015 (axial)	278

[a] Signs reversed compared to original paper.
[b] ^{18}O marked with •.
[c] Compounds with both oxygens enriched show an effect of +0.060 ppm.
[d] pH 9.
[e] pH 8.2.
[f] Per ^{18}O.
[g] α, β and γ refer to the position of ^{31}P; α indicates the one nearest the sugar moiety.
[h] Isotope effect caused by a bridge ^{18}O.
[i] One ^{18}O.
[j] A long-range effect of 0.0055 ppm is reported in reference 271.
[k] AMP is an abbreviation for adenosine mononucleotide phosphate; Ad is an abbreviation of adenosine.
[l] Shown on spectra, but not given explicitly.
[m] Equatorial refers to the position of the O—P bond.

incorporation and in enzymatic studies. Most studies are performed at frequencies equal to, or greater than, 150 MHz to ensure good resolution of the lines.

Cohn and Hu[260] have suggested several ways of using ^{18}O isotope effects on ^{31}P nuclear shieldings. The advantage of the NMR method is, in this case as in many other cases, its non-destructive nature, which makes continuous monitoring possible without perturbing the system.

The phosphate unit plays an important role in biochemistry, and labelling of this group with ^{18}O makes it possible to follow both the fate of the phosphate group and in some cases also the scrambling of oxygens within the phosphate group, provided that exchange with solvent oxygen does not take place.

A study of ^{18}O-enriched ADP and ATP compounds reveals a difference between the ^{31}P shieldings of α and β phosphorus, when the ^{18}O is incorporated in a bridge or in a non-bridge position [28]. Furthermore, the γ phosphorus shows a different isotope effect from those of the α and β phosphorus nuclei[261,262] (Table 41). N.B.: the effects are given per ^{18}O.

$$
\underset{\text{Bridge}}{\overset{\displaystyle\parallel\qquad\parallel}{-\text{P}-\text{O}-\text{P}-}} \qquad\qquad \underset{\underset{\text{Non-bridge}}{\displaystyle\text{O}}}{\overset{\displaystyle\text{O}}{-\text{P}-\text{O}}}
$$

[28]

A difference due to bridge and non-bridge oxygens was likewise observed in ATP α S-α,β-^{18}O and ATP α S-α-^{18}O.[263] The difference between bridging and non-bridging oxygens is a very important one for their use in configurational studies (*vide infra*). The magnitudes of the $^1\Delta P(^{18}O)$ data correlate well with the amount of double-bond character of the bond.[261] A linear relationship between $^1\Delta P(^{18}O)$ and the square of the frequency of the A_1 stretching mode of the phosphate group has also been observed.[262]

A single example of a $^2\Delta$ value has been reported in ATP.[264]

1. Uses

Of the many reactions studied, $P_i \rightleftharpoons$ ADP exchange catalysed by polynucleotide phosphorylase may be mentioned. This study also reveals that the cleavage occurs as shown in [29].[260] Inorganic orthophosphate does not exchange with water at room temperature, but the reaction may be speeded up by the addition of inorganic pyrophosphatase[260] or other enzymes.

Bock and Cohn[265] have studied the mechanism of exchange of ^{18}O between pyrophosphate and water catalysed by alkaline phosphatase of

$$
\underset{\substack{|\\O}}{\overset{\substack{O\\||}}{A-O-P}}-\underset{\substack{|\\O}}{\overset{\substack{O\\||}}{P}}-O+\dot{O}-\underset{\substack{|\\\dot{Q}}}{\overset{\substack{\dot{O}\\|}}{P}}-\dot{O} \rightleftharpoons \underset{\substack{|\\O}}{\overset{\substack{O\\||}}{A-O-P}}-\dot{O}-\underset{\substack{|\\\dot{O}}}{\overset{\substack{\dot{O}\\||}}{P}}-\dot{O}+\underset{\substack{|\\O}}{\overset{\substack{O\\||}}{O-P}}-O
$$

[29]

Escherichia coli. The native zinc metalloenzyme differs from its cobalt analogue according to the isotope distribution.

The isotope method was used to establish that human acid phosphatase catalyses phosphate oxygen–water exchange.[266] Exchange between $P^{18}O_4^{3-}$ and H_2O catalysed by myosin subfragment 1 in the presence of MgADP was studied in the process of elucidating muscle contraction.[267]

Hackney *et al.*[256,268] have studied the photoreaction between ADP and $P^{18}O_4^{3-}$. The γ-phosphoryl group of the ATP formed is transferred to glucose by hexokinase to form glucose 6-phosphate. To avoid overlap between α- and β-glucose 6-phosphate, this is oxidized into 6-phosphogluconate, which is then analysed (Table 41).

Cohn[269] has reported preliminary results on the use of ^{18}O-labelled ATP in the study of the reaction mechanism of formation of valylacyl-tRNA, the goal being to prove the existence of an enzyme–aminacyl-AMP intermediate.

By equilibrating a mixture of ^{18}O-enriched orthophosphate, α-D-ribosephosphate-^{16}O, inosine and hypoxanthine by means of calf spleen purine nucleoside phosphorylase, it is demonstrated that the C—O bond of the α-D-ribose 1-phosphate is cleaved in the enzymatic reaction.[263]

Sharp and Benkovic[270] have studied the fructose 1,6-biphosphatase catalysis of fructose 1,6-biphosphate to fructose 6-phosphate and inorganic phosphate and conclude that mass spectrometric methods are $10^2–10^3$ times more sensitive and considerably more precise in the determination of low abundance species than the observation of one-bond isotope effects. By studying the randomization of adenosine 5′-triphosphate-α,β-^{18}O-β-$^{18}O_2$ caused by pyruvate kinase and pyruvate with and without inhibitor, a dissociative S_N1 (P) mechanism was arrived at.[271]

Synthesis of enantiomeric phosphates and the determination of the absolute configuration is of interest in mechanistic studies of enzymes.

Jarvest and Lowe[272] have been able to determine the absolute configuration of adenosine 5′-phosphorothioate-^{18}O by taking advantage of the difference in isotopic effects of bridging and non-bridging oxygens in the hydrolysed products.

Starting with phospho (S)-1,2-diol-1-$^{16}O,^{17}O,^{18}O$, cyclization followed by methylation leads to 2,2-dioxo-1,3,2-dioxaphopholanes. As ^{17}O causes line-broadening, only those isomers without ^{17}O give rise to observable ^{31}P signals. The *syn* and *anti* compounds have different nuclear shieldings. The **R** and **S** isomers are thus identified because they show different isotope effects[273] (Table 41). A stereospecific method is available to cyclize D-glucose 6-phosphate-^{16}O,^{17}O,^{18}O and adenosine 5′-phosphate-^{16}O,^{17}O,^{18}O into phosphate diesters;[272] these may be further esterified and the configuration has been determined by means of isotope effects.[274]

Webb and Trentham[275] have determined the absolute configuration of inorganic thiophosphate-^{16}O,^{17}O,^{18}O synthesized from ADP-^{16}O,^{17}O,^{18}O taking advantage of the two previously mentioned findings, i.e. ^{17}O broadens the ^{31}P signal and ^{18}O gives different isotope effects when in a bridging or in a non-bridging position.

Coderre and Gerlt[276] have determined the configuration of ADP-^{18}O by complexing it with [Co(III)(NH$_3$)$_4$] ions, leading to a diastereomeric mixture of an α,β-bidentate complex. After separation the position of ^{18}O may be determined since the ^{18}O complexed with Co(III) has the smaller isotope effect and the screw senses have been assigned previously. Slightly different $^1\Delta$ values are observed in the cobalt complex compared to the uncomplexed material due to different bond orders in the two (Table 41).

Gorenstein and Rowell[277] have used the isotope effect to follow the stereochemistry of hydrolysis in phosphate triesters. Diastereomers of cyclic 2′-deoxy-AMP-^{18}O have been prepared and the absolute configuration at phosphorus has been determined.[278]

B. ^{18}O isotope effects on other nuclei

Such effects are only reported in a few cases. Van Etten and Risley[281] have observed $^1\Delta^{15}$N(^{18}O) effects in nitrite and nitrate ions (Table 42). This large effect is used to monitor nitrite oxygen–water exchange. ^{18}O effects on ^{55}Mn[282,283] and ^{95}Mo[283] are also given in Table 42. Again, oxygen exchange may be studied.[283]

TABLE 42

^{18}O isotope effects on nuclear shieldings of ^{15}N, ^{55}Mn and ^{95}Mo (in ppm).

	$^1\Delta$	Ref.
Silver nitrite	0.138^b	281
Sodium nitrite	0.138^b	281
Nitrate	0.056^b	281
KMnO$_4$	0.599^b	282
Na$_2$MoO$_4$	+0.25^b	283

^{a} Sign reversed compared to original paper.
^{b} Per oxygen atom.

XIV. MISCELLANEOUS ISOTOPE EFFECTS

A. $\Delta C(B)$ and $\Delta F(B)$ effects

The ^{11}B isotope effect on C-1 of methylpentaborane-9 is smaller than 0.0012 ppm.[227]

Gillespie et al.[284,285] have used the collapse of splittings due to isotope effects as a monitor in the exchange reactions of F^- in $AgBF_4$ and in ketone–BF_3 complexes.

B. $\Delta F(^{34}S)$, $\Delta C(^{34}S)$ and $\Delta Mo(^{34}S)$ effects

A $^2\Delta F(^{34}S)$ value of 0.014 ppm is observed in 2-fluorothiophene.[223] A $^1\Delta C(S)$ value is observed under very high resolution conditions in CS_2.[286] The value of $^1\Delta Mo(S)$ is somewhat larger and is found in MoS_4^{2-} ions[287] (Table 43).

TABLE 43

^{34}S isotope effects on ^{13}C and ^{19}F nuclear shieldings (in ppb).

	$^1\Delta$	$^2\Delta$	Ref.
$\Delta^{13}C(^{34}S)$			
CS_2-$^{34}S_1$	9		75
$\Delta^{19}F(^{34}S)$			
2-Fluorothiophene		14	223
$\Delta^{95}Mo(^{34}S)$			
MoS_4^- ^{b}	90^c		287

^{a} Sign reversed compared to original paper.
^{b} Aqueous solutions of the potassium or ammonium ions.
^{c} Per ^{34}S.

C. $\Delta F(^{37}Cl)$, $\Delta C(^{37}Cl)$ and $\Delta Pt(^{37}Cl)$ effects

A $^2\Delta F(^{37}Cl)$ effect is observed in chlorofluoroalkanes. Marco and Gatti[288] have observed no difference between CF and CF_2 signals. Carey et al.[289] have observed better resolved isotope splittings at low temperatures. Brey et al.[220] have studied $^2\Delta F(^{37}Cl)$ results at two different temperatures and report that $^2\Delta F(^{37}Cl)$ decreases at higher temperatures (Table 44).

A $^1\Delta C(^{37}Cl)$ value has been resolved in polychlorinated hydrocarbons.[290]

^{195}Pt spectra of chloro- and bromoplatinates show nicely split signals due to isotope splittings.[291]

TABLE 44

^{35}Cl isotope effects on ^{19}F, ^{13}C and ^{195}Pt nuclear shieldings (in ppb).

	$^2\Delta$	Ref.
^{19}F		
Fluorotrichloromethane-^{37}Cl$_1$ [b]	8.5[c]	220
	7.1[d]	
	5.6[f]	
Fluorotrichloromethane-^{37}Cl$_2$	14.3[d]	220
	11.3[f]	
Fluorotrichloromethane-^{37}Cl$_3$	21.3[d]	220
1,1-Difluoro-1,2,2,2-tetra-chloroethane-1-^{37}Cl$_1$	8.8[c]	220
	7.5[d]	
	6.3[f]	
1,2,2,3-Tetrafluoro-1,1,3-trichloro-3-bromopropane-1-^{37}Cl$_1$	7.9[d]	220
	6.7[e]	
	6.4[f]	
1,2,2,3-Tetrafluoro-1,1,3-trichloro-3-bromopropane-1-^{37}Cl$_2$	14.3[d]	220
	13.0[e]	
1,2,2,3-Tetrafluoro-1,1,3-trichloro-3-bromopropane-3-^{37}Cl$_1$	6.5[f]	220
	5.5[f]	
1,1,2,3,3-Pentafluoro-1-chloro-2-bromobutyroni-trile-1-^{37}Cl$_1$	5.8[f]	220
^{13}C		
1,2-Dichloroperfluoro-propane[g]	+6.7	288
Hexachloropropene	+4[h]	290
Hexachlorobutadiene	+4[h]	290
^{195}Pt		
Pt[Cl$_6$]$^{2-}$ [j]	+167[h]	291

[a] Sign reversed compared to original paper.
[b] A splitting due to CF^{35}Cl^{37}Cl$_2$ is observed in reference 289.
[c] Temperature, 193 K.
[d] Temperature, 230 K.
[e] Temperature, 295 K.
[f] Temperature, 305 K.

D. $\Delta^{77}Se(^nSe)$ effects

For $n = 76, 78, 80$ and 82 the isotope effect is 0.50 Hz. For $n = 77$ the Δ value is slightly different. This difference is suggested to be related to differences in the vibrational modes for the symmetric isotopomer with $n = 77$ and for those with $n = 76, 78, 80$ and 82.[173]

E. $^1\Delta F(^{29}Si)$ effects

One-bond isotope effects for a series of compounds with the formula $RSiF_3$ are given;[292] the range is from 9 to 36 ppb. The isotope effects are believed to depend strongly on the two silicon–fluorine stretching modes.

XV. COMMON FEATURES

A. Additivity

One of the most important characteristics of isotope effects is that they are usually additive. If not, it would be impossible to separate out the various effects observed in systems with more than one isotopic replacement, e.g. biological enriched materials.

Additivity has been observed for most of the types of isotope effects treated; examples are $^1\Delta C(D)$ of methane[13] and $\Delta F(D)$ in cyclo-hexylfluoride-2,2,6,6-D_4.[154] $^1\Delta P(D)$ of phenylphosphine,[170] $^1\Delta P(^{18}O)$ data of PO_4^{2-} ions and $^1\Delta C(^{18}O)$ values of acetic acid and of orthocarbonate;[244] $^1\Delta^{55}Mn(^{18}O)$ in $KMnO_4$;[282] $^1\Delta Se(D)$ data in H_2Se[172] and many other examples may be found in the accompanying tables.

Baldry and Robinson[96] have observed additivity in dimethylcyclohexane. The equilibrium isotope effect increases linearly as the number of deuterium atoms increases from one to three. A similar effect is found for the ^{19}F shielding of trans-1-trifluoromethyl-3-methylcyclohexane-3'-D_1, D_2 and D_3.

Batiz-Hernandez and Bernheim[1] have suggested that the additivity of isotope effects could be directly related to the additivity of zero-point vibrational energies of isotopically substituted molecules which result in the so-called low of the mean. It is interesting to note that small deviations

---

Footnotes to Table 44 continued

[g] Both CF and CF_2 groups show this effect.

[h] Per ^{37}Cl.

[i] Approximately the same effect is reported for all carbons except C-4, which is too small to be resolved. However, the effect at an sp^3-hybridized carbon is slightly larger than at an sp^2-hybridized carbon.

[j] $Pt[Br_6]^{2-}$ shows an isotope effect of 28 ppb, per ^{81}Br.

from isotope effect additivity have their counterparts in deviations from the law of the mean.

Non-additivity is reported in the following cases: $^2\Delta H(D)$ of methane,[119] $^2\Delta H(D)$ of lactic acid,[121] $^1\Delta C(D)$ of toluene-D,[100] $^1\Delta P(D)$ of phosphine.[169] Non-additivity is also reported for $\Delta F(D)$ of p-fluorophenyldimethyl- and p-fluorophenylmethyl-α-D_n carbonium ions[97] (Section V.B). The cause of non-additivity in this case is probably that deuterium perturbs the rotamer populations of the methyl groups.

The non-additivity observed in methane and phosphine is possibly related to deviations from the law of the mean. The non-additivity in lactic acid and toluene is more likely to be related to pertubations of rotamer populations.

B. Hyperconjugation

Isotope effects are caused by differences in vibrational zero-point energy. Halevi et al.[293,294] have expressed the vibronic phenomena in terms of substituent parameters, such as inductive and hyperconjugative effects. Along these lines, isotope effects are related to the greater electron-donating tendency of a deuterium atom compared with a proton and the reduced hyperconjugative ability of a C—D compared to that of a C—H bond.

Transplanting these ideas to isotope effects on nuclear shielding and to toluene and 4-fluorotoluene in particular, Maciel et al.[100,157] have proposed an explanation for these long-range isotope effects. They caution, however, against an uncritical use of these qualitative ideas.

A study of toluene-d_1, -d_2 and -d_3 clearly reveals that $^n\Delta C(D)$ effects are additive, but more interestingly that both $^3\Delta C(D)$ and $^5\Delta C(D)$ effects are negative, whereas $^4\Delta C(D)$ is zero and in addition that $^5\Delta C(D) > {}^3\Delta C(D)$. The negative isotope effects are observed for positions at which hyperconjugative interactions might contribute. That $^3\Delta C(D) < {}^5\Delta C(D)$ is due to the intrinsic, positive effect, adding to the value of $^3\Delta C(D)$. That hyperconjugation plays a role is strongly supported from studies on ethylbenzene-α-D_1 and isopropylbenzene-α-D_1 in which the value of $^3\Delta(C-2)$ becomes increasingly positive because the rotamer, especially of the latter, in which the C—D bond is in the plane of the ring, dominates [4]. For indane-α,α-D_2, $^3\Delta C-2(D) = -0.037$ ppm, whereas $^3\Delta C-6(D) = -0.061$ ppm, which is equal to the sum of -0.037 and 2×-0.012 ppm; -0.012 ppm is the intrinsic effect observed in cyclopentane (Table 1).[30] However, why $^3\Delta C-2(D)$ of the indane should be larger than in toluene-D_2 is not quite obvious.

Ernst et al.[31] have observed negative isotope effects in 4-methylpara-cyclophane-4'-D (Table 5) and ascribed them to hyperconjugation. In addition they propose hyperconjugation as the explanation of isotope effects in acetone,[32] 3,3-dideuteriocamphor derivatives,[33] on C-β in Ndeuterioenamones,[295] on C-4 in O-deuteriophenols[58] and on C-1 in 3-deuteriocyclobutene.[20]

^{19}F nuclear shielding is very sensitive to isotope effects. This feature is exploited in a study of hyperconjugation.[157,158] Deuterium isotope effects over six bonds, on fluorine, are observed in p-fluorophenylmethyl carbonium ions-β-D$_3$ [30]. Isotope effects are only observed in those cases in

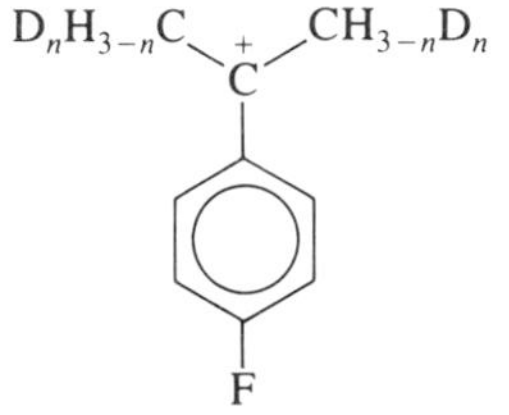

$$D_nH_{3-n}C \quad \overset{+}{\underset{C}{}} \quad CH_{3-n}D_n$$

F

[30]

which methyl group hyperconjugation is expected to take place[158] (Table 25). The effect observed for carbonium ions is an order of magnitude larger than for the p-fluorotoluene system.[157] A relation to the electronic properties of the X group substituted is also found, but not to the size of the X substituent [31].

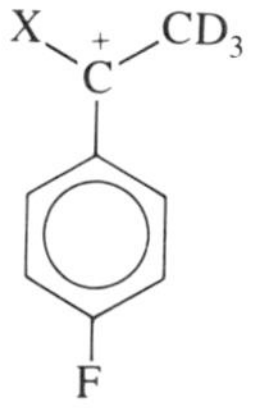

$$X \quad \overset{+}{\underset{C}{}} \quad CD_3$$

F

[31]

Forsyth *et al.*[97] have investigated hyperconjugation effects of deuterium substitution both by ^{13}C and ^{19}F NMR spectroscopy. Phenyldimethyl carbonium ion shows[35] (Table 5) high frequency shifts for the *ortho* and *para* carbons in accord with a hyperconjugative effect. α-Methyl-α-neopentyl-benzyl anion, which should exhibit negative hyperconjugation, shows a positive isotope effect for the *para* carbon (Table 5). The possibility that the unusual effects should be caused by equilibrium isotope effects is ruled out. ^{19}F NMR, which is more sensitive, was used to study the effect of varying numbers of deuterium atoms in the methyl groups of p-fluoro-phenyldimethyl carbonium ion (Table 25) and in p-fluorophenylmethyl carbonium ion (Table 25). The effect of one trideuteriomethyl group is very close to half the isotope effect observed for two trideuteriomethyl groups. However, for successive deuterium substitution within a methyl group, the long-range isotope effects on ^{19}F are not additive (Table 25), whereas the short-range effect on C is additive. The non-additivity is attributed to unequal rotamer populations of the partially deuteriated methyl groups. The preferred conformers are those in which the C—H bonds are better aligned for hyperconjugation than the C—D bonds. The long-range isotope effects should hence show a temperature dependence and show better additivity at higher temperatures, which is actually observed.

The long-range effects all lead to deshielding (as seen in Table 5), except for the compound in which negative hyperconjugation is thought to take place. Furthermore, it is the *ortho* and *para* positions that are perturbed. All these facts indicate that hyperconjugation is responsible.[158] The isotope effects on nuclear shielding are a ground state property. The NMR isotope effects nevertheless mimic the direction of kinetic isotope effects as though an electronic substituent effect was responsible. The apparent "electronic" effect is simply a reflection of the fact that observed NMR shieldings are vibrationally and rotationally averaged.[97] Forsyth *et al.*[97] find that the anharmonicity should be taken into account for the calculation of isotope effects.

C. Temperature and solvent effects

1. *Solvent effects*

Solvent effects have been reported in a few cases. Crecely *et al.*[191] have observed a small difference in the $^1\Delta H(^{13}C)$ value of *p*-benzoquinone in the solvents DMSO and CDCl$_3$. Pendleburg and Philips[159] have found, in *N*-deuterated fluoroamides, that $^4\Delta F(ND)$ varies from 12.4 Hz in methanol to 13.5 Hz in dichloromethane. The solvent dependence of $^1\Delta P(ND)$ has been investigated in alkylphosphates.[168,171] The largest effect is reached in a dilute solution of cyclohexane in which inter-molecular hydrogen bonding should be minimized. In neat liquids, and in methanol, a somewhat lower value is ascribed to the effect of hydrogen bonding, as seen in Table 28. Vederas[145] has found no solvent effects on the $^1\Delta C(^{18}O)$ value for aldehydes and ketones, whereas solvent effects on the $^1\Delta C(^{18}O)$ data of alcohols are substantial. Solvent effects are also believed to have an effect on the $^2\Delta C(OD)$ value of carbohydrates.[49] The low value of $^1\Delta C(^{18}O)$ of 1-(*p*-aminophenyl)ethanol[80] may be due either to hydrogen bonding or to solvent effects.

The case in which solvent effects may be expected to be important are those in which inter- or intra-molecular hydrogen bonding takes place. Intra-molecular hydrogen bonding is treated in Section III.K.1. The extent of intra-molecular hydrogen bonding is clearly solvent dependent.

Equilibrium isotope effects often show a solvent dependence.

2. Temperature effects

The temperature variation of the shielding of various isotopes of hydrogen has been calculated by several authors recently.[296,297] Beckett and Carr[298] performed experiments over a wide range of temperatures and gas pressures. These refined experimental data, in the zero pressure limit, are compared with improved theoretical data. A temperature dependence is observed, but it is not as strong as that predicted.[296] The trend agrees well with that predicted by Reid,[297] but numerically the experimental and the theoretical values disagree by 10%.

Brey et al.[220] have investigated $^1\Delta(F,^{13}C)$ and $^2\Delta(F,^{13}C)$ data at two different temperatures, 233 and 303 K. The largest effects are observed at the lower temperature. For $^1\Delta$ the difference is ~0.30 Hz (at 94.1 MHz), but for $^2\Delta$ the variation is much larger, 0.08–0.56 Hz.

Brey et al.[220] have also investigated the effect of ^{37}Cl on ^{19}F nuclear shielding over a wide temperature interval, 193–305 K, and find much smaller isotope effects at higher temperatures, which is in agreement with the hypothesis of Gutowsky[299] that isotope effects are related to the nature of molecular vibrations.

Lambert and Greifenstein[154] have observed no temperature effects on $\Delta F(D)$ values in the temperature range between −85 and +25 °C.

D. Correlations

Very few correlations have been proposed, and some of those have since been refuted. Grishin et al.[21] have noticed a parallelity between the values of $^1\Delta C(D)$ and $^2\Delta H(D)$ for a series of methanes. An exception is observed for methyl iodide.

Batiz-Hernandez and Bernstein[1] have noted that the value of $^2\Delta H(D)$ decreases with an increase in the s character of the carbon directly bonded to the deuterium. No such correlation is apparent[22] for $^1\Delta C(D)$. A relationship which is regularly cited[1,15,140,300] is

$$\Delta F(^{13}C) = 0.007 - 4.36 \times 10^{-4} \times J(C-F), \tag{1}$$

linking $\Delta F(^{13}C)$ and the one-bond carbon–fluorine coupling. Aydin and Günther[140] have found a relationship between $^1J(C-H)$ and $^1\Delta$ as well as between $^1\Delta$ and $\delta(C-1)$ in the series cyclopropane to cycloheptane. Similarly, for $^1\Delta P(D)$, Borisenko et al.[170] have found

$$^1\Delta P(D) = 1.50 - 1.80 \times 10^{-3} \times {}^1J(P-H). \tag{2}$$

Using an equation derived by Manatt,[300] i.e. $^1J(P-H) = 29.30\ S_P$, where S_P is the s character of the P—H bond, they arrived at

$$^1\Delta P(D) = 1.50 - 5.3 \times 10^{-2} \times S_P. \tag{3}$$

A plot of $\Delta P(^{18}O)$ against the square of the frequency of the A_1 stretching mode of the phosphates shows a good linear relationship, indicating that the magnitude of the isotope shift is related to the force constant of the phosphorus–oxygen bond.[262]

$^1\Delta P(^{18}O)$ values correlate with the amount of double-bond character of the O—P bond.[261] In general, reasonable correlations may only be observed for closely related compounds. Both molecular structure and symmetry must be similar.

E. Assignments

The purpose of this section is not to quote all those cases in which isotope effects are mentioned, but rather to give a very short overview of how isotope effects may be used for spectral assignment purposes, e.g. ^{13}C NMR spectra, by means of $^n\Delta C(D)$ data. In ^{13}C spectra the two-bond isotope effect is very useful.[302,303] $\Delta C(D)$ values are usually too small to be useful except in cases where *peri* interactions are present [32]. Deuterium may be used to distinguish OH-bearing aromatic carbons from other oxygen-bearing carbons. This effect is, furthermore, useful, as it may be made to disappear either by heating or by the addition of H_2O.

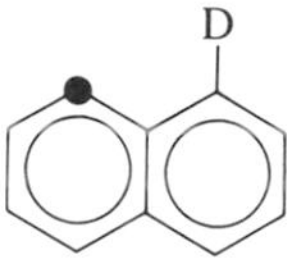

[32]

The trends and characteristic ranges may of course be taken into account. An example of the use of isotope effects for structural assignment is the great importance of the bond order dependence of values of $^1\Delta P(^{18}O)$. This property is a fundamental feature of the use of $^1\Delta P(^{18}O)$ for the determination of chirality and for use in enzyme studies.

XVI. EQUILIBRIUM ISOTOPE EFFECTS

One of the most exciting features of isotope effects on nuclear shielding is the equilibrium effect observed in systems in which the presence of, for example, deuterium rather than ^1H may shift the position of the equilibrium or, in other words, change the equilibrium constant. The observation of equilibrium isotope effects tells us nothing about the nature of the intrinsic effect which is often masked by the equilibrium effects and which must be subtracted from the observed effect to get the true equilibrium effect. However, it may lead to exciting information about the nature of carbonium ions, the presence of tautomeric equilibria, acid–base equilibria and conformational equilibria.

The magnitude of equilibrium isotope effects depends on three factors: (i) the nuclei observed; (ii) the magnitude of the change in the equilibrium constant; (iii) the difference of the nuclear shielding of the two equilibrating nuclei. For this reason isotope effects on carbon have been used to monitor equilibrium effects and the perturbing isotope has in most cases been deuterium, but the first studies were of ^1H with deuterium as the perturber and the first occurrence was evident from a study of dioxane-2,3,5,6-D$_4$.[304]

Equilibrium isotope effects may, quite naturally, have either sign. The observation of a negative isotope effect may thus be an indicator of the presence of an equilibrium situation. However, as discussed in Section XV.B, hyperconjugation often leads to negative isotope effects.

A. Carbonium ions

1. ^1H spectra

The ^1H spectrum of the 2,3-dimethylbutyl-2-ium ions shows a singlet for the methyl groups as the fast hydride shift causes averaging. The spectrum of the 3-deuterio-2,3-dimethylbutyl-2-ium ion [33] shows, however, line

$$
\begin{array}{ccc}
\text{CH}_3 \quad \text{CH}_2\text{D} & & \text{H}_3\text{C} \quad \text{CH}_2\text{D} \\
\text{H}-\text{C}-\text{C}^+ & \rightleftharpoons & {}^+\text{C}-\text{C}-\text{H} \\
\text{CH}_3 \quad \text{CH}_3 & & \text{H}_3\text{C} \quad \text{CH}_3 \\
\text{I} & & \text{II}
\end{array}
$$

[33]

doublings due to a perturbation of the equilibrium by deuterium. The carbonium ion prefers to be substituted by an unlabelled methyl group or, in other words, is stabilized less by a β-deuterium, as in [33, I], than by a β-hydrogen, as in [33, II]. The hydride shift that interconverts [33, I] and [33, II] is very fast, so the individual shieldings are estimated from model compounds, leading to a value of 1.132 at $-56\,^\circ$C[305] for the equilibrium constant.

If more hydrogens are replaced by deuterium, a linear increase of the splitting between the two lines is observed.

In a mixture of 2,2,3-trimethylbutyl-3-ium ion and 1,1,1-trideuterio-2,2,3-trimethylbutyl-3-ium ion a temperature-dependent splitting of the methyl signal is observed. The methide shift that interconverts [34, I] and [34, II] is fast, but [34, I] is favoured over [34, II]. No equilibrium isotope effect is observed for the tetradeuteriocyclopentyl cation.[306]

In the 2,3-dimethylbutyl-2-ium-D$_6$ ion the interchange of nonequivalent methyl groups has been followed.[307]

[34]

2. ^{13}C spectra

For investigations of equilibria ^{13}C chemical shifts are highly favoured when compared to those of ^1H, and quite a number of studies have been performed. In this section isotope perturbation of resonance is also treated. Compounds showing equilibrium isotope effects in ^1H spectra have also been investigated by ^{13}C NMR.

Deuteration of one of the methyl groups in 1,2-dimethylcyclopentyl carbonium ion lifts the degeneracy of C-1 and C-2. Two peaks are observed, split symmetrically about the averaged resonance of the protiated compound. If the separation between the two signals is called δ, this can be expressed as

$$\delta = \{(a\delta_1 + b\delta_2) - (a\delta_2 + b\delta_1)\}/(a+b). \qquad (4)$$

Now let $K = b/a$, and δ_1 and δ_2 are the shieldings of the averaging carbons. This can be simplified by the substitution of $D = \delta_2 - \delta_1$. Therefore

$$K = (D + \delta)/(D - \delta). \qquad (5)$$

For fast reactions D must be estimated from suitable model compounds. Values of 261 and 202 ppm are estimated in the case of the dimethylnorbornyl cation and for the dimethylcyclopentyl cation, respectively.[308] These very large differences make the method very sensitive and it is thus possible to observe a small ^{13}C isotope effect in 1,2-dimethylcyclohexane-1-^{13}C.[308]

A further study of ^{13}C isotope effects has been performed on the 2,3-dimethyl-2-butylium ion-2-^{13}C. The dilabelled compound serves as a reference. Splittings as large as 91.8 Hz (67.89 MHz) are observed at $-135\,°C$. No intrinsic isotope effects larger than 4 Hz are reported.[309]

Saunders and Kates[310] have observed a different type of isotope effect, isotopic perturbation of resonance. In the cyclohexenyl-1-D and cyclopentenyl-1-D cations, both of which are strongly delocalized ions, small deshielding isotope effects are observed. The canonical form in which the positive charge is next to the hydrogen is preferred.

The ratio δ/Δ is suggested as a criterion for distinguishing between symmetrical, delocalized and rapidly equilibrating systems.[310]

The isotopic perturbation of resonance is not an equilibrium effect, but it is treated in this section as it is observed in the same type of system as the equilibrium effect and because isotope effects may be used to distinguish between the perturbation of equilibrium and the perturbation of resonance.

Isotopic perturbation of resonance is also observed in the 2-bicyclo[2.1.1]hexyl-2-D cation. This points towards a σ-delocalized, and probably symmetrically bridged and not interconverting, ion.[311]

The non-classical nature of the norbonyl cation has been discussed in many contexts. From isotope effect arguments a static structure is proposed.[312]

Another carbonium ion that has attracted much attention is the cyclobutyl-cyclopropylcarbinyl cation. The cyclopropylcarbinyl-α-D_1 cation has been studied by ^1H and ^{13}C NMR. The ^1H nuclear shielding varies with temperature in a fashion expected for equilibrium isotope effects.[313] In the ^{13}C spectrum both positive and negative isotope effects are observed. The difference in magnitude points towards a structure such as [35]. The rapid degenerate equilibrium may occur via B or C.[313] The deuterium isotope effect method has also been applied to a pyramidal dication.[314] The small isotope effect, ~0.4 ppm, is in agreement with a symmetrical, bridged structure.[314]

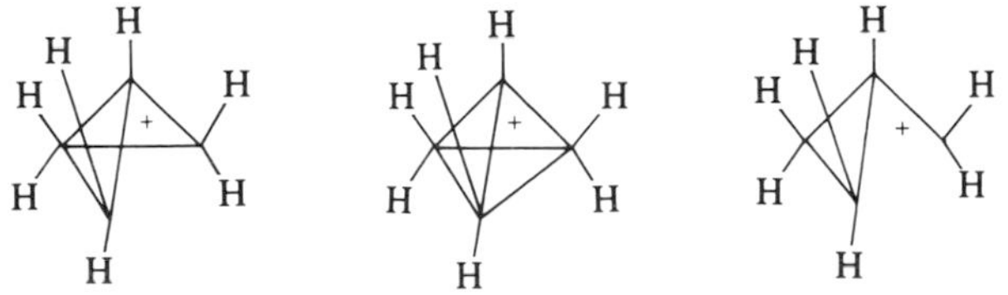

[35]

1,5- and 1,6-dimethylcyclododecyl cations show μ-hydrido bridging as judged from the shielding of the bridge proton and from the small couplings to this hydrogen. The isotope effects are likewise small, confirming a static structure.[315]

Ahlberg et al.[316] have investigated the nature of the 9-barbaralyl (tricyclo[3.3.1.0^{2,8}]nona-3,6-dien-9-yl) cation using the 3-^{13}C-1,2,4,5,6,7,8,9-D_8 derivative. They observe a deshielding of 6 ppm and suggest a structure such as [36]. The value of 6 ppm is fairly small for an equilibrating structure.

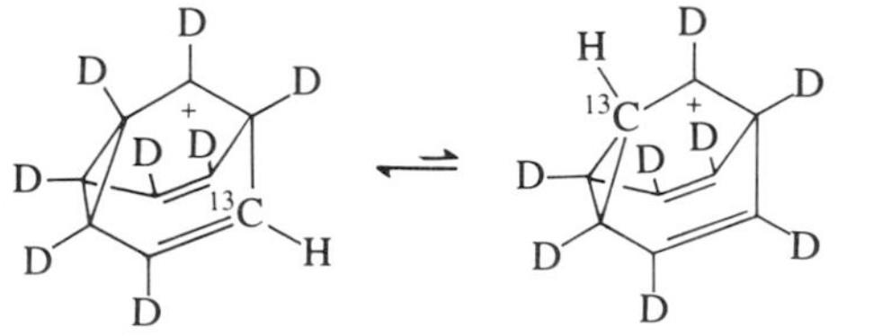

[36]

Some intermediate values are observed in 1,2-dimethylnorbornyl-1'-D$_3$ cations[308] and in 1,2-dimethyl-2-bicyclo[2.1.1]hexyl cations. The intermediate values have led to the suggestion that the bonding in some carbocations precludes a ready classification into "classical" or "non-classical".[317] Typical values of isotope effects in carbonium ions are given in [37].

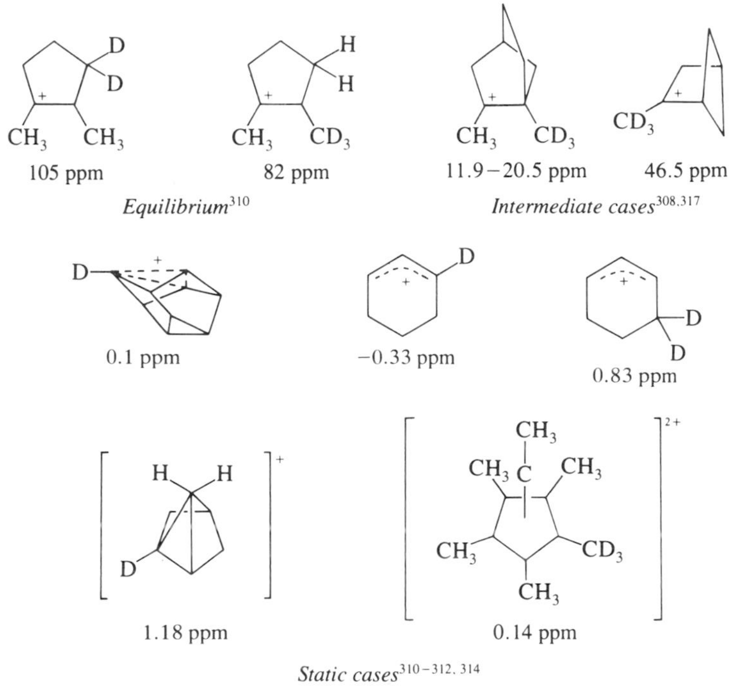

105 ppm	82 ppm	11.9–20.5 ppm	46.5 ppm
Equilibrium[310]		*Intermediate cases*[308,317]	

0.1 ppm	−0.33 ppm	0.83 ppm

1.18 ppm	0.14 ppm

Static cases[310–312, 314]

Splittings caused by deuterium substitution

[37]

Isotope effects are also used to distinguish between the η^1-C$_5$H$_5$ and η^5-C$_5$H$_5$ rings in (η^1-C$_6$H$_5$)Sn(CH$_3$)$_3$. The isotope effect between the monodeuteriated and the protiated compound is temperature dependent. No such effect is observed in ferrocene, (η^5-C$_5$H$_5$)$_2$Fe.[318]

Deuterium isotope effects are also used to study the structure of magnesium, lithium and potassium allyls. The magnesium derivative shows a much larger effect than the others.[319] Isotope effects provide a means of assignment for HOs$_3$(CO)$_{10}$CH$_3$.[103]

B. Valence isomerization

The valence isomerization of 1,5-dimethylsemibullvalene has been studied by deuterium substitution at C-2.[320] Deuterium prefers the cyclopropane position. This is another case of isotopic perturbation of equilibrium. The largest effect is 4.6 ppm and is strongly temperature dependent.

C. Conformational studies

Conformational changes, caused by deuterium substitution, have been observed in ring systems. Six-ring systems have been studied in some detail with deuterium at the ring carbons or on side-chain methyl groups. The reason for the effect is thought to be the slightly shorter $C-D$ bond compared to a $C-H$ bond.

1. ^1H spectra

Observation of deuterium isotope effects on ^1H shieldings in conformational studies has been described for cis-3,5-dideuteriocyclopentene[321] and for cycloheptatriene-7-D.[322]

2. ^{13}C spectra

Variable temperature studies of 1-deuteriocyclohexane have revealed that $^1\Delta(C-D_{eq}) < {}^1\Delta(C-D_{ax})^{16}$ based on the finding[17] that $^1J(C-D_{eq}) > {}^1J(C-D_{ax})$ and also in agreement with the finding from cyclohexane-D_{11}[17] (Table 1). From coalescence temperature measurements a ΔG value of 42.68 kJ mol^{-1} is determined. Furthermore, a small difference between the observed average $^1\Delta$ value at 20 °C and the mean value of $^1\Delta_{eq}$ and $^1\Delta_{ax}$ points (Table 1) towards an equilibrium constant different from 1. In addition, measurements on cyclohexane-1,1-D$_2$ at 20 °C and -80 °C show that the magnitude of $^1\Delta$ is temperature dependent. By extrapolating the mean value from -80 to $+20$ °C, a value of 1.100 for the equilibrium constant is obtained which favours the conformer with deuterium in an equatorial position. This value has been corroborated by studying a mixture of cyclohexane-1-D and cyclohexane-1,1-D$_2$.

A study of t-butylcyclohexane-4-D shows a much smaller difference between $^1\Delta(C-D_{ax})$ and $^1\Delta(C-D_{eq})^{19}$ (Table 1).

Alkylcyclohexanes with deuteriated alkyl groups have been studied in some detail.[29,96,323] Trans-1,3-dimethylcyclohexane[96] shows, upon deuteriation of one methyl group, small positive and negative isotope effects, as do the corresponding 2- and 5-cyclohexanones. The equilibrium effect at the non-deuteriated methyl group (assuming no direct isotope effect) is used to derive the ratio

$$\frac{K^D}{K^H} = 1 - \left(2 + K^H + \frac{1}{K^H}\right)\Delta\delta(\delta E - \delta A), \qquad (6)$$

giving the K^D/K^H ratio for the three mentioned compounds values of 1.018, 1.032 and 1.016. The equilibrium favours the conformer in which the CD$_3$ group is in an axial position.[96] A study of *cis*-1-ethyl-4-methyl-cyclohexane-4'-D$_3$ gives a similar result.[29]

In 1,1,3,3-tetramethylcyclohexane-1'-D$_3$ a splitting of 0.184 ppm of the signal of the 3'-CH$_3$ group occurs and the CD$_3$ group spends more time in the axial position. An equilibrium constant of 1.042 at 18 °C is arrived at. The 1,3-diaxial interaction increases the vibrational frequency and hence the zero-point energy of the methyl group in the axial position. The CD$_3$ group has a lower zero-point energy and is thus raised less in energy than the CH$_3$ group. In the equatorial position no real difference between the CH$_3$ and CD$_3$ groups is likely.[324] The authors claim that the measurement of isotope effect is a more precise method of obtaining equilibrium constants than freezing out the isomers and subsequent measurement of the relevant signal areas due to difficulties in accurate area measurement.[323]

Incorporation of ^{13}C into one of the methyl groups in 1,4-dimethyl-cyclohexane has no perturbing effect.[211] Small deuterium isotope effects are observed in *N,N*-dimethylpiperidinium-C'-D$_3$ ions.[325]

An early report of a negative isotope effect and of effects over as many as six bonds in cyclodecanone[326] has since been ascribed to a conformational change. This probably also covers 2-methylcyclohexanone[327] and possibly the unexplained missing effect in androstanone.[328]

The case of cyclodecanone has been discussed from a general point of view by Andrews *et al.*[329] They propose two rules to explain isotope shifts due to the lifting of conformational degeneracies:

(1) Any nucleus positioned on a molecular symmetry element will have the same chemical shift in any member of a set of conformational equivalents about that element, and hence will experience no net shift as a result of unequal weighting.

(2) The *average* shift of any set of nuclei interchanged by symmetry operations, on the average structure, will not change as the result of weighting.

3. ^{19}F spectra

^{19}F shielding differences are very large and thus sensitivity to isotope effects is high. Fluorine is thus a potentially very good gauge of isotope effects, except that it is not an unperturbing probe. The size of fluorine may play a role in equilibrium studies. Furthermore, fluorine is highly electronegative and is in addition strongly mesomerically active in conjugated systems. A minor drawback may be that synthesis of appropriate fluorinated compounds may be difficult.

Equilibrium isotope effects may also be studied by means of ^{19}F chemical shieldings. *Trans*-trifluoromethyl-3-methylcyclohexane-3-D$_3$ shows an isotope effect of 0.0294 ppm and *trans*-trifluoromethylhydroxymethylcyclohexane-2′-D$_2$ shows a similar effect,[96] leading to an equilibrium constant of 1.02. The isotope effects observed in the latter are solvent independent.

Conformational effects on ^{19}F nuclear shielding are observed in methyl(p-fluorophenyl) carbenium ions,[97] as discussed in Section V.

D. Acids and bases

Led and Petersen[64] have studied isotope effects on the ^{13}C shielding of amino acids as a function of pH. The isotope effects observed for the H$_2$A and HA species agree reasonably well with those predicted from studies of amino acids in trifluoroacetic acid.[63] However, for the A$^-$ species a negative isotope effect is observed for many of the resonances. A plot of the isotope effects against pH reveals that in the pH region close to the pK_a value of the ammonium group, a negative effect of 0.9 ppm for the COOH carbon of lysine is observed. An effect of ~ -0.4 ppm is observed for C-δ at a pH value approximately one unit higher. These negative isotope values are primarily caused by a change in the pK_a values due to deuteriation and are, as such, mainly equilibrium isotope effects.[64]

The magnitude of the equilibrium isotope effect depends on the difference between the shielding of the carbons in the charged and the non-charged species (titration shifts) and should in principle also be observable for the carboxylic acid carbons in the low pH range, near the pK_a value of the carboxylic acid. No effect is observed. The titration shift is, in the case of the carboxylic acid carbon, smaller and the equilibrium effect is not, perhaps for that reason, discernible.

The dramatic effect observed causes the authors[64] to warn that such effects should be taken into account when nuclear shieldings of amino acids in H$_2$O and D$_2$O are compared.

The unusual effect observed in the value of $^1\Delta P(D)$ of glucose 6-phosphate[49] may possibly also be related to equilibrium effects.

E. Tautomeric systems

In tautomeric systems an equilibrium isotope effect may be observed as in β-thioxoketones [16]. The magnitude and sign of the equilibrium effect depend on (i) the value of $\delta C_I - \delta C_{II}$; (ii) the position of the equilibrium; (iii) the nature of the heteroatoms, as shown in Table 20. A collection of compounds in which R and R^1 are either aliphatic or aromatic species has been investigated. The observed effects in CDCl$_3$ are as large as -7.98 ppm.[74] Point (ii) is demonstrated by the pair [38] and [39]. In [38] a large isotope effect is observed, whereas in [39] a very small effect is observed as the equilibrium is strongly shifted towards the thiol tautomer, as is also indicated by the shielding of the C=S and C=O carbons.

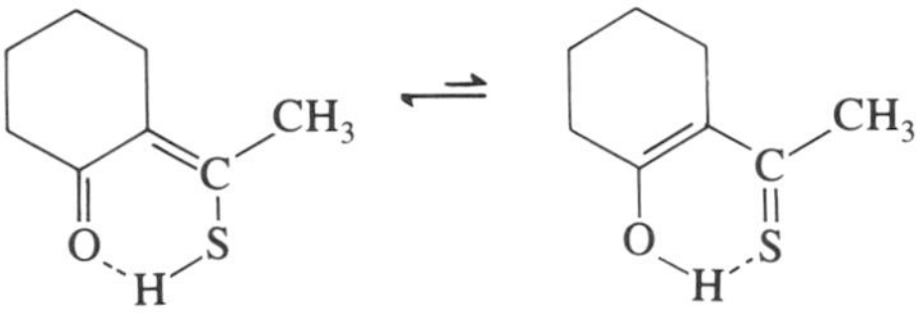

[38]

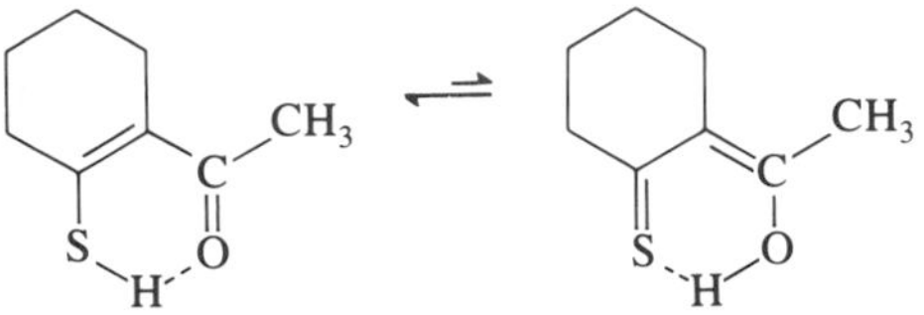

[39]

In Schiff's bases, in which tautomeric equilibria may also exist, the few examples known show much smaller equilibrium effects.[74] A recent case of deuterium isotope effects on enaminoketones is explained on the basis of averaged isotope effects.[295] It was later suggested that the effects may be due to a shift in the tautomeric equilibrium.[74] An explanation including both of these elements is probably closer the truth.

Systems of type [40] have also been investigated. The possibility of tautomerism was discussed but discarded.[79] In the present author's opinion the isotope effects observed in [40, A] are strongly indicative of a tautomeric equilibrium, although the effect on C-3 is unusually large. Tautomeric equilibrium effects may possibly play a role in the long-range effects observed in hydrogen-bonded phenols (Section III.K.1).

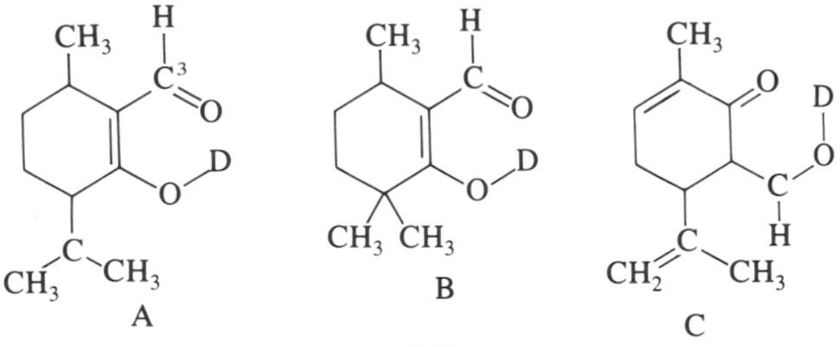

[40]

F. Complex formation

1. Charge transfer complexes

Complex formation between benzene, toluene and p-xylene and fluoranil may be studied by abstracting the difference in ^{19}F shielding when the protiated and the perdeuteriated compounds are complexed with fluoranil.[330]

2. Lanthanide-induced shifts

The observation of line-doubling upon the addition of Eu(DPM)$_3$-2-py (a pyridine adduct of tridipivalomethanatoeuropium (III)) to a mixture of *trans*-verbenol and its 4-D derivative is explained by assuming a greater association constant between the deuterium-substituted compound and the metal complex than between the protiated compound and the metal complex. The effect is also observed in the case of other alcohols and an aldehyde.[331] Several other shift reagents have been investigated, such as Eu(DMP)$_3$, Pr(DMP)$_3$, tris(dibenzoylmethanato)europium (III) (Eu(DPM)$_3$), tris(dibenzoylmethanato)europium (III) (Eu(DBM)$_3$), and tris(1-benzoylacetonato)europium (III) (Eu(BAT)$_3$).[332] Effects are also observed for amines[332] and ethers.[333] A prerequisite is normally that the deuterium substitution is close to the basic coordination site.[333] Substitution at a β position generally gives an effect.[333,334] Lanthanide-induced shifts have also been studied in glymes.[335,336]

Long-range effects are observed[337] in 1-methoxy-2,3-dimethylbutane-3-D, in which the deuterium and the methoxy group may be close in space. In the 3,3'-D$_2$ derivative no extra effect is observed. That spatial closeness is necessary is demonstrated by compounds such as [41, A] and [41, B], which show isotope effects, while [41, C] does not.[337] Allenic esters also show long-range effects. Lang and Hansen[338] have observed a change coupled to a shift in the S–*cis*/*trans* conformer equilibrium upon deuteriation.

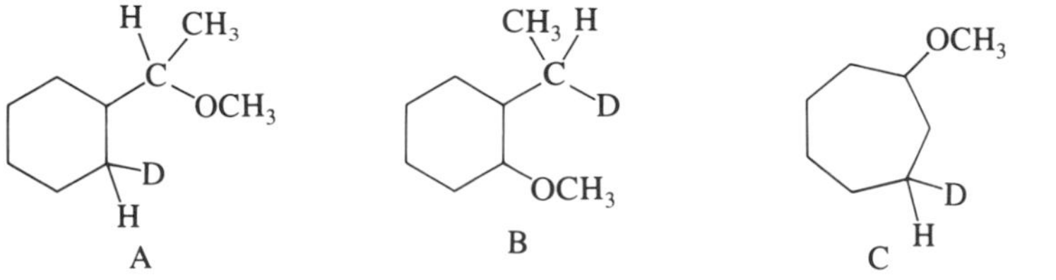

[41]

Deuteriation of the 2- and 6-methyl groups of 2,4,6-trimethylpyridine gives a shift to H-3 and H-5, but deuteriation of the 4-methyl group has no effect.[339,340] 2,6-Ethyl-4-methylpyridine-α-D$_4$ shows effects similar to that of 2,4,6-trimethyl, whereas the 2,6-diisopropyl-4-methylpyridine-α-D$_2$ shows no effect, neither do the *t*-butyl derivatives.[339] Steric effects can prevent complex formation.[340]

G. Conclusions

Equilibrium isotope effects are the reason for the observation of a large number of apparent long-range isotope effects of either sign. As isotope effects depend upon the difference in the nuclear shielding of the two equilibrating species, the observation of a proportionality with such differences may be a strong indication for the occurrence of equilibrium isotope effects.

Several of the unusual isotope effects reported may possibly be ascribed to equilibrium isotope effects.

XVII. PRIMARY ISOTOPE EFFECTS

Primary isotope effects (p.i.e.) are defined as the difference in magnetic shielding of the isotopes of nuclei of the same element.

Tupćiauskas et al.[341] have measured p.i.e. of the pair ^{117}Sn, ^{119}Sn and find that the effect is beyond experimental uncertainty, especially for $(C_2H_5)_4$Sn. However, McFarlane et al.[342] refute this finding on experimental grounds and state that the p.i.e. is less than 0.1 ppm.

In the case of ^{14}N, ^{15}N no significant effect has been detected.[343] The same conclusion is reached for ^{10}B, ^{11}B;[344] ^{85}Rb, ^{87}Rb[345] and ^{111}Cd, ^{113}Cd.[345]

Bloxsidge et al.[346] have discussed the difficulties of measuring the primary isotope effects of the pair T/^{1}H in organic compounds.

Primary isotope effects $^P\Delta(H, D)$ or $^P\Delta(H, T)$ are usually small in organic compounds. Values of $^P\Delta(H, T)$ are less than 0.03 ppm for tritium bound to sp^2- or sp^3-type carbons.

The most useful p.i.e. is probably that of the D/^{1}H pair of hydrogen-bonded hydrogens.

Exceptionally large primary isotope effects are observed in hydrogen-bonded systems. Chan et al.[347] have ascribed this to the existence of fast equilibrium between a symmetrical and an unsymmetrical structure, [42],

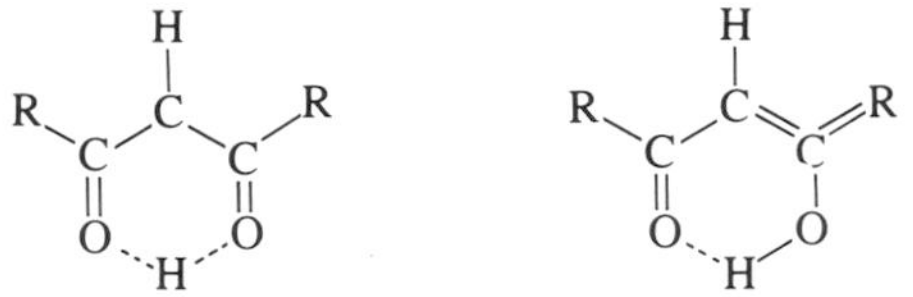

[42]

with different nuclear shieldings for the bridge hydrogen in the two structures. Temperature measurements showing that both ^{1}H and D shieldings vary with temperature support the idea that deuterium isotope effects have the same origin as the temperature dependence. This is furthermore corroborated by a correlation between anomalous isotope effects and temperature effects.

An explanation based on a double minimum vibrational potential is considered difficult to fit to all of the experimental findings.

More recent studies[78,348] reveal a correlation between the shielding of the OH proton and the isotope effect, $^P\Delta(H, D)$. Based on a simple linear model and the assumption that a lengthening of the O—H bond leads to deshielding, it is possible to predict that $^P\Delta(^1H, D) = -0.7$ ppm for a single minimum potential and $^P\Delta(H, D) = 0.4$ ppm for a double minimum potential. In the case of weak hydrogen bonds the potential minimum should be deep, with a low anharmonicity and a negligible isotope effect. For stronger hydrogen bonds the potential minima draw together, a lower barrier and increased anharmonicity of the potential leading to different equilibrium bond lengths for O—H and O—D, the former being the longer and hence leading to a positive $^P\Delta$ value. For extremely strong hydrogen bonds a single-minimum potential is expected, giving rise to a negative shielding effect.

From a theoretical treatment[349] it is furthermore predicted that $^P\Delta(H, T) = 1.44 \, ^P\Delta(H, D)$ for the case of an anharmonic cubic term in the potential. For those cases in which unusually large positive primary isotope effects are present (Table 45), a ratio of between 1.34 and 1.4 is actually observed.

The observation of tritium or deuterium isotope effects may be used to assess the shape of the hydrogen-bond potential energy function, in particular to distinguish between single- and double-minimum potentials.

Robinson et al.[60] have debated the solution suggested by Chan et al.[347] and proposed a slightly unsymmetrical structure of the hydrogen bridge of the enolic form of β-diketones. The oxygen–oxygen distance is suggested to play a decisive role and deuteriation is supposed to lead to a lengthening of this distance. Ab initio calculations show that the build up of excessive charge on oxygen (negative) and hydrogen (positive) increases with decreasing O···O distance. Assuming a double potential minimum and a lengthening of $r(O···O)$ upon deuteriation, a shielding of OD in the deuterium compound, compared to OH, is predicted, in agreement with experimental findings.

Shapetko et al.[79] have explained the p.i.e. observed in β-diketones by a tunnelling process. Replacement of the bridging proton with a deuterium leads to a reduction of the tunnelling effects, which brings about an increase of the O···O distance. An increase in distance causes a higher shielding for the deuterium. The conclusion is thus in line with that reached by Robinson et al.[60] Shapetko et al.[79] predict, furthermore, a higher shielding for the carbonyl carbon upon deuteriation. When an unsymmetrical double minimum potential occurs as in the alicyclic β-ketoaldehydes of [40, A]–[40, C], different effects are expected for the two carbon atoms involved.

Egan et al.[350] have discussed the temperature dependences of the nuclear shielding of the enolic proton of pentane-2,4-dione and suggested, following the line of Muller and Reiter,[351] that increasing the temperature brings

TABLE 45

Primary isotope effects[a] in hydrogen-bonded systems.

	H/D	H/T	Ref.
Acetylacetone[b]	+0.58		347
Acetylacetone[c]	+0.55		347
Acetylacetone[d]	+0.58		347
Acetylacetone[e]	+0.61	0.83	78
Acetylacetone	+0.50		79[n]
3-Methyl-2,4-pentanedione[f]	+0.45		347
1,3-Diphenyl-1,3-propanedione[g]	+0.45		347
1,3-Diphenyl-1,3-propanedione[e]	+0.72	1.01	78
1-Phenyl-1,3-butanedione[b]	+0.42		347
1-Phenyl-1,3-butanedione[e]	+0.67	+0.93	78
1-Phenyl-1,3-butanedione	+0.64		79
Methylpentan-2,4-dione-oate	+0.61		79
2-Carboethoxypentane-1,3-dione	+0.64		79
2,2,6,6-Tetramethyl-3,5-heptanedione	+0.59		79
4-Phenyl-2,4-butanedione	+0.64		79
Ethyl acetoacetate[i]	−0.02		347
Ethyl acetoacetate[e]	+0.16	+0.21	78
Ethyl acetoacetate	+0.01		79
Salicylaldehyde	+0.06[j]	+0.008[e]	348
Salicylaldehyde	−0.03		79
o-Hydroxyacetophenone	+0.10[j]	0.074[e]	348
Malondialdehyde	+0.42		78
1,1,1,5,5,5-Hexafluoropentane[b]	+0.30		348
Benzyl alcohol[j]	−0.02	−0.016	78
Methanol[b]	0.00		347
Acetic acid[b]	+0.07		347
Hydrogen furan-3,4-dicarboxylate ion[k]	+0.11		348
Hydrogen maleate ion[k]	−0.03	−0.07[e]	348
Hydrogen phthalate ion[k]	−0.15	−0.25[e]	348
Hydrogen difluoride ion[l]	−0.30		348
1-Dimethylamine-8-dimethylammonium naphthalene cation[m]	+0.66	+0.915	78

[a] $^P\Delta\delta(^1H)-\delta(^2H)$.

[b] Neat liquid.

[c] 0.16 mole fraction in cyclohexane.

[d] 0.12 mole fraction in n-butyl ether.

[e] ~0.25 M in C_6H_6.

[f] 0.5 mole fraction in cyclohexane.

[g] 0.15 mole fraction in CCl_4.

[h] 0.3 mole fraction in CCl_4.

[i] 0.2 mole fraction in CCl_4.

[j] Solvent, methylene chloride.

[k] Temperature −55 °C. Solvent, methylene chloride.

[l] Temperature −36 °C. Neat liquid.

[m] Tetraphenylborane as counterion in methylene chloride.

[n] Data for compounds [40, A, B and C] are as follows: +0.44, +0.60 and +0.60 ppm, respectively.

about a redistribution of the population between ground and low-lying excited states of the vibrational modes, which will lengthen the O—O distance and hence lead to an increase in shielding. Additionally, $\Delta\delta/T$ for the deuteriated compound should be different from that of the hydrogenated compound.

The p.i.e. has been investigated in acetylacetone in the presence of nitrogen bases. As the mole fraction of triethylamine is increased to 0.6, the p.i.e. decreases from 0.62 to 0.10 ppm. In the acetylacetone–pyrrole system the p.i.e. remains almost constant. This is because the keto form is involved in hydrogen bonding with pyrrole, leaving the intramolecular bond intact.[352]

Tunnel effects have further been studied in aqueous solution[353] in the trifluoroacetate–trifluoroacetic acid systems in methylene chloride[354] and in the trifluoroacetic acid–N,N-dimethylaniline system.[355]

XVIII. TECHNIQUES

As mentioned in the introduction, high fields are often useful for the observation of isotope effects, especially over several bonds. Closely related to high fields is, of course, high NMR signal resolution.

A. $^n\Delta C(D)$ effects

Observation of deuterium isotope effects in a mixture of deuteriated and non-deuteriated materials depends on the exchange rate. This problem is most important for alcohols, phenols, amines and amides with easily exchangeable protons.

In amides and peptides $\Delta C(ND)$ effects may be observed by dissolution in a 50:50 mixture of H_2O and D_2O, as originally shown by Feeney et al.[65] DMSO is known as a solvent that slows down exchange. Nemark and Hill[47] have derived a simple method for obtaining the spectra of mixtures of deuteriated and non-deuteriated alcohols and phenols. They add sufficient D_2O to a solution of the alcohol or phenol in DMSO to give a 1:1 mixture of deuteriated and non-deuteriated species. The excess water is dried off by addition of 200 mg $CaSO_4$ to the NMR sample tube. Molecular sieves (3A), $CaCl_2$, alumina and sodium methoxide prove less suitable as they increase the exchange rate. The method is not universally applicable since phenol, some halogen-substituted phenols, phenol-containing amino and carbonyl groups and some others show no doublets or line-broadenings, probably because of exchange. The addition of 5% D_2O to a DMSO solution has been practised.[55,56,59] Addition of CH_3OD or CD_3OD may also be used.[71]

Pfeffer et al.[48,49] have used concentric tubes, one containing the deuteri-ated, the other the undeuteriated species; a method they termed deuterium-induced differential isotope shift (DIS). By this method one avoids the problems of exchange, but different concentrations in the two compartments may influence the results. Data obtained by this method, and those obtained by other methods, may be compared with the results given in Table 13. The use of the method for compounds in which the nuclear shielding depends on pH is difficult, as pointed out by the authors.[49] To minimize bulk negative susceptibility contributions to the observed differential shifts, the H_2O solvent is placed in the inner tube and the D_2O in the outer tube.[48]

Gagnaire and Vincendon[51] chose a dipolar solvent and a low temperature (287 K) to limit the exchange rate of carbohydrate hydroxyl protons.

Bock, Gagnaire and Vincendon[53] have recorded the protiated analogues of carbohydrates in $CDCl_3$, after which 0.2 ml D_2O is added, the tube shaken vigorously, excess water decanted off and a Vertex stopper inserted to separate the chloroform phase from traces of water.

In order to avoid the triplet splitting at the α-carbon, deuterium decoup-ling may be used simultaneously with ^1H noise decoupling. In a study of bile acids,[84] switching on deuterium decoupling gives sharp signals of the deuteriated carbon, whereas leaving it off gives broad signals. This simple technique provides a sensitive method for identifying the deuteriated carbons.

B. $^n\Delta H(^{13}C)$ and $^n\Delta F(^{13}C)$ effects

Isotope effects, or differences thereof, may be obtained from a complete analysis of the coupled spectra.

1. Coupled spectra

The effects of ^{13}C on ^1H shieldings have been introduced in the analysis of ^{13}C single frequency spectra;[183,185] for some carbons they give an improved root mean square error.

2. $^{13}C–H$ satellite spectra

The inclusion of isotope effects is necessary to fit ^{13}C satellite spectra.[181,191] Englert et al.[190] introduced the difference between the directly bonded hydrogen shielding and that of all the other hydrogens in the analysis of the ^1H spectrum of benzene-1-^{13}C.

Goldstein et al.[181] have observed $^1\Delta$ data as well as the differences $^2\Delta - ^3\Delta$ and $^2\Delta - ^4\Delta$ from the analysis of 1,4-dihalobenzene (Table 30). For p-dimethoxybenzene, hydroquinone and p-benzoquinone no long-range isotope effects are observed. Differences are obtained in the analysis of the satellite spectrum of pyrene and 1,3,6,8-tetradeuteropyrene (Table 30).

The large effect observed in the latter is most likely to be due to broad lines preventing an exact analysis.[192] Schaefer *et al.*[180] have reported absolute values for 1,3-dithiole-2-thione, 1,3-dithiole-2-one, 1,3-dioxole and 1,3-dioxolyl.

Likewise ^{13}C isotope effects on ^{19}F nuclear shielding may be obtained from ^{13}C–^{19}F satellite spectra.[186–188,218,219,221,222]

C. $^1\Delta C(^{18}O)$ effects

The observation of oxygen isotope effects on the ^{13}C shielding of ^{13}C and ^{18}O simultaneously in biosynthetically enriched materials may be hampered by the presence of signals due to ^{13}C–^{13}C couplings. This difficulty may be circumvented by the use of a spin-echo pulse sequence $[90°–\tau–180°–\tau–\text{acquisition}–T]_N$ with $\tau = \frac{1}{2}J(C—C)$ and T being the delay between one acquisition and the next pulse sequence. The coupled carbons are phase-modulated by this procedure and their signals are out of phase with the normal spectrum.[246,254]

ACKNOWLEDGMENTS

I wish to express my gratitude to a number of authors for supplying me with reprints and preprints. I am indebted to Mrs Ella M. Larsen and to Søren Møller for the typing of the manuscript. The author is grateful to Dr. K. Schaumburg for helpful comments.

REFERENCES

1. H. Batiz-Hernandez and R. A. Bernheim, *Progr. NMR Spectr.*, 1967, **3**, 63.
2. W. T. Raynes, in *Specialist Periodical Report on NMR*, Vols 2 and 3 (R. K. Harris, ed.), The Chemical Society, London, 1973, 1974.
3. R. B. Mallion, in *Specialist Periodical Report on NMR*, Vol. 4 (R. K. Harris, ed.), The Chemical Society, London, 1975.
4. R. Ditchfield, in *Specialist Periodical Report on NMR*, Vol. 5 (R. K. Harris, ed.), The Chemical Society, London, 1976.
5. W. T. Raynes, in *Specialist Periodical Report on NMR*, Vols 7 and 8 (R. J. Abraham, ed.), The Chemical Society, London, 1978, 1979.
6. C. J. Jameson and J. Mason, in *Specialist Periodical Report on NMR*, Vol. 9 (G. A. Webb, ed.), The Chemical Society, London, 1980.
7. C. J. Jameson, in *Specialist Periodical Report on NMR*, Vol. 10 (G. A. Webb, ed.), The Chemical Society, London, 1981.
8. N. F. Ramsey, *Phys. Rev.*, 1952, **87**, 1075.
9. C. J. Jameson, *Bull. Magn. Reson.*, 1981, **3**, 3.
10. C. J. Jameson, *J. Chem. Phys.*, 1977, **66**, 4983.
11. A. D. Buckingham and W. Urland, *Chem. Rev.*, 1975, **75**, 113.
12. E. Breitmaier, G. Jung, W. Voelter and L. Pohl, *Tetrahedron*, 1973, **29**, 2485.
13. M. Alei and W. E. Wageman, *J. Chem. Phys.*, 1978, **68**, 783.

14. A. P. Tulloch and M. Mazurek, *J. Chem. Soc., Chem. Comm.*, 1973, **1973**, 692.
15. R. Aydin and H. Günther, *J. Amer. Chem. Soc.* 1981, **103**, 1301.
16. R. Aydin and H. Günther, *Angew. Chem.* 1981, **93**, 1000.
17. V. A. Chertkov and N. M. Sergeev, *J. Amer. Chem. Soc.*, 1977, **99**, 6750.
18. R. Aydin and H. Günther, *Z. Naturforsch.*, 1979, **34b**, 528.
19. D. Doddrell, W. Kitching, W. Adcock and P. A. Wiseman, *J. Org. Chem.*, 1976, **41**, 3036.
20. P. E. Hansen and J. J. Led, *Org. Magn. Reson.*, 1981, **15**, 288.
21. Yu. K. Grishin, N. M. Sergeev and Yu. A. Ustynyuk, *Mol. Phys.*, 1971, **22**, 711.
22. D. Doddrell and I. Burfitt, *Aust. J. Chem.*, 1972, **25**, 2239.
23. A. S. Serianni and R. Barker, *Canad. J. Chem.*, 1979, **57**, 3160.
24. Ch. Grathwohl, R. Schwyzer, A. Thun-kyi and K. Wüthrich, *FEBS Lett.*, 1973, **29**, 271.
25. P. A. J. Gorin, *Canad. J. Chem.*, 1974, **52**, 458.
26. K. G. R. Pachler, P. L. Wessels, J. Dekker, J. J. Dekker and P. G. Dekker, *Tetrahedron Lett.*, 1976, **1976**, 3059.
27. A. P. Tulloch, *Canad. J. Chem.*, 1977, **55**, 1135.
28. A. P. Tulloch, *Org. Magn. Reson.*, 1978, **11**, 109.
29. H. Booth and J. R. Everett, *Canad. J. Chem.* 1980, **58**, 2714.
30. J. R. Wesener and H. Günther, *Tetrahedron Lett.*, 1982, **23**, 2845.
31. L. Ernst, S. Eltamany and H. Hopf, *J. Amer. Chem. Soc.*, 1982, **104**, 299.
32. G. E. Maciel, P. D. Ellis and D. C. Hofer, *J. Phys. Chem.*, 1967, **71**, 2160.
33. D. G. Morris and A. M. Murray, *J. Chem. Soc., Perkin Trans. II*, 1976, **1976**, 1579.
34. L. Ernst, *Liebigs Ann. Chem.*, 1981, **1981**, 376.
35. K. L. Servis and F.-F. Shue, *J. Amer. Chem. Soc.*, 1980, **102**, 7233.
36. J. Dabrowski, D. Griebel and H. Zimmermann, *Spectr. Lett.*, 1979, **12**, 661.
37. H. Günther and G. Jikeli, *Chem. Ber.*, 1973, **106**, 1863.
38. R. H. Martin, J. Moriau and N. Defay, *Tetrahedron*, 1974, **30**, 179.
39. S. Berger, private communication.
40. R. A. Bell, C. L. Chan and B. G. Sayer, *J. Chem. Soc., Chem. Comm.* 1972, **1972**, 67.
41. R. W. Taft, in *Steric Effects in Organic Chemistry*, (M. S. Newman, ed.), Wiley, New York, 1956, p. 601.
42. H. Günther, H. Seel and M.-E. Günther, *Org. Magn. Reson.*, 1978, **11**, 97.
43. R. Aydin, H. Günther, J. Runsink, H. Schmickler and H. Seel, *Org. Magn. Reson.*, 1980, **13**, 210.
44. W. Kitching, M. Bullpitt, D. Doddrell and W. Adcock, *Org. Magn. Reson.*, 1974, **6**, 289.
45. G. L. Lebel, J. D. Laposa, B. G. Sayer and R. A. Bell, *Anal. Chem.*, 1971, **43**, 1500.
46. H. Günther, private communication.
47. R. A. Newmark and J. R. Hill, *Org. Magn. Reson.*, 1980, **13**, 40.
48. P. E. Pfeffer, K. M. Valentine and F. W. Parrish, *J. Amer. Chem. Soc.*, 1979, **101**, 1265.
49. P. E. Pfeffer, F. W. Parrish and J. Unruh, *Carbohydrate Res.* 1980, **84**, 13.
50. S.-C. Ho, H. J. Koch and R. S. Stuart, *Carbohydrate Res.*, 1978, **64**, 251.
51. D. Gagnaire and M. Vincendon, *J. Chem. Soc., Chem. Comm.*, 1977, **1977**, 509.
52. D. Gagnaire, D. Mancier and M. Vincendon, *Org. Magn. Reson.*, 1978, **11**, 344.
53. K. Bock, D. Gagnaire and M. Vignon, *Compt. Rend. Acad. Sci. Paris, Ser. C*, 1979, **289**, 345.
54. J. R. Andersen, W. Batsberg, L. Carlsen, H. Egsgaard, E. Larsen and A. Senning, *Biomed. Mass. Spectrometry*, 1980, **7**, 205.
55. F. W. Wehrli, *J. Chem. Soc., Chem. Comm.*, 1975, **1975**, 663.
56. E. G. Sundholm, *Acta Chem. Scand. B*, 1978, **32**, 177.
57. I. Kurobane, L. C. Vining and A. G. McInnes, *Tetrahedron Lett.*, 1978, 4633.
58. D. H. O'Brien and R. D. Stipanovic, *J. Org. Chem.*, 1978, **43**, 1105.
59. G. G. I. Moore, A. R. Kirk and R. A. Newmark, *J. Heterocyclic Chem.*, 1979, **16**, 789.
60. M. J. T. Robinson, K. M. Rosen and J. D. B. Workman, *Tetrahedron*, 1977, **33**, 1655.

61. P. E. Hansen and F. Duus, to be published.
62. E. Oldfield, R. S. Norton and A. Allerlhand, *J. Biol. Chem.*, 1975, **250**, 6381.
63. K. H. Ladner, J. J. Led and D. M. Grant, *J. Magn. Reson.*, 1975, **20**, 530.
64. J. J. Led and S. B. Petersen, *J. Magn. Reson.*, 1979, **33**, 603.
65. J. Feeney, P. Partington and G. C. K. Roberts, *J. Magn. Reson.*, 1974, **13**, 268.
66. R. A. Newmark and J. R. Hill, *J. Magn. Reson.*, 1976, **21**, 1.
67. L. A. Fonina, G. Ya. Avotin', T. A. Balashova, N. V. Starovoitova, L. B. Senyaviva, I. S. Savelov, V. F. Bystrov, V. T. Ivanov and Ya. A. Ovchinnikov, *Bioorg. Khim.*, 1980, **6**, 1285.
68. G. E. Hawkes, E. W. Randall, W. E. Hull, D. Gattegno and F. Conti, *Biochemistry*, 1978, **17**, 3986.
69. G. Schlingmann, B. Dresow, V. Koppenhagen and L. Ernst, *Liebigs Ann. Chem.*, 1980, **1980**, 1186.
70. V. Wray, A. Gossauer, B. Grüning, G. Reifenstahl and H. Zilch, *J. Chem. Soc., Perkin Trans. II*, 1979, **1979**, 1558.
71. P. E. Hansen, H. Thiessen and R. Brodersen, *Acta Chem. Scand. B*, 1979, **33**, 281.
72. T. Bundgaard, H. J. Jakobsen and E. J. Rahkama, *J. Magn. Reson.*, 1975, **19**, 345.
73. P. E. Hansen, H. J. Jakobsen and R. Brodersen, to be published.
74. P. E. Hansen, F. Duus and P. Schmitt, *Org. Magn. Reson.*, 1982, **18**, 58.
75. T. J. Simpson, *J. Chem. Soc., Perkin Trans. I*, 1979, **1979**, 1233.
76. P. S. Steyn, R. Vleggar, P. I.. Wessels, R. J. Cole and D. B. Scott, *J. Chem. Soc., Perkin Trans. I*, 1979, **1979**, 451.
77. C. P. Gorst-Allman, K. G. R. Pachler, P. S. Steyn, P. L. Wessels and D. B. Scott, *J. Chem. Soc., Perkin Trans. I*, 1977, **1977**, 2181.
78. L. J. Altman, D. Laungani, G. Gunnarson, H. Wennerstrøm and S. Forsén, *J. Amer. Chem. Soc.*, 1978, **100**, 8264.
79. N. N. Shapetko, Yu. S. Bogachev, I. L. Radushnova and D. N. Shigorin, *Dokl. Akad. Nauk SSSR*, 1976, **231**, 409.
80. J. Diakur, T. T. Nakashina and J. C. Vederas, *Canad. J. Chem.*, 1980, **58**, 1311.
81. C. Verchère, D. Rouselle and C. Viel, *Org. Magn. Reson.*, 1978, **11**, 395.
82. G. W. Gribble, R. B. Nelson, J. L. Johnson and G. C. Levy, *J. Org. Chem.*, 1975, **40**, 3720.
83. F. H. Köhler and W. Prössdrof, *J. Amer. Chem. Soc.*, 1978, **100**, 5970.
84. D. M. Wilson, A. L. Burlingame, T. Cronholm and J. Sjövall, *Biochem. Biophys. Res. Comm.*, 1974, **56**, 828.
85. A. G. McInnes, J. A. Walter and J. L. C. Wright, *Tetrahedron Lett.*, 1979, **1979**, 3245.
86. (a) U. Sankawa, H. Shimada, T. Sato, T. Kinoshita and K. Yamasaki, *Tetrahedron Lett.*, 1977, 483; (b) U. Sankawa and H. Shimada, *Tetrahedron Lett.*, 1978, 3375; (c) U. Sankawa, H. Shimada and K. Yamasaki, *Chem. Pharm. Bull.*, 1981, **25**, 3601.
87. (a) M. J. Garson, R. A. Hill and J. Stautnon, *J. Chem. Soc., Chem. Comm.*, 1977, **1977**, 624; (b) M. J. Garson, R. A. Hill and J. Staunton, *J. Chem. Soc., Chem. Comm.*, 1977, **1977**, 921.
88. M. J. Garson and J. Staunton, *J. Chem. Soc., Chem. Comm.*, 1981, **1981**, 708.
89. P. W. Bradshaw and J. R. Hanson, *J. Chem. Soc., Chem. Comm.*, 1981, **1981**, 1169.
90. R. Hunther and G. Mellows, *Tetrahedron Lett.*, 1978, **1978**, 5051.
91. C. Abell and J. Staunton, *J. Chem. Soc., Chem. Comm.*, 1981, **1981**, 856.
92. R. E. Wasylishen and T. Schaefer, *Canad. J. Chem.*, 1974, **52**, 3247.
93. H. N. Colli, V. Gold and J. E. Pearson, *J. Chem. Soc., Chem. Comm.*, 1973, **1973**, 408.
94. G. C. Levy and J. D. Cargioli, *J. Magn. Reson.*, 1972, **6**, 143.
95. Y. Fujikura, K. Aigami, N. Takaishi, H. Ikeda and Y. Imamoto, *Chem. Lett.*, 1976, **1976**, 507.
96. K. W. Baldry and M. J. T. Robinson, *Tetrahedron*, 1977, **33**, 1663.
97. D. A. Forsyth, P. Lucas and R. M. Burk, *J. Amer. Chem. Soc.*, 1982, **104**, 240.

98. R. A. Archer, D. W. Johnson, E. W. Hagaman, L. N. Moreno and E. Wenkert, *J. Org. Chem.*, 1977, **42**, 489.

99. J. M. A. Al-Rawi, J. A. Elvidge, J. R. Jones and E. A. Evans, *J. Chem. Soc., Perkin Trans. II*, 1975, **1975**, 449.

100. D. Lauer, E. L. Motell, D. D. Traficante and G. E. Maciel, *J. Amer. Chem. Soc.*, 1972, **94**, 5335.

101. F. W. Wehrli and T. Wirthlin, *Interpretation of Carbon-13 NMR Spectra*, Heyden, London, 1976, p. 108.

102. F. A. L. Anet and A. H. Dekmezian, *J. Amer. Chem. Soc.*, 1979, **101**, 5449.

103. R. B. Calvert and J. R. Shapley, *J. Amer. Chem. Soc.*, 1978, **100**, 7726.

104. C. R. Lassigne and E. J. Wells, *J. Magn. Reson.*, 1978, **31**, 195.

105. N. Gurudata, *Canad. J. Chem.*, 1972, **50**, 1956.

106. P. Diehl, H. Bösiger and J. Jokisari, *Org. Magn. Reson.*, 1979, **12**, 282.

107. R. S. Ozubko, G. W. Buchanan and I. C. P. Smith, *Canad. J. Chem.*, 1974, **52**, 2493.

108. P. E. Hansen, *Org. Magn. Reson.*, 1979, **12**, 109.

109. J. Giraud and C. Marzin, *Org. Magn. Reson.*, 1979, **12**, 647.

110. A. Lýcka, D. Šnobl, V. Macháček and M. Večeřa, *Org. Magn. Reson.*, 1981, **15**, 390

111. J. Elguero, A. Fruchier and M. del C. Pardo, *Canad. J. Chem.*, 1976, **54**, 1329.

112. R. J. Cushley, R. J. Sykes, C.-K. Shaw and H. H. Wasserman, *Canad. J. Chem.*, 1975, **53**, 148.

113. S. Gronowitz, I. Johnson and A.-B. Hörnfeldt, *Chem. Scripta*, 1975, **7**, 76.

114. P. E. Hansen, Thesis, University of Aarhus, 1974.

115. R. Wehner and H. Günther, *Chem. Ber.*, 1974, **107**, 3152.

116. R. Martino, P. Mison, F. W. Wehrli and T. Wirtlin, *Org. Magn. Reson.*, 1975, **7**, 175.

117. M. E. Jung, T. J. Shaw, R. R. Fraser, J. Banville and K. Taymaz, *Tetrahedron Lett.*, 1979, **1979**, 4149.

118. R. A. Bernheim and B. J. Lavery, *J. Chem. Phys.*, 1965, **42**, 1464.

119. O. S. Tee and J. Warkentin, *Canad. J. Chem.*, 1965, **43**, 2424.

120. A. L. Allred and W. D. Wilk, *J. Chem. Soc., Chem. Comm.*, **1969**, **1969**, 273.

121. W. Saur, H. L. Crespi and J. J. Katz, *J. Magn. Reson.*, 1970, **2**, 47.

122. J. W. Marsman, J. Lubach and W. Drenth, *Recuil*, 1969, **88**, 1969.

123. E. Bengsch, M. Corval and G. L. Martin, *Org. Magn. Reson.*, 1974, **6**, 195.

124. H. Jensen and K. Schaumburg, *Acta Chem. Scand.*, 1971, **25**, 663.

125. V. R. Haddon and L. M. Jackman, *Org. Magn. Reson.*, 1973, **5**, 333.

126. B. A. Énglin, Yu. V. Belov, A. E. Zolotarev, T. A. Onishchenko, I. K. Shmyrev and R. Kh. Friedlina, *Dokl. Akad. Nauk SSSR*, 1973, **210**, 1973.

127. R. J. Jablonski and E. I. Snyder, *J. Amer. Chem. Soc.*, 1968, **90**, 2316.

128. M. Buza and E. I. Snyder, *J. Amer. Chem. Soc.*, 1966, **88**, 1161.

129. M. C. Baird, *J. Magn. Reson.*, 1974, **14**, 117.

130. G. M. Ford, L. G. Robinson and G. B. Savitsky, *J. Magn. Reson.*, 1970, **4**, 109.

131. G. B. Savitsky, L. G. Robinson, W. A. Tallon and L. R. Womble, *J. Magn. Reson.*, 1969, **1**, 139.

132. T. Yonemitsu and K. Kubo, *J. Chem. Soc., Chem. Comm.*, 1981, **1981**, 309.

133. T. Yonemitsu and K. Kubo, *Chem. Lett.*, 1981, **1981**, 1061.

134. T. Yonemitsu, S. Fujino and K. Kubo, *Kyushu Sangy Daigaku Kogakubo Kenkyu Hokoku*, 1980, **17**, 30.

135. M. L. Crespi, U. Smith and J. J. Katz, *Biochemistry*, 1968, **7**, 2232.

136. L. Canuel and M. St.-Jacques, *Canad. J. Chem.*, 1974, **52**, 3581.

137. G. Fraenkel, Y. Asahi, H. Batiz-Hernandez and R. A. Bernheim, *J. Chem. Phys.*, 1966, **44**, 4647.

138. M. Shporer and A. Loewenstein, *Mol. Phys.*, 1968, **15**, 9.

139. D. E. Leyden and J. M. McCall, *Anal. Chim. Acta*, 1970, **49**, 77.

140. H. Jensen and K. Schaumburg, *Mol. Phys.*, 1971, **22**, 1041.
141. U. Eichhoff, W. E. Hull, V. A. Scherbakov and V. G. Khlopin, in *Magnetic Resonance and Related Phenomena, Proceedings of the 20th AMPERE Congress*, (E. Kundla, E. Lipmaa and T. Saluvere, eds), Springer, Berlin, 1979, p. 460.
142. F. R. Taravel and M. R. Vignon, *Nouveau J. Chim.*, 1982, **6**, 37.
143. M. Rossi, F. Coletta and G. Rigatti, *Lincei Rend. Sci. Fis. Mat. Nat.*, 1972, **12**, 80.
144. B. D. James, B. E. Smith and R. H. Newman, *J. Chem. Soc., Chem. Comm.*, 1974, **1974**, 294.
145. B. E. Smith, B. D. James and R. M. Peachey, *Inorg. Chem.*, 1977, **16**, 2057.
146. W. M. Litchman, M. Alei and A. E. Florin, *J. Chem. Phys.*, 1969, **50**, 1897.
147. G. Angelini, M. Speranza, A. L. Segre and L. J. Altman, *J. Org. Chem.*, 1980, **45**, 3291.
148. K. Schaumburg and C. Deverell, *J. Amer. Chem. Soc.*, 1968, **90**, 2495.
149. C. Deverell, K. Schaumburg and H. J. Bernstein, *J. Chem. Phys.*, 1968, **49**, 1276.
150. D. K. Hindermann and C. D. Cornwell, *J. Chem. Phys.*, 1968, **48**, 2017.
151. K. Radley and L. W. Reeves, *J. Chem. Phys.*, 1971, **54**, 4509.
152. V. A. Sherbakov, B. N. Chernyshov and R. L. Davidovich, *Zhur. Struk. Khim.*, 1973, **14**, 1109.
153. J. B. Lambert and L. G. Greifenstein, *J. Amer. Chem. Soc.*, 1973, **95**, 6150.
154. J. B. Lambert and L. G. Greifenstein, *J. Amer. Chem. Soc.*, 1974, **96**, 5120.
155. L. Phillips and V. Wray, *J. Chem. Soc., Perkin Trans. II.*, 1972, **1972**, 223.
156. P. J. Mitchell and L. Phillips, *J. Chem. Soc., Chem. Comm.*, 1975, 1975, 908.
157. D. D. Traficante and G. E. Maciel, *J. Amer. Chem. Soc.*, 1965, **87**, 4917.
158. J. W. Timberlake, J. A. Thomson and R. W. Taft, *J. Amer. Chem. Soc.*, 1971, **93**, 274.
159. M. H. Pendleburg and L. Phillips, *Org. Magn. Reson.*, 1972, **4**, 529.
160. H. Akiyama, F. Yamanchi and K. Ouchi, *J. Chem. Soc., Chem. Comm.*, 1970, **1970**, 355.
161. I. B. Golovanov, V. N. Gazlow, I. A. Soboleva and V. V. Snolyaninov, *Zhur. Obs. Khim.*, 1973, **43**, 905.
162. J. T. Gerig, D. T. Loehr, K. F. S. Luk and D. C. Roe, *J. Amer. Chem. Soc.*, 1979, **101**, 7482.
163. W. E. Hull and B. D. Sykes, *Biochemistry*, 1976, **15**, 1535.
164. (a) P. C. Lauterbur, private communication mentioned in reference 163; (b) P. C. Lauterbur, B. V. Kaufman and M. K. Crawford, in *Biomolecular Structure and Function* (P. F. Agris, ed.), Academic Press, New York, 1978, p. 329.
165. (a) D. S. Hagen, J. H. Weiner and B. D. Sykes, *Biochemistry*, 1979, **18**, 2007; (b) D. S. Hagen, J. H. Weiner and B. D. Sykes, in *NMR and Biochemistry, A Symposium Honoring Mildred Cohn* (S. J. Opella and P. Lu, eds), Marcel Dekker, New York, 1979, p. 51.
166. B. Kaufman, MS thesis, State University of New York at Stony Brook, and private communication mentioned in reference 164.
167. R. E. London, T. E. Walker, T. W. Whaley and N. A. Matwiyoff, *Org. Magn. Reson.*, 1977, **9**, 598.
168. W. J. Stec, N. Goddard and J. R. Van Wazer, *J. Phys. Chem.*, 1971, **75**, 3549.
169. A. K. Jameson and C. J. Jameson, *J. Magn. Reson.*, 1978, **32**, 455.
170. A. A. Borisenko, N. M. Sergeev and Yu. A. Ustynyuk, *Mol. Phys.*, 1971, **22**, 715.
171. W. McFarlane and D. S. Rycroft, *Mol. Phys.*, 1972, **24**, 893.
172. H. J. Jakobsen, A. J. Zozulin, P. D. Ellis and J. D. Odom, *J. Magn. Reson.*, 1980, **38**, 219.
173. H. J. Jakobsen and R. S. Hansen, *J. Magn. Reson.*, 1978, **30**, 397.
174. O. Lutz and H. Oehler, *Z. Naturforsch.*, 1977, **32a**, 131.
175. M. R. Bendall and D. M. Doddrell, *Aust. J. Chem.*, 1978, **31**, 1141.
176. O. Lutz and G. Stricker, *Phys. Lett.*, 1971, **35A**, 397.
177. B. W. Epperlein, H. Krüger, O. Lutz and A. Schwenk, *Z. Naturforsch.*, 1974, **29a**, 1553.
178. C. Hassler, J. Kronenbitter and A. Schwenk, *Z. Phys. A*, 1977, **280**, 117.
179. O. Lutz, *Phys. Lett.*, 1969, **29A**, 58.

180. T. Schaefer, K. Chum, D. McKinnon and M. S. Chauhan, *Canad. J. Chem.*, 1975, **53**, 2734.
181. J. M. Read, R. W. Crecely and J. H. Goldstein, *J. Mol. Spectr.*, 1968, **25**, 107.
182. L. Ernst, V. Wray, V. A. Chertkov and N. M. Sergeev, *J. Magn. Reson.*, 1977, **25**, 123.
183. M. Hansen and H. J. Jakobsen, *J. Magn. Reson.*, 1973, **10**, 74.
184. M. Hansen, R. S. Hansen and H. J. Jakobsen, *J. Magn. Reson.*, 1974, **13**, 386.
185. M. Hansen and H. J. Jakobsen, *J. Magn. Reson.*, 1975, **20**, 520.
186. V. Wray and D. N. Lincoln, *J. Magn. Reson.*, 1975, **18**, 374.
187. L. Ernst, D. N. Lincoln and V. Wray, *J. Magn. Reson.*, 1976, **21**, 115.
188. V. Wray, L. Ernst and E. Lustig, *J. Magn. Reson.*, 1977, **27**, 1.
189. V. A. Chertkov and N. M. Sergeev, *J. Magn. Reson.*, 1976, **21**, 159.
190. G. Englert, P. Diehl and W. Niederberger, *Z. Naturforsch.*, 1971, **26a**, 1829.
191. R. W. Crecely, R. S. Butler and J. H. Goldstein, *J. Mol. Spectr.*, 1968, **26**, 489.
192. P. E. Hansen and A. Berg, *Acta Chem. Scand.*, 1971, **25**, 3377.
193. J. Runsink and H. Günther, *Org. Magn. Reson.*, 1980, **13**, 249.
194. T. Wamsler, J. T. Nielsen, E. J. Pedersen and K. Schaumburg, *J. Magn. Reson.*, 1978, **31**, 177.
195. J. M. Briggs, E. Rahkama and E. W. Randall, *J. Magn. Reson.*, 1973, **12**, 40.
196. V. A. Chertkov and Yu. K. Grishin, *Zhur. Struk. Khim.*, 1977, **18**, 776.
197. N. J. Koole and M. J. A. de Bie, *J. Magn. Reson.*, 1976, **23**, 9.
198. G. Schrumpf, G. Becher and W. Lüttke, *J. Magn. Reson.*, 1973, **10**, 90.
199. D. M. McKinnon and T. Schaefer, *Canad. J. Chem.*, 1971, **49**, 89.
200. V. A. Chertkov, Yu. K. Grishin and N. M. Sergeev, *J. Magn. Reson.*, 1976, **24**, 275.
201. T. Tokuhiro, L. Menefra and H. H. Szmant, *Rev. Latinoamer. Quim.*, 1977, **8**, 36.
202. E. W. Garbisch and M. G. Griffith, *J. Amer. Chem. Soc.*, 1968, **90**, 6543.
203. K. Bachman and W. von Philipsborn, *Org. Magn. Reson.*, 1976, **8**, 648.
204. G. Schrumpf, *Chem. Ber.*, 1973, **106**, 246.
205. G. Schrumpf, *Chem. Ber.*, 1973, **106**, 266.
206. V. Wray, L. Ernst, T. Lund and H. J. Jakobsen, *J. Magn. Reson.*, 1980, **40**, 55.
207. P. Sándor and L. Radics, *Org. Magn. Reson.*, 1981, **16**, 148.
208. P. S. Nielsen, R. S. Hansen and H. J. Jakobsen, *J. Organomet. Chem.*, 1976, **114**, 145.
209. E. Bengsch and M. Ptak, *Stable Isotopes in Life Sciences*, International Atomic Energy Agency, Vienna, 1977, p. 197.
210. F. J. Weigert and J. D. Roberts, *J. Amer. Chem. Soc.*, 1972, **94**, 6021.
211. H. Booth and J. R. Everett, *Canad. J. Chem.*, 1980, **58**, 2709.
212. J. Jokisari, *Org. Magn. Reson.*, 1978, **11**, 157.
213. S. M. Cohen, S. Ogawa and R. G. Shulman, *Proc. Natl Acad. Sci.*, 1976, **76**, 1603.
214. S. Mohanty, *Chem. Phys. Lett.*, 1973, **18**, 581.
215. P. E. Hansen, O. K. Poulsen and A. Berg, *Org. Magn. Reson.*, 1976, **8**, 632.
216. P. E. Hansen, O. K. Poulsen and A. Berg, *Org. Magn. Reson.*, 1979, **12**, 43.
217. P. E. Hansen, O. K. Poulsen and A. Berg, *Org. Magn. Reson.*, 1975, **7**, 23.
218. J. F. Hinton and L. W. Jacques, *J. Magn. Reson.*, 1975, **17**, 95.
219. L. H. Sutcliffe and B. Taylor, *Spectrochim. Acta*, 1972, **28A**, 619.
220. W. S. Brey, K. H. Ladner, R. E. Block and W. A. Tallon, *J. Magn. Reson.*, 1972, **8**, 406.
221. L. Ernst and V. Wray, *J. Magn. Reson.*, 1977, **28**, 373.
222. V. A. Chertkov and N. M. Sergeev, *Zhur. Strukt. Khim.*, 1977, **19**, 613.
223. S. Rodmar, B. Rodmar, M. K. Sharma, S. Gronowitz, H. Christiansen and M. Rosén. *Acta Chem. Scand.* 1968, **22**, 907.
224. G. V. D. Tiers, *J. Chem. Phys.*, 1958, **29**, 963.
225. J. Jokisari, K. Räisänen, L. Lajunen, A. Passoja and P. Pyykkö, *J. Magn. Reson.*, 1978, **31**, 121.
226. J. Jokisari, *Mol. Phys.*, 1978, **36**, 113.

227. A. J. Zozulin, H. J. Jakobsen, T. F. Moore, A. R. Garber and J. D. Odom, *J. Magn. Reson.*, 1980, **41**, 458.
228. S. Aime and R. K. Harris, *J. Magn. Reson.*, 1974, **13**, 236.
229. R. Pasonen, J. Enquist, M. Karhu, E. Rahkama, M. Sundberg and R. Uggle, *Org. Magn. Reson.*, 1978, **11**, 42.
230. A. Loewenstein and M. Shporer, *Mol. Phys.*, 1965, **10**, 293.
231. L. N. Kurkovskaya and N. N. Shapet'ko, *Zhur. Strukt. Khim.*, 1975, **16**, 682.
232. L. N. Kurkovskaya, *Zhur. Strukt. Khim.*, 1978, **19**, 946.
233. J. P. Jacobsen, K. Schamburg and J. T. Nielsen, *J. Magn. Reson.*, 1974, **13**, 372.
234. R. E. Wasylishen, D. H. Muldrew and K. J. Friesen, *J. Magn. Reson.*, 1980, **41**, 341.
235. J. M. Ostrander, L. H. Hurley, A. G. McInnes, D. G. Smith, J. A. Walter and J. L. C. Wright, *J. Antibiotics*, 1980, **33**, 1167.
236. S. J. Gould, K. J. Martinkus and C.-H. Tann, *J. Amer. Chem. Soc.*, 1981, **103**, 4639.
237. E. Leete and J. A. McDonell, *J. Amer. Chem. Soc.*, 1981, **103**, 658.
238. C. Bremard, B. Monchel and S. Sneur, *J. Chem. Soc., Chem. Comm.*, 1982, **1982**, 300.
239. H. Pikman and S. Pinchas, *J. Inorg. Nucl. Chem.*, 1970, **32**, 2441.
240. S. Pinchas and E. Meshulem, *J. Chem. Soc., Chem. Comm.*, 1970, **1970**, 1147.
241. B. Sredni and S. Pinchas, *J. Magn. Reson.*, 1972, **7**, 289.
242. J. M. Risley and R. L. Van Etten, *J. Amer. Chem. Soc.*, 1979, **101**, 252.
243. J. M. Risley and R. L. Van Etten, *J. Amer. Chem. Soc.*, 1980, **102**, 4609.
244. J. M. Risley and R. L. Van Etten, *J. Amer. Chem. Soc.*, 1980, **102**, 6699.
245. J. C. Vederas, *J. Amer. Chem. Soc.*, 1980, **102**, 374.
246. T. T. Nakashima and J. C. Vederas, *J. Chem. Soc., Chem. Comm.*, 1982, **1982**, 206.
247. J. C. Vederas and T. T. Nakashima, *J. Chem. Soc., Chem. Comm.*, 1980, **1980**, 183.
248. S. A. Benner, J. E. Maggio and H. E. Simmons, *J. Amer. Chem. Soc.*, 1981, **103**, 1581.
249. C. R. Hutchinson, M. M. Sherman, J. C. Vederas and T. T. Nakashima, *J. Amer. Chem. Soc.*, 1981, **103**, 5953.
250. J. M. Risley and R. L. Van Etten, *J. Amer. Chem. Soc.*, 1981, **103**, 4389.
251. D. J. Darensbourg, *J. Organomet. Chem.*, 1979, **174**, C70.
252. D. J. Darensbourg and B. J. Baldwin, *J. Amer. Chem. Soc.*, 1979, **101**, 6447.
253. D. J. Darensbourg, B. J. Baldwin and J. A. Froelich, *J. Amer. Chem. Soc.*, 1980, **102**, 4688.
254. M. P. Lane, T. T. Nakashima and J. C. Vederas, *J. Amer. Chem. Soc.*, 1982, **104**, 913.
255. D. E. Cane, H. Hasler and T. C. Liang, *J. Amer. Chem. Soc.*, 1981, **103**, 5960.
256. D. D. Hackney, J. A. Sleep, G. Rosen, R. L. Hutton and P. D. Boyer, in *NMR and Biochemistry* (S. J. Opella and P. Lu, eds), Marcel Dekker, New York, 1979, p. 285.
257. B. Brandänge, O. Dahlman and L. Mörch, *J. Chem. Soc., Chem. Comm.*, 1980, **1980**, 555.
258. D. E. Cane, T. C. Liang and H. Hasler, *J. Amer. Chem. Soc.*, 1981, **103**, 5962.
259. J. C. Vederas, T. T. Nakashima and J. Diakur, *Planta Med.*, 1980, **39**, 201.
260a. M. Cohn and A. Hu, *Proc. Natl Acad. Sci.*, 1978, **75**, 200.
260b. M. Cohn and A. Hu, *Bull. Magn. Reson.*, 1978, **1**, 38.
261. M. Cohn and A. Hu, *J. Amer. Chem. Soc.*, 1980, **102**, 913.
262. G. Lowe, B. V. L. Potter, B. S. Sproat and W. E. Hull, *J. Chem. Soc., Chem. Comm.*, 1979, **1979**, 733.
263. F. Jordan, J. A. Patrick and S. Salomone, *J. Biol. Chem.*, 1979, **254**, 2384.
264. G. Lowe and B. S. Sproat, *J. Chem. Soc., Perkin Trans. I*, 1978, **1978**, 1622.
265. J. L. Bock and M. Cohn, *J. Biol. Chem.*, 1978, **253**, 4082.
266. R. L. Van Etten and J. M. Risley, *Proc. Natl Acad. Sci.*, 1978, **75**, 4784.
267. M. R. Webb, G. G. McDonald and D. R. Trentham, *J. Biol. Chem.*, 1978, **253**, 2908.
268. D. D. Hackney, G. Rosen and P. D. Boyer, *Proc. Natl Acad. Sci.*, 1979, **76**, 3646.
269. M. Cohn, in *NMR in Biochemistry* (S. J. Opella and P. Lu, eds), Marcel Dekker, New York, 1979, p. 7.

270. T. R. Sharp and S. J. Benkovic, *Biochemistry*, 1979, **18**, 2911.

271. G. Lowe and B. S. Sproat, *J. Chem. Soc., Perkin Trans. II*, 1978, **1978**, 1622.

272. R. L. Jarvest and G. Lowe, *J. Chem. Soc., Chem. Comm.*, 1979, **1979**, 364.

273. S. L. Buchwald and J. R. Knowles, *J. Amer. Chem. Soc.*, 1980, **102**, 6601.

274. R. L. Jarvest, G. Lowe and B. V. L. Potter, *J. Chem. Soc., Perkin Trans. I*, 1981, **1981**, 3186.

275. M. R. Webb and D. R. Trentham, *J. Biol. Chem.*, 1980, **255**, 1775.

276. J. A. Coderre and J. A. Gerlt, *J. Amer. Chem. Soc.*, 1980, **102**, 6597.

277. G. Gorenstein and R. Rowell, *J. Amer. Chem. Soc.*, 1980, **102**, 6165.

278. J. A. Gerlt and J. A. Coderre, *J. Amer. Chem. Soc.*, 1980, **102**, 4531.

279. G. Lowe and B. S. Sproat, *J. Chem. Soc., Chem. Commun.*, 1978, **1978**, 565.

280. O. Lutz, A. Nolle and D. Staschewski, *Z. Naturforsch.*, 1978, **33a**, 380.

281. R. L. Van Etten and J. M. Risley, *J. Amer. Chem. Soc.*, 1981, **103**, 5633.

282. A. R. Haase, O. Lutz, M. Müller and A. Nolle, *Z. Naturforsch.*, 1976, **31a**, 1427.

283. K. U. Buckler, A. R. Haase, O. Lutz, M. Müller and A. Nolle, *Z. Naturforsch.*, 1977, **32a**, 126.

284. R. J. Gillespie, J. S. Hartman and M. Parekh, *Canad. J. Chem.*, 1968, **46**, 1601.

285. R. J. Gillespie and J. S. Hartman, *Canad. J. Chem.*, 1968, **46**, 2147.

286. S. Aa. Linde and H. J. Jakobsen, *J. Magn. Reson.*, 1975, **17**, 411.

287. O. Lutz, A. Nolle and P. Kronek, *Z. Phys. A*, 1977, **282**, 157.

288. A. DeMarco and G. Gatti, *J. Magn. Reson.*, 1972, **6**, 200.

289. P. R. Carey, H. W. Kroto and M. A. Turpin, *J. Chem. Soc., Chem. Comm.*, 1969, **1969**, 188.

290. W. Büchner and D. Scheutzow, *Org. Magn. Reson.*, 1975, **7**, 615.

291. I. M. Ismail, S. J. S. Kerrison and P. J. Sadler, *J. Chem. Soc., Chem. Comm.*, 1980, **1980**, 1175.

292. J. Dyer and J. Lee, *Spectrochim. Acta*, 1970, **26A**, 1045.

293. E. A. Halevi, *Progr. Phys. Org. Chem.*, 1963, **1**, 109.

294. E. A. Halevi, M. Nussin and A. Ron, *J. Chem. Soc.*, 1963, **1963**, 866.

295. G. M. Coppola, R. Damon, A. D. Kahle and M. J. Shapiro, *J. Org. Chem.*, 1981, **46**, 1221.

296. W. T. Raynes, A. M. Davis and D. B. Cook, *Mol. Phys.*, 1971, **21**, 123.

297. R. V. Reid, *Phys. Rev. A*, 1975, **11**, 403.

298. J. R. Beckett and H. Y. Carr, *Phys. Rev. A*, 1981, **24**, 144.

299. H. S. Gutowsky, *J. Chem. Phys.*, 1959, **31**, 1683.

300. S. G. Frankiss, *J. Phys. Chem.*, 1963, **67**, 752.

301. S. L. Manatt, G. L. Juvinall, R. I. Wagner and D. D. Elleman, *J. Amer. Chem. Soc.*, 1966, **88**, 2689.

302. D. H. Hunther, A. L. Johnson, J. B. Stothers, A. Nickon, J. L. Lambert and D. F. Covey, *J. Amer. Chem. Soc.*, 1972, **94**, 8581.

303. S. H. Grover, D. H. Marr, J. B. Stothers and C. T. Tan, *Canad. J. Chem.*, 1975, **53**, 1351.

304. F. R. Jensen and R. A. Neese, *J. Amer. Chem. Soc.*, 1971, **93**, 6329.

305. M. Saunders, M. H. Jaffe and P. Vogel, *J. Amer. Chem. Soc.*, 1971, **93**, 2558.

306. M. Saunders and P. Vogel, *J. Amer. Chem. Soc.*, 1971, **93**, 2559.

307. M. Saunders and P. Vogel, *J. Amer. Chem. Soc.*, 1971, **93**, 2561.

308. M. Saunders, L. Telkowski and M. R. Kates, *J. Amer. Chem. Soc.*, 1977, **99**, 8070.

309. M. Saunders, M. R. Kates and G. E. Walker, *J. Amer. Chem. Soc.*, 1981, **103**, 4623.

310. M. Saunders and M. R. Kates, *J. Amer. Chem. Soc.*, 1977, **99**, 8071.

311. M. Saunders, M. R. Kates, K. B. Wiberg and W. Pratt, *J. Amer. Chem. Soc.*, 1977, **99**, 8072.

312. M. Saunders and M. R. Kates, *J. Amer. Chem. Soc.*, 1980, **102**, 6867.

313. M. Saunders and H. U. Siehl, *J. Amer. Chem. Soc.*, 1980, **102**, 6868.

314. H. Hogeveen and E. M. G. A. van Kruchten, *J. Org. Chem.*, 1981, **46**, 1350.

315. R. P. Kirchen, K. Ranganayakulu, A. Rauk, B. P. Singh and T. S. Sorensen, *J. Amer. Chem. Soc.*, 1981, **103**, 588.
316. P. Ahlberg, C. Engdahl and G. Jonsäll, *J. Amer. Chem. Soc.*, 1981, **103**, 1583.
317. L. R. Smitz and T. S. Sorensen, *J. Amer. Chem. Soc.*, 1980, **102**, 1645.
318. J. W. Faller, H. H. Murray and M. Saunders, *J. Amer. Chem. Soc.*, 1980, **102**, 2306.
319. M. Schlosser and M. Stähle, *Angew. Chem.*, 1980, **92**, 497.
320. R. Askani, H. O. Kalinowski and B. Weuste, *Org. Magn. Reson.*, 1982, **18**, 176.
321. F. A. L. Anet and F. Leyendecker, *J. Amer. Chem. Soc.*, 1973, **95**, 156.
322. F. R. Jensen and L. A. Smith, *J. Amer. Chem. Soc.*, 1964, **86**, 956.
323. F. A. L. Anett, V. J. Bassus, A. P. W. Hewett and M. Saunders, *J. Amer. Chem. Soc.*, 1980, **102**, 3945.
324. M. Wolfsberg, *Ann. Rev. Phys. Chem.*, 1969, **20**, 449.
325. K. W. Baldry and M. J. T. Robinson, *J. Chem. Res.* (*M*), 1977, **1977**, 1001.
326. F. W. Wehrli, D. Jeremíc, M. Lj. Mihailovic and S. Milosavljevíc, *J. Chem. Soc., Chem. Comm.*, 1978, **1978**, 302.
327. F. W. Wehrli and T. W. Wirthlin, *Interpretation of Carbon-13 NMR Spectra*, Heyden, New York, 1976, p. 108.
328. H. Eggert and C. Djerassi, *J. Org. Chem.*, 1973, **38**, 3788.
329. G. C. Andrews, G. N. Churmny and E. B. Whipple, *Org. Magn. Reson.*, 1981, **15**, 324.
330. J. A. Chudek and R. Foster, *Tetrahedron*, 1978, **34**, 2209.
331. G. V. Smith, W. A. Boyd and C. C. Hinckley, *J. Amer. Chem. Soc.*, 1971, **93**, 6319.
332. C. C. Hinckley, W. A. Boyd and G. V. Smith, *Tetrahedron Lett.*, 1972, **1972**, 879.
333. C. C. Hinckley, W. A. Boyd and G. V. Smith, in *Nuclear Magnetic Shift Reagents* (R. E. Sievers, ed.), Academic Press, New York, 1973, p. 1.
334. J. K. M. Sanders and D. H. Williams, *J. Chem. Soc., Chem. Comm.*, 1972, **1972**, 436.
335. A. M. Grotens, J. Smid and E. de Boer, *Tetrahedron Lett.*, 1971, **1971**, 4863.
336. A. M. Grotens, C. W. Hilbers, E. de Boer and J. Smid, *Tetrahedron Lett.*, 1972, **1972**, 2067.
337. C. M. DePuy, P. C. Fünfschilling and J. M. Olson, *J. Amer. Chem. Soc.*, 1976, **98**, 276.
338. R. W. Lang and H.-J. Hansen, *Helv. Chim. Acta*, 1980, **63**, 1215.
339. A. T. Balaban, I. I. Stanoiu, F. Chiraleu, *Rev. Roumaine Chim.*, 1978, **23**, 187.
340. A. T. Balaban, I. I. Stanoiu and F. Chiraleu, *J. Chem. Soc., Chem. Comm.*, 1976, **1976**, 984.
341. A. Tuóciauskas, N. M. Sergeev and Yu. A. Ustynyuk, *Mol. Phys.*, 1971, **21**, 179.
342. H. C. E. McFarlane, W. McFarlane and C. J. Turner, *Mol. Phys.*, 1979, **37**, 1639.
343. E. D. Becker and R. B. Bradley, *J. Magn. Reson.*, 1971, **4**, 136.
344. W. McFarlane, *J. Magn. Reson.*, 1973, **10**, 98.
345. H. Krüger, O. Lutz, A. Nolle, A. Schwenk and G. Stricker, *Z. Naturforsch.*, 1973, **28a**, 484.
346. J. P. Bloxsidge, J. A. Elvidge, J. R. Jones, R. B. Mane and M. Saljoughian, *Org. Magn. Reson.*, 1979, **12**, 574.
347. S. I. Chan, L. Lin, D. Clutter and P. Dea, *Proc. Natl Acad. Sci.*, 1970, **65**, 316.
348. G. Gunnarson, H. Wennerström, W. Egan and S. Forsén, *Chem. Phys. Lett.*, 1976, **38**, 96.
349. T. W. Marshall, *Mol. Phys.*, 1961, **4**, 61.
350. W. Egan, G. Gunnarson, T. E. Bull and S. Forsén, *J. Amer. Chem. Soc.*, 1977, **99**, 4568.
351. N. Muller and R. C. Reiter, *J. Chem. Phys.*, 1965, **42**, 3265.
352. T. K. Leipert, *Org. Magn. Reson.*, 1977, **9**, 157.
353. A. A. Samoilenko, N. P. Makoeeva and N. N. Shapet'ko, *Dokl. Akad. Nauk SSSR*, 1980, **252**, 652.
354. A. A. Samoilenko, N. P. Makoeeva, Yu. S. Bogachev and N. N. Shapet'ko, *Zhur. Strukt. Khim.*, 1979, **20**, 928.

355. A. A. Samoilenko, A. I. Serebryanskaya, Yu. S. Bogachev, N. N. Shapet'ko and A. I. Shatenstein, *Zhur. Obsch. Khim.*, 1980, **50**, 1379.
356. I. I. Stanoiu, F. Chiraleu, E. Gard and A. T. Balaban, *Rev. Roumaine Chim.*, 1977, **22**, 117.
357. I. A. Oxton, A. G. McInnes and J. A. Walter, *Canad. J. Chem.*, 1979, **57**, 503.
358. H. Lackner, *Chem. Ber.*, 1971, **104**, 3653.
359. E. Breitmaier, *Chimia*, 1974, **28**, 120.
360. P. Junkes, M. Baudler, J. Dobbers and D. Rackwitz, *Z. Naturforsch.*, 1972, **27b**, 1451.
361. J. Eberling, M. A. Leva, H. Stary and A. Schmidpeter, *Z. Naturforsch.*, 1971, **26b**, 650.

Recent Advances in Silicon-29 NMR Spectroscopy

E. A. WILLIAMS

*General Electric Company, Corporate Research and Development, Schenectady,
New York 12301, USA*

I. INTRODUCTION

As early as 1962 the promise of ^{29}Si NMR as a structural probe in
organosilanes and siloxanes was demonstrated in Lauterbur's review[1] of
the reports of direct observation of ^{29}Si in liquids. A few papers appeared
in the late 1960s further suggesting its potential for studies of organic[2] and
inorganic[3] silicon compounds. The tremendous growth of ^{29}Si NMR papers
since the early 1970s has been due largely to the development and application of the pulsed Fourier transform (FT) method[4] in commercial spectrometers. More recently, better sensitivity and the availability of high
resolution solids capability on newer spectrometers have further increased
the scope of ^{29}Si NMR.

ANNUAL REPORTS ON NMR SPECTROSCOPY
VOLUME 15   ISBN 0-12-505315-0

The interest in this field is multifaceted. The importance of silicon-containing materials to the electronics and chemical industries in the form of substrates, catalysts, adhesives, rubbers, fillers, additives and consumer products has necessitated the development of better characterization methods. For theoretical chemists and physicists, ^{29}Si chemical shifts, couplings and relaxation data provide information about bonding, molecular interactions and motions.

Finally, since silicon lies immediately below the well-understood carbon and at the beginning of the heavier elements in the periodic table, understanding the factors involved in determining its chemical shifts and couplings will play a significant role in the development of a unified theory. As a result of this interest, by 1979 several comprehensive reviews had been published.[5-8] Since there are extensive compilations of chemical shift data available,[8,9] there will be no attempt to repeat that here. It is the purpose of this chapter to survey the literature since 1978 and to present the more recently obtained results in ^{29}Si NMR spectroscopy.

II. PRACTICAL CONSIDERATIONS

^{29}Si is the only naturally occurring isotope of silicon with non-zero spin ($I = 1/2$). Its low natural abundance (4.7%) and low NMR sensitivity (7.84×10^{-3}) at constant field relative to ^{1}H places it with ^{13}C in the group of "rare" spin nuclei. Fortunately the recent use of FT NMR spectrometers eliminates the problem of low sensitivity. The theory of FT NMR has been described in detail[10] and is not included in this discussion. Further difficulties arise from the long spin–lattice relaxation times (T_1) frequently associated with ^{29}Si and its negative magnetogyric ratio (γ). The latter means that under proton decoupling conditions the nuclear Overhauser effect (NOE) is negative, producing reduced signal intensities, null or negative signals. The long T_1 values (> 20 s) necessitate long delays between pulses in an FT experiment. Both of these problems can be overcome with the addition of a small amount (0.01 M) of a paramagnetic relaxation reagent[11] such as the transition metal complex Cr(acac)$_3$. This serves to replace other relaxation mechanisms with the highly efficient electron–nuclear dipole–dipole interaction which reduces T_1 to a few seconds. Since the proton–nuclear dipole–dipole contribution to spin–lattice relaxation is eliminated, the NOE is removed as well. Short T_1 values and the absence of any NOE produces an absorption spectrum obtained with the same relative ease as a ^{13}C spectrum. For cases in which it is undesirable to add a relaxation reagent, pulse-modulated decoupling ("gated decoupling")[12] may be used to eliminate the NOE. This does not eliminate the problem of long T_1 values, however, and long recovery times may be needed in an FT experiment.

Other techniques have been used for sensitivity enhancement. Selective population transfer (SPT) for ^{29}Si[13,14] takes advantage of Si–H coupling. It involves the application of a weak irradiating field, B_2 to a single ^{1}H transition for a time τ. Under the condition $\gamma_H(B_2)\tau = \pi$, where γ_H is the gyromagnetic ratio of ^{1}H and $\tau \ll T_1$, the line intensities in ^{29}Si spectra can be affected by a factor of $|\gamma_H/\gamma_{Si}| = 5$, where γ_{Si} is the gyromagnetic ratio of ^{29}Si. This technique provides coupling information in addition to sensitivity improvement. Cross-polarization in liquids (J cross-polarization, JCP)[15] has been used for silanes[16] to measure Si–H couplings. This also takes advantage of scalar coupling and can effect enhancements of up to five (γ_H/γ_{Si}). Substantial sensitivity improvements have been achieved by Doddrell $et\ al.$[17] utilizing Freeman and Morris's nonselective polarization transfer pulse sequence (insensitive nuclei enhanced by polarization transfer, or "INEPT"[18]). Very recently Helmer and West have extended this to a whole series of silanes[19] for which enhancement factors (E_d) from 3 to > 9 have been achieved. The only requirement for using this technique is that there be sufficiently strong coupling between Si and H. Several representative examples are collected in Table 1.

Substantial advances in high resolution NMR of solids, specifically the combination of cross-polarization (CP)[184] and magic angle sample spinning[185,186] (MASS), have led to many applications of ^{29}Si CP/MASS within the last four years (Section III.K). For a detailed discussion of these techniques see reference 187 and references therein.

The present known ^{29}Si chemical shift range is > 550 ppm, although most silicon nuclei appear within a 250 ppm region. As in ^{13}C NMR, the tetraiodo

TABLE 1

Enhancement of ^{29}Si resonances in various silicon compounds by proton polarization transfer.[19]

	E_d	δ (ppm)	J (Hz)
Me$_4$Si	9.3	0	6.6
Me$_3$SiCl	7.5	29.87	6.8
t-BuSiCl$_3$	3.6	17.75	10.4
Ph$_2$SiCl$_2$	3.2	6.47	6.6
PhSiH$_3$	5.7	-59.75	199.9
			6.4 (*ortho*), 1.1 (*meta*)
(MeO)$_4$Si	5.5	-78.25	3.7
(Me$_3$\underline{Si}OSiMe$_2$)$_2$O	6.5	7.17	6.7
(Me$_3$SiOSi\underline{Me}$_2$)$_2$O	6.2	-21.66	7.3
(Me$_3$Si)$_2$\underline{Si}C$_2$	8.9	-19.13	7.1

derivative SiI$_4$ is the most shielded resonance[20] found to date, while the high frequency (low field) end of the scale is presently defined by the three-membered metallocycle bis(tetracarbonyliron)dimethylsilane ($\delta_{Si} = 173$ ppm; determined by the INDOR technique).[21] Since the number of ^{29}Si chemical shifts reported to date includes more than 2000 compounds, it is becoming increasingly important to define the ^{29}Si chemical shift scale relative to an accepted reference standard. The convention for selecting a standard reference compound for most nuclei has been to use a compound that has a single resonance line, has good solubility in most solvents yet remains unreactive, and appears at the low frequency end of the spectrum so that all other resonance lines can be reported in ppm as positive numbers to the high frequency side of the standard.[22] Most researchers currently use Me$_4$Si (TMS) as a reference compound for ^{29}Si as well as for ^{13}C and ^{1}H. It has the advantage of reasonable solubility in most organic solvents, low solvent shifts,[23] stability in most chemical systems and a relatively short T_1 at ambient temperatures.[2,24] Its volatility can either be an advantage for removing it from the solution after use or a disadvantage if high temperature studies are desired. A major disadvantage is the occurrence of its resonance at relatively high frequency, which means that most chemical shifts are negative. It also occurs in the middle of the shift range for a large number of organosilanes. This can be problematic in making chemical shift assignments, particularly in systems in which structures of the type $(CH_3)_3SiCH_2X$ are involved (where X is virtually any substituent).

A number of other reference compounds have been suggested including polydimethylsiloxane,[1] tetramethoxysilane ($(MeO)_4Si$),[25] tetraethoxysilane ($(EtO)_4Si$),[23] tetrafluorosilane (SiF_4)[3,26] and octamethylcyclotetrasiloxane ($(Me_2SiO)_4$).[23] Tetramethoxysilane is quite toxic. It is also chemically reactive, as are $(EtO)_4Si$ and SiF_4. Both $(EtO)_4Si$ and $(MeO)_4Si$ have long relaxation times.[28,24] Very recently, tetrakis(trimethylsilyl)methane has been suggested as a good reference compound for ^{1}H, ^{13}C and ^{29}Si NMR ($\delta_{Si} = 3.06$ ppm, CHCl$_3$).[29] In studies of aqueous silicate solutions, the monosilicate anion has been used as an internal reference.[30] Obviously the best reference will depend on the application. Harris and Kimber have suggested the avoidance of reference compounds altogether through the indirect calculation of ^{29}Si chemical shifts by the measurement of absolute resonance frequencies.[31] In spite of all suggestions, TMS is still the most widely accepted internal and external standard. All chemical shifts in this chapter are reported in ppm relative to TMS with positive values to high frequency (low field) and negative values to low frequency (high field). Figure 1 shows the range of silicon chemical shifts and some representative types of compounds. Possible reference compounds and their chemical shifts are also indicated.

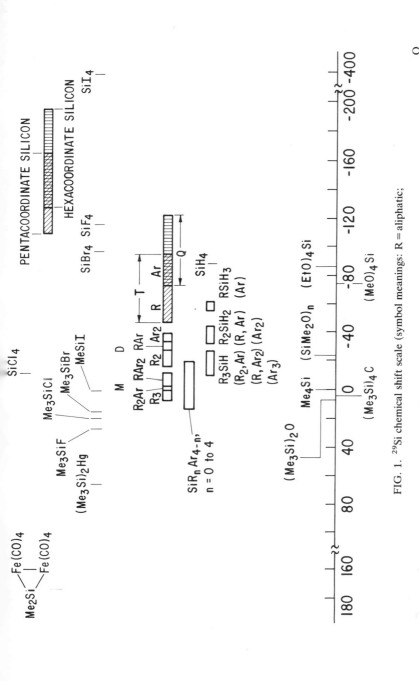

FIG. 1. ^{29}Si chemical shift scale (symbol meanings: R = aliphatic;

M = monofunctional —Si—O;   D = difunctional O—Si—O;   T = trifunctional O—Si—O;   Q = tetrafunctional O—Si—O;

Me = CH₃). Reference compounds are shown immediately above and below the scale.

III. CHEMICAL SHIFTS

A. Theoretical background

The magnetic shielding of a nucleus is a function of the electron distribution within the molecule and can be described as a sum of local and long-range effects:[32]

$$\sigma = \sigma_d + \sigma_p + \sigma' \tag{1}$$

The local contribution represents the electron density and motion of the electrons around the nucleus in question and is comprised of a diamagnetic term σ_d and a paramagnetic term σ_p. The long-range effects (σ') include contributions such as magnetic anisotropy and electric field effects from other atoms in the molecule. It is usually assumed that for ^{29}Si local effects are dominant. The diamagnetic term is given by Lamb's formula:[33]

$$\sigma_d = \frac{\mu_0 e^2}{12\pi m} \sum_i P_{ii} \langle r_i^{-1} \rangle \tag{2}$$

where P_{ii} is the charge density in atomic orbital i at an average distance of r_i from the nucleus. The paramagnetic term, according to the Jameson–Gutowsky formulation, is[34]

$$\sigma_p = -\frac{\mu_0 e^2 \hbar^2}{6\pi m^2 \Delta E} [\langle r^{-3} \rangle_p P_u + \langle r^{-3} \rangle_d D_u] \tag{3}$$

where ΔE is a mean excitation energy, $\langle r^{-3} \rangle_p$ and $\langle r^{-3} \rangle_d$ are the mean inverse cubes of the distances of the valence p and d electrons from the nucleus, and P_u and D_u represent the p and d electron imbalance about the nucleus. These last terms include elements of the charge density and bond order matrix.

Most semi-empirical calculations of ^{29}Si chemical shifts consider variations only in the paramagnetic term. Ambiguities arise because of uncertainties regarding the magnitude of substituent effects on the ΔE, $\langle r^{-3} \rangle$ and P_u terms. It is also not clear if the d-orbital term may be ignored, as is frequently done, or whether (p–d) π bonding is important. Using this approach and assuming an average excitation energy, ΔE, Engelhardt et al.[35] calculated a paramagnetic term relative to a hypothetical situation in which electrons are evenly distributed between the bonds. Electronegativity differences between silicon and the ligand are used as a measure of bond polarities and tetrahedral bond angles on silicon are assumed. They are able to predict successfully the trends in chemical shifts, although an empirical correction factor is required. Similar results involving CNDO/2 calculations were obtained by Roelandt, van der Vondel and van den Berghe.[36] Further extensions of the work by Radeglia and coworkers[37–43] include a recent study of substituted disilanes,[43] considerations of shielding contributions by heavier elements (to account for the deviations from calculated values for molecules containing bromine and iodine),[40] corrections for departure from tetrahedral bond angles,[37] use of CNDO/2 calculations instead of electronegativity arguments in determining the electron distribution about silicon[38] and a comparison of CNDO/2 calculations with and without the assumption of a constant value of ΔE.[39] In the last study it was concluded that in compounds of general structure $Me_{4-n}SiX_n$ (X = electronegative substituent) the averaged excitation energies cannot be considered to be constant. Lyubimov and Ionov[44] previously demonstrated that variations in ΔE are sufficient to account for deviations in additivity in the series of silanes F_nSiCl_{4-n} and F_nSiBr_{4-n}. In summary, it is known in a qualitative sense that substituent electronegativities and the coordination number at silicon[45] have a major effect on silicon chemical shifts. Other factors which must be considered are steric interactions, variations from tetrahedral bond angles, the possibility of (p–d) π bonding and magnetic anisotropy of neighbouring groups. For a comprehensive and quantitative model, some of the simplifying assumptions (constant ΔE, neglect of d-orbitals) which have been used up to the present may have to be revised.

Empirical approaches to predicting silicon shielding have included relating the chemical shifts to the sum of the substituent electronegativities to give a U-shaped relation,[48] and the use of substituent parameters. Both methods must be used with caution, however, since large deviations can occur. The substituent effect of Cl, for example, varies greatly depending upon the other substituents bonded to the silicon atom. Substituent parameters are discussed under the appropriate compound types.

Some success has been obtained with the use of pairwise interaction parameters.[27] Deviations are significant for fluorine or substituents with oxygen bonded directly to silicon.

B. Substituted methyl and aryl silanes

A large portion of the literature on ^{29}Si chemical shifts concerns substituted methyl silanes, $Me_{4-n}SiX_n$ and $Me_{3-n}X_nSi(CH_2)_mY$. The simplest case is that of the monosubstituted trimethylsilyl derivatives, Me_3SiX. In an early study Hunter and Reeves[2] noted that electronegative substituents produced a deshielding of the silicon nucleus which was greatest for the most electronegative substituents. Later work[31,46] on 14 trimethylsilyl compounds shows that although a general trend may be seen, the correlation between δ_{Si} and the electronegativity of the substituent, χ_X is quite poor. Likewise, no correlation between the ^{29}Si and ^{13}C chemical shifts in the series is found. If the structural changes are more removed from the silicon atom, however, other factors are eliminated and the silicon shielding is governed by the electronic effect of the substituent. McFarlane and Seaby[47] found a good linear correlation between the pK_a of the acid RCO_2H and the silicon chemical shift of substituted methylsilylcarboxylates $Me_{4-n}Si(O_2CR)_n$. This is shown graphically for $n = 1$ in Fig. 2. The *magnitude* of this dependence on the electronic effect of R depends upon the nature of the other substituents on silicon, however, and the slope of the lines for $n = 1, 2$ and 3 decreases from 1.0 to 0.88 and 0.42 respectively. This is typical for silicon shieldings and is most dramatically shown for the aryl silanes[48] $p\text{-}XC_6H_5SiY_3$, where the correlation between the silicon chemical shift and Hammett σ constants actually reverses slope from Y = F, OEt and Cl to Y = H and CH_3. (Fig. 3).

In other studies, an oxygen atom between the phenyl ring and the silicon atom has been shown to double the substituent effect in aryl substituted phenoxytrimethyl silanes compared with phenyltrimethyl silanes, even though the substituent is one bond further removed from silicon.[49,50] Similar results have been previously found in a qualitative sense for the same

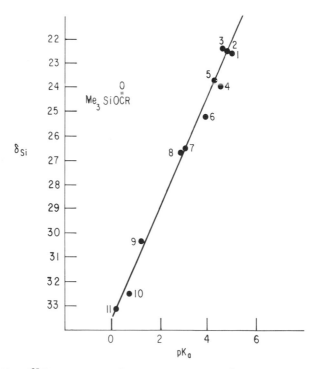

FIG. 2.   Plot of ^{29}Si chemical shifts (in ppm relative to Me_4Si) of the methylsilyl carboxylates Me_3SiO_2CR versus the pK_a of the corresponding acid RCO_2H,[47] where R = (1) $t\text{-}Bu$, (2) Me, (3) CH=CHMe, (4) CH_2Ph, (5) Ph, (6) H, (7) CH_2Br, (8) CH_2Cl, (9) $CHCl_2$, (10) CCl_3, and (11) CF_3.

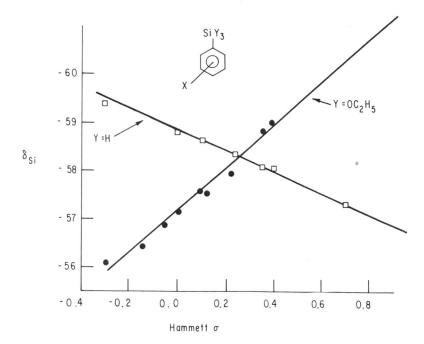

FIG. 3. Hammett plots of ^{29}Si chemical shifts (in ppm from TMS) of arylsilanes.

compounds,[25] for the two series of trimethyl silanes Me_3SiOR and Me_3SiR,[51] and in some organosilicon derivatives of salicylic acid.[86]

In a recent study,[52] Schraml et al. have confirmed this enhanced sensitivity to substituent effects for the silicon chemical shift in phenoxysilanes relative to phenylsilanes. Since the effects of substituents X are transmitted equally well within $XC_6H_4—$ fragments in both systems (as shown by ^{13}C chemical shifts and CNDO/2 net atomic charges), the authors attribute this to a higher sensitivity of the silicon shielding to electron density changes in the phenoxysilanes rather than to a better transmission of electronic effects in these compounds.

The greatest effect on the silicon shielding is, of course, caused by substitution directly on silicon. Multiple α substitution at silicon by electro-negative substituents causes a sagging pattern, as shown in Fig. 4 for $X = OMe$ and Cl. In contrast, the other substituents show unidirectional changes, although the substituent effect is not necessarily a constant value and may be either shielding or deshielding (see $X = Et$ and $X = H$ in Fig. 4).

In some recent studies[53,54] the ^1H, ^{13}C and ^{29}Si NMR data for methylallyl silanes and methylvinyl silanes were examined. In the allyl-substituted compounds $Me_{4-n}Si(CH_2CH=CH_2)_n$ the very slight shielding observed for silicon from $n = 1$ to $n = 4$ (+0.08 to -1.98 ppm) is attributed to the

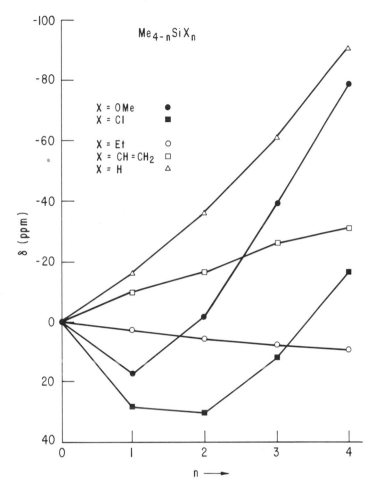

FIG. 4. ^{29}Si chemical shifts (in ppm relative to Me$_4$Si) for the methylsilanes Me$_{4-n}$SiX$_n$. Data are from reference 6 and references therein.

anisotropy effect of the π system.[53] In the case of the vinyl silanes, Me$_{4-n}$Si(CH=CH$_2$)$_n$ the silicon becomes progressively more shielded by about 7 ppm per vinyl substituent. This is ascribed to a (d, σ^*–π) hyperconjugation in accordance with theoretical discussions.[55]

A series of methylpropenyl derivatives of silicon, tin and lead, Me$_n$M(C$_3$H$_5$)$_{4n}$, has been examined.[56] In these systems approximate chemical shift additivity is observed and a regression analysis gives the values -12.4, -8.5 and -4.8 ppm respectively for the *cis, trans* and isopropenyl substituent effects at silicon. The differences between estimated and observed chemical shifts are attributed to steric effects since the greatest deviations occur in those compounds where steric crowding would be expected, such as in $Si(cis\text{-}C_3H_5)_4$ (-49.6 calc., -51.2 exp.) for example. The longer Sn—C and Pb—C bond lengths preclude steric interactions in these compounds and the agreement between the estimated and observed shifts is much better (deviation $<5\%$).

The effect of changing a remote substituent on the trends observed for silicon shielding has been the subject of many papers.[57–69] A series of derivatives of general structure $Me_{3-n}X_nSi(CH_2)_mY$ was examined and in most cases compared with the corresponding methyl derivatives, Me_4SiX_n. In all cases of β substitution ($m = 1$) the general trend with increasing substitution of X is maintained. This is not surprising since the changes at the more remote β position are not expected to dominate the silicon shielding. Only very slight changes in the shielding trends for the same X are observed upon changing substituent $(CH_2)_mY$ from methyl to allyl to vinyl ($Y = H$, $m = 1$; $Y = CH{=}CH_2$, $m = 1$; $Y = CH{=}CH_2$, $m = 0$, respectively).[60] Likewise, very little change is observed for the sequence $(CH_2)_mY = $ methyl, benzyl and phenyl.[60] If Y is an electronegative element, however, the β substituent effect is in general found to be greater ($m = 1$). As n increases from 1 to 3 the difference in ^{29}Si chemical shift between the compound $Me_{3-n}X_nSiCH_2Y$ and the model compound $Me_{3-n}X_nSiCH_3$ (defined as $\Delta\delta$) also increases.[57,60]

C. Perhalosilanes

Most of the data on halosilanes concerns methyl silanes. Unlike the methylhalosilanes, the mixed silicon halides do not show the "sagging pattern" characteristic of sequential α substitution by an electronegative element. The data for some of these halosilanes are shown graphically in Fig. 5. The anomalously strong shielding effect of iodine is similar to that observed for the analogous carbon derivatives and is not well understood.[71]

The data for some chloro and fluoro polysilanes have been reported,[3,70,72,73] as well as information for a series of perchlorosilanes.[73] Table 2 shows the data for the perchlorosilanes. The terminology

$$M = {-}SiCl_3 \text{ (monosubstituted)} \qquad D = {>}SiCl_2 \text{ (disubstituted)}$$

$$T = {\equiv}SiCl \text{ (trisubstituted) and } Q = {-}\overset{\displaystyle |}{\underset{\displaystyle |}{Si}}{-} \text{ (quaternary)}$$

is used in this discussion. Some trends may be deduced from Table 2. The T [7, B] and Q [6, C], [8, B] silicon atoms appear in distinctly unique regions of the spectrum which are characteristic of these functionalities. A progressively larger deshielding effect is apparent for M silicon atoms as a β chlorine is replaced by successive $SiCl_3$ substituents $[2] \rightarrow [3, A] \rightarrow [7, A] \rightarrow [8, A]$,

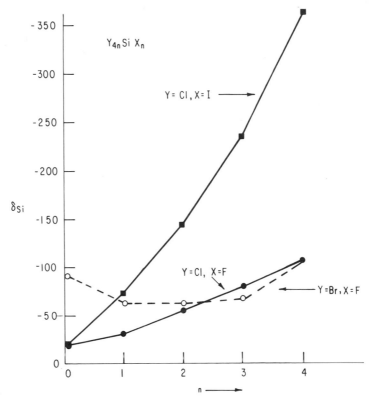

FIG. 5. Chemical shifts trends for some mixed halosilanes $Y_{4-n}SiX_n$.[26,68] Shifts are in ppm relative to Me_4Si.

TABLE 2

^{29}Si chemical shifts for oligomeric perchlorosilanes.[73]

	$\delta_{Si}{}^a$			
Compound	A	B	C	D
[1] $SiCl_4$	−18.45			
[2] $Cl_3SiSiCl_3$	−6.13			
[3] $(Cl_3Si^A)_2Si^BCl_2$	−3.70	−7.44		
[4] $(Cl_3Si^ASi^BCl_2)_2$	−3.91	−4.15		
[5] $(Cl_3Si^ASi^BCl_2)_2Si^CCl_2$	−3.82	−3.82	−1.15	
[6] $Cl_3Si^ASi^BCl_2Si^C(Si^DCl_3)_3$	−3.50	+6.27	−79.48	+4.13
[7] $(Cl_3Si^A)_3Si^BCl$	0.0	−31.83		
[8] $(Cl_3Si^A)_4Si^B$	+3.95	−80.41		

^{a} In ppm relative to Me_4Si.

$\Delta\delta = 2.4$, 3.7, 4.0 ppm, respectively). The chemical shift of [6, D] is very close to that of [8, A] (both are attached to O silicon atoms) even though [6, D] has an additional substituent in the γ position. The terminal $SiCl_3$ silicon (M) changes only slightly as the chain length increases in the linear species [3, A] → [4, A] → [5, A]. It is interesting to note that in the n-Si_5Cl_{12} the chemical shifts of the terminal and second silicon atom in the chain are identical.

D. Silicon hydrides, alkylsilanes, silaethylenes and disilenes

Most silicon hydrides appear at relatively low frequency compared with their alkyl analogs. A few representative compounds are collected in Table 3. Although, to a very rough approximation, the silicon trihydrides ($RSiH_3$) contain the most shielded silicon atoms (-30 to -70 ppm), followed by the dihydrides (R_2SiH_2, -20 to -40 ppm) and monohydrides (R_3SiH, 0 to -30 ppm), there is considerable overlap among the three types of compounds and the nature of the other substituents obviously plays a major role. The chemical shifts of $(MeO)_3SiH$ (-54.9 ppm)[75] and $PhCH_2SiH_3$ (-56.0 ppm[76]) are quite similar, for example. Likewise the two monohydrides, triethylsilane (Et_3SiH) and tris(trimethylsilyl)silane [$(Me_3Si)_3Si^*H$] have dramatically different chemical shifts of $+0.15$ ppm[77] and -117.40 ppm,[6] respectively.

Harris and Kimber have examined a series of homologous silanes R_3SiH where R = Me, Et, n-Pr, n-Bu and n-hexyl.[77] The approach of Grant and Paul[78] is used to evaluate the additivity of alkyl substituent effects in these silanes. The silicon chemical shift is assumed to be represented by

$$Si = B + \sum_i A_i n_i \tag{4}$$

TABLE 3

^{29}Si chemical shifts for some silicon hydrides and the analogous methylsilanes.

Compound	δ_{Si} [a]	Ref.
SiH_4	-91.9	70
$SiMe_4$	0.0	
$HSiMe_3$	-16.34	31
H_3SiSiH_3	-104.8	74
$Me_3SiSiMe_3$	-19.78	31
$Me_3SiSi^*Me_2H$	-39.1	72
Ph_2SiH_2	-34.5	2
Ph_2SiMe_2	-8.19	75

[a] In ppm relative to Me_4Si.

where B is the chemical shift of silane, n_i is the number of carbon atoms at the ith position from silicon and A_i are constants. A series of parameters is derived from the chemical shifts of the silanes R_3SiH and $Me_{4-n}SiH_n$ ($n=0$ to 4). These include $B = -91.9$ ppm, $A_\alpha = +25.2$ ppm, $A_\beta = +5.50$ ppm (A_β includes effect of chain branching), $A_\gamma = -2.75$ ppm, $A_\delta = +0.51$ ppm and $4°(1°) = -2.2$ ppm. All of the values except A_α are very close to those calculated for carbon,[78] which suggests a common origin for the effects. It is not clear why the value for A_α is so different for silicon ($+25.2$ ppm as opposed to $+9.09$ ppm for carbon). Remarkably, the estimated value of B, which is supposed to represent the chemical shift of SiH_4, is identical to the experimental value.[70] In a recent investigation of 30 trialkylsilanes and trialkylbromosilanes the importance of the other substituents on silicon is re-emphasized with the observation that these β, γ and δ substituent parameters are different for Br versus H.[79]

The ^{29}Si chemical shifts for two new classes of compounds containing a silicon atom double bonded to carbon or silicon have been reported.[80,201,202] As would be expected by analogy with ^{13}C chemical shifts of alkenes, these silicon atoms are strongly deshielded relative to the saturated silicon compounds and have chemical shifts in the range 40–65 ppm. The data for the silaethylenes are collected in Table 4. The silicon atoms B, C and D which are not in the double bond have "normal" shifts when compared with model systems. Si^D, for example, appears at about 13 ppm compared with Me_3SiOEt which has a chemical shift of 13.5 ppm.[2] Likewise, Si^B and Si^C appear in the same region as the analogous silicon atoms in $(Me_3Si^*)_2 SiMe_2$ ($\delta_{Si} = -15.9$ ppm[25]). The recently reported tetramesityldisilene [8, A] has a ^{29}Si chemical shift of 63.6 ppm.[202] More examples of these types of

TABLE 4

^{29}Si chemical shifts of some silaethylenes.[80]

$$Me_3Si^B \diagdown \qquad \diagup OSi^DMe_3$$
$$Si^A = C$$
$$Me_3Si^C \diagup \qquad \diagdown R$$

		δ_{Si} [a]	
R	A	B, C	D
CMe_3	41.44	$-12.18, -12.64$	13.34
CEt_3	54.26	$-11.79, -12.19$	12.74
Adamantyl	41.80	$-12.31, -12.62$	13.39

[a] In ppm relative to Me_4Si.

$$\left(Me - \underset{Me}{\overset{Me}{\bigodot}} \right)_2 Si = Si \left(\underset{Me}{\overset{Me}{\bigodot}} - Me \right)_2$$

[8, A]

compounds are needed before specific trends with substitution can be established.

E. Cyclic and polysilanes

A number of ^{29}Si chemical shifts are now available for cyclic silanes and disilanes.[6,25,80–85,91] From the data in Tables 5 and 6, some trends of chemical shift with ring size are apparent. In the monosilanes incorporating the silicon atom in a smaller ring causes a strong deshielding, of 14–20 ppm, for a five-membered ring relative to the six-membered ring (compare [9] and [10], [11] and [12], [17] and [18]). Another 2 ppm deshielding occurs when the silicon is incorporated in a four-membered ring as in [13]. The effect is much smaller for the disilyl compounds in Table 6.[19–24] A deshielding of 10 ppm is found for the four-membered disilane [22] relative to the six-membered [19]–[21],[84] and only 8 ppm separates [23] and [24]. The cyclic silanes [25] and [26] show even less change in chemical shift (only 3 ppm), although a six- and a four-membered ring are again being compared. Finally, the polysilanes [27]–[31] show a high frequency shift only for the four-membered homolog [31], whereas the five- to eight-membered rings all have silicon resonances within 3.3 ppm of each other. A similar effect for four-membered rings relative to the five- and 6-membered compounds is noted for perchloro and perbromo cyclic silanes.[83]

In the disilacyclohexene, [19], it is noted that the chemical shift more closely resembles divinyldisilane (-24.8 ppm) than diallyldisilane (-18.1 ppm).[84] Based upon a comparison of their studies with those previously published for 1-silacyclopent-3-ene the authors conclude that the ring strain in these systems is very weak, and a double "β" effect arises from $\sigma(CH_2-Si)-\pi(C=C)$ conjugation, which is a result of the nonplanarity of the rings.

In all the examples discussed above, the chemical shift decreases with smaller ring size. The silacyclopropanes and silacyclopropenes form a unique class of compounds with unusually shielded silicon resonances.[87–90] The tetraalkylsilacyclopropane resonances occur between -50 and -60 ppm from Me$_4$Si (see [14] in Table 5 for example), in marked contrast to normal cyclic and alicyclic tetraalkylsilanes which usually appear between $+5$ and -20 ppm. A very low frequency silicon resonance is noted[88] for the silacyclopropene [32]. The possibility of a ring current has been considered to

E. A. WILLIAMS

TABLE 5

[29]Si chemical shifts of cyclic silanes with one silicon atom incorporated in the ring.

	Compound	δ_{Si}	Ref.
[9]	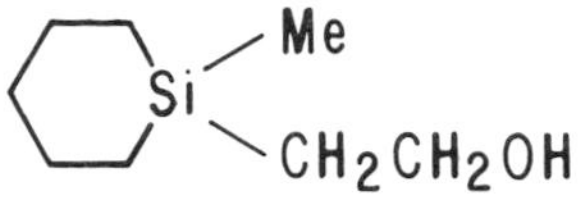 Si / Me \ CH$_2$CH$_2$OH	−4.11	25
[10]	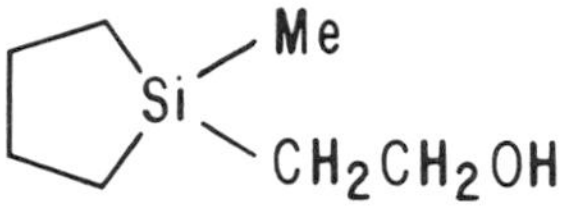 Si / Me \ CH$_2$CH$_2$OH	16.33	25
[11]	 SiMe$_2$	−1.2	84
[12]	 SiMe$_2$	16.77	25
[13]	 SiMe$_2$	18.90	25
[14]	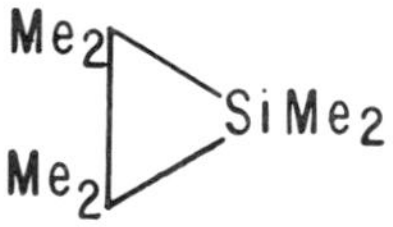 Me$_2$ Me$_2$ SiMe$_2$	−49.31	89
[15]	 SiMe$_2$	16.5	85
[16]	 SiMe$_2$	17.4	85
[17]	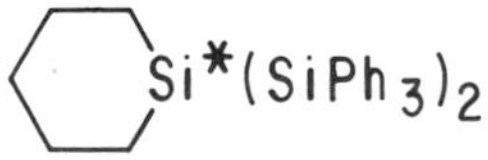 Si*(SiPh$_3$)$_2$	−48.6	82
[18]	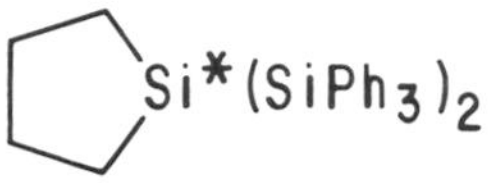 Si*(SiPh$_3$)$_2$	−34.8	82

[a] In ppm relative to Me$_4$Si. Data from reference 25 converted using δ_{Si} (MeO)$_4$Si = −78.5 ppm.

TABLE 6

^{29}Si chemical shifts of cyclic polysilanes.[a]

Compound	δ_{Si}	Ref.
[19] (ring with two SiMe$_2$)	−22.3	84
[20] (Me-substituted ring with two SiMe$_2$)	−22.1, −22.0	84
[21] (diMe-substituted ring with two SiMe$_2$)	−22.1	84
[22] (Me$_3$Si, Me$_3$Si four-membered ring with two SiMe$_2$)	−12.5	84
[23] (six-membered ring with two SiMe$_2$)	−19.5	84
[24] (Me$_2$, Me$_2$ ring with two SiMe$_2$)	−11.6	84
[25] (Me$_2$Si, Me$_2$Si, SiMe$_2$ ring)	0.3	81
[26] (SiMe$_2$, Me$_2$Si ring)	2.73	25
[27] Si_8Me_{16}	−39.9	91
[28] Si_7Me_{14}	−41.7	91
[29] Si_6Me_{12}	−41.81	83
[30] Si_5Me_{10}	−42.12	83
[31] Si_4Me_8	−27.56	83

[a] In ppm relative to Me$_4$Si. Data from reference 25 converted using δ_{Si} (MeO)$_4$Si $= -78.5$ ppm. Data from reference 81 converted using δ_{Si} (Me$_2$SiO)$_4 = -19.51$ ppm.

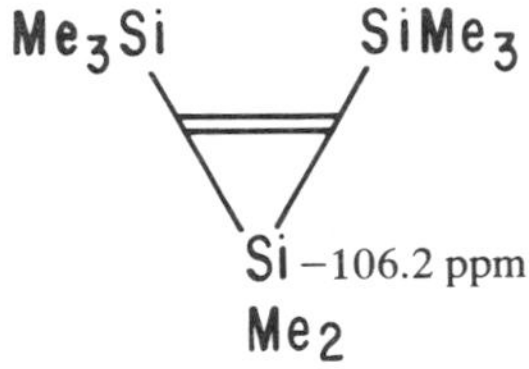

$$Me_3Si \qquad SiMe_3$$

$$Si \; -106.2 \; ppm$$

$$Me_2$$

[32]

explain the unusual shielding in the ^{13}C and ^1H resonances for cyclopropane[92] and might also be contributing to the chemical shifts in the silacyclopropanes. The extreme shielding in [32] may be due in part to π bonding with involvement of the silicon d-orbitals.[88]

Information is now available on a variety of polysilanes.[3,25,43,70,72,82,91,93–96] Data for linear and branched permethylpolysilanes are collected in Table 7. Stanislawski and West have used the approach of Grant and Paul[78] to relate the ^{29}Si chemical shifts in the linear permethylpolysilanes to substituent parameters. Equation (5) describes the chemical shift of the kth silicon nucleus,

$$\delta_{Si(k)} = B + \sum_l A_l n_{kl} \qquad (5)$$

in terms of an additive chemical shift parameter for the lth atom, A_l, times the number of silicon atoms (n_{kl}) in the lth position relative to the kth silicon atom. B is a constant (8.5 ppm) and the parameters which arise out of a regression analysis of 11 permethylpolysilanes are $\alpha = -28.5$, $\beta = +3.9$, $\gamma = +1.2$ and $\delta = +0.2$ ppm.[91] It is apparent that additional correction factors will have to be included for branched and cyclic polysilanes. The good correlation found for the linear permethylpolysilanes does not extend to the cyclic silanes.[91] Likewise, the branched polysilanes have chemical shifts which deviate considerably from values calculated using equation (5); deviations are greatest for the silicon atoms with the most branching. As with the cyclic silanes, the observed chemical shifts are at lower frequency than the calculated values. The range of chemical shifts for the different types of silicon atoms is:

$$Me_3Si-, \; -9 \; to \; -20 \; ppm \qquad Me_2Si\lessgtr, \; -26 \; to \; -50 \; ppm$$

$$MeSi\lessgtr, \; -79 \; to \; -81 \; ppm \qquad -\overset{|}{\underset{|}{Si}}-, \; -118 \; to \; -136 \; ppm$$

An empirical relationship between the silicon chemical shift and the number of neighbouring silicon atoms in α, β or γ positions has been

TABLE 7

^{29}Si chemical shifts for permethylated polysilanes.[93] ^{a}

Compound	δ_{Si}					
	A	B	C	D	E	F
Me$_3$Si^A—SiMe$_3$	−19.7					
(Me$_3$Si^A)$_2$—Si^{B}Me$_2$	−16.1	−48.6				
(Me$_3$Si^A—Si^{B}Me$_2$)$_2$	−15.2	−44.9				
(Me$_3$Si^A—Si^{B}Me$_2$)$_2$—Si^{C}Me$_2$	−15.1	−43.6	−40.9			
(Me$_3$Si^A—Si^{B}Me$_2$—Si^{C}Me$_2$)$_2$	−15.1	−43.4	−39.5			
(Me$_3$Si^A—Si^{B}Me$_2$—Si^{C}Me$_2$)$_2$Si^{D}Me$_2$	−15.0	−43.3	−39.3	−38.2		
(Me$_3$Si^A—Si^{B}Me$_2$—Si^{C}Me$_2$—Si^{D}Me$_2$)$_2$	−15.0	−43.3	−39.2	−37.9		
(Me$_3$Si^A—Si^{B}Me$_2$—Si^{C}Me$_2$— Si^{D}Me$_2$)$_2$Si^{E}Me$_2$	−15.0	−43.3	−39.1	−37.8	−37.4	
(Me$_3$Si^A—Si^{B}Me$_2$—Si^{C}Me$_2$— Si^{D}Me$_2$—Si^{E}Me$_2$)$_2$	−15.0	−43.3	−39.1	−37.8	−37.6	
(Me$_3$Si^A—Si^{B}Me$_2$—Si^{C}Me$_2$—Si^{D}Me$_2$ —Si^{E}Me$_2$)$_2$Si^{F}Me$_2$	−15.0	−43.2	−39.1	−37.8	−37.5	−37.5
(Me$_3$Si^A—Si^{B}Me$_2$—Si^{C}Me$_2$—Si^{D}Me$_2$ —Si^{E}Me$_2$—Si^{F}Me$_2$)$_2$	−15.0	−43.1	−39.0	−37.6	−37.4	−37.3
(Me$_3$Si^A)$_3$—Si^{B}Me	−12.5	−88.1				
(Me$_3$Si^A)$_4$Si^B	−9.8	−135.5				
(Me$_3$Si^A)$_2$MeSi^{B}SiMe(SiMe$_3$)$_2$	−11.7	−81.4				
(Me$_3$Si^A)$_3$Si^{D}Si^{C}Me$_2$Si^{B}Me$_3$	−9.5	−14.7	−39.8	−131.9		
(Me$_3$Si^A)$_3$Si^{D}Si^{C}Me(Si^{B}Me$_3$)$_2$	−9.2	−11.7	−79.8	−129.4		
(Me$_3$Si^A)$_3$Si^{E}Si^{C}Me$_2$Si^{D}Me(Si^{B}Me$_3$)$_2$	−9.5	−11.1	−29.2	−78.8	−126.0	
(Me$_3$Si^A)$_3$Si^{B}Si(SiMe$_3$)$_3$	−9.5	−130.0				
(Me$_3$Si^A)$_3$Si^{F}Si^{C}Me$_2$Si^{D}Me$_2$Si^E Me(Si^{B}Me$_3$)$_2$	−9.4	−11.7	−31.8	−33.2	−81.2	−128.6
(Me$_3$Si^A)$_3$Si^{C}Si^{B}Me$_2$Si(SiMe$_3$)$_3$	−9.2	−26.0	−118.2			
Me$_3$Si^A)$_3$Si^{C}Si^{B}Me$_2$SiMe$_2$Si(SiMe$_3$)$_3$	−9.2	−29.0	−126.5			
[(Me$_3$Si^A)$_3$Si^{D}Si^{B}Me$_2$Si^{C}Me$_2$]$_2$	−9.5	−31.4	−35.7	−128.0^b		

^{a} Data obtained in CDCl$_3$ and reported in ppm from Me$_4$Si.
^{b} In benzene.

developed for linear and branched silanes:[96]

$$\delta_{Si} = -96.02 - 2.43a - 3.73a^2 + B_a b + C_a \qquad (6)$$

where a is the number of silicon atoms in the α position and b is the number of silicon atoms in the β position, B_a and C_a are incremental chemical shifts produced by β and γ silicon atoms. The resonances observed for the polysilanes range from −90 to −103 ppm for SiH$_3$, −103 to −115 ppm for SiH$_2$, −126 to −136 ppm for SiH and −165 ppm for Si(SiH$_4$)$_4$.[96]

Several studies of polysilanes substituted with substituents of widely varying electronegativity have been made.[43,72,82,94,95] In the substituted disilanes $Me_3Si^ASi^BMe_2X$, a low frequency shift analogous to that observed for the trimethylsilyl compounds, Me_3SiX, is observed for the directly substituted silicon (Si^B) when X is an electronegative substituent. Similar results are obtained for the isotetrasilanes $(Y_3Si^A)_3Si^BX$, regardless of the nature of Y (Y = Me, OMe, Cl).[95]

The substituent effects observed for tris(trimethylsiloxy)silanes $(Me_3Si^AO)_3Si^BX$[97] are opposite of those observed for the pentamethyldisilanyl and trimethylsilyl compounds. In this case electronegative substituents (such as $-CCl_3$) at Si^B have a shielding effect relative to X = CH_3. As is the case for the pentamethyldisilanes and the isotetrasilanes, the substituent effects are opposite for Si^A and Si^B. Although both δ_{Si^A} and δ_{Si^B} give linear correlations with Taft σ^* constants, the slopes have different signs.[97] The effects of increasing chain length and chain branching in alkyl substituents are shown for $Ph_3Si^ASi^BMe_2R$,[82] $Me_3Si^ASi^BMe_2OR$[43] and $Me_2(OR)SiSiMe_2OR$[43] in Table 8. The shieldings observed for Si^B are consistent with the trends observed for the trialkylsilanes with increasing chain length.[77] The chemical shift of Si^A is quite insensitive to the changes of Si^B, virtually no chemical shift differences are observed for Si^A in the series [33], and only very slight changes in [34] and [35]. In the alkoxy derivatives [34] and [35] branching causes shifts to lower frequency (shielding effect), whereas the opposite is true when the alkyl substituent is attached directly to silicon as in [33].

A series of chloro- and fluoro-terminated linear permethylpolysilanes were examined by Stanislawski and West.[94] Chlorine and fluorine substituent parameters are obtained which gave good calculated values for all chemical shifts except for the dihalodisilanes. The values of the α, β and γ

TABLE 8

Effect of chain branching on ^{29}Si chemical shifts on some alkyl and alkoxy substituted silanes.[a]

R	$Ph_3Si^ASi^BMe_2R$[82] [33]		$Me_2(OR)SiSiMe_2OR$[43] [34]	$Me_3Si^ASi^BMe_2OR$[43] [35]	
	δ_{Si^A}	δ_{Si^B}	δ_{Si}	δ_{Si^A}	δ_{Si^B}
Me	−20.4	−18.4	+11.70	−22.41	+15.68
Et	—	—	+8.39	−22.49	+12.98
n-Pr	−20.4	−17.0	+9.10	−22.74	+12.80
n-Bu	—	—	+9.05	−22.44	+13.12
n-Oct	−20.4	−16.8	—	—	—
i-Pr	−20.6	−11.9	+6.56	−22.87	+10.09
t-Bu	−20.4	−8.0	−0.19	−22.67	+3.24

[a] In ppm relative to Me_4Si.

substituent constants for F (+53.0, −2.6 and −0.7 ppm, respectively) and Cl (+42.2, +2.5 and −0.3 ppm) demonstrate that the effect of halogen substitution is rapidly attenuated down the chain.

F. Silicon-containing transition metal complexes

One of the areas of ^{29}Si NMR that has grown rapidly as a result of improvements in technique and instrumentation is the study of silyl metal complexes.[3,14,21,98-107] Early work by Ebsworth et al.[98] and later work by Mitchell and Marsmann[99] demonstrate the marked shift to high frequency (30–60 ppm) which occurs in organosilyl mercurials. A similar deshielding effect had been previously noted for the transition metal derivative $F_3SiCo(CO)_4$ relative to F_3SiMe.[3] In the past three years several papers have appeared reporting chemical shifts of various types of silicon-containing transition metal complexes including metal carbene and carbyne complexes $(L_n^1L_m^2M\!=\!C\!\!\stackrel{Y}{\underset{SiR_3}{\leqslant}}$ and $L_n^1L_m^2M\!\equiv\!CSiR_3)$[103] and those with silicon directly bonded to the transition metal.[14,105] Several examples are collected in Table 9. In each case a silicon atom directly bonded to a transition metal [36]–[39], [42], Si^A in [40] and [41] Si^A and Si^B in [44], exhibits a strong shift to high frequency relative to the analogous methysilane (10–50 ppm). Silicon atoms β to the metal such as Si^B in [40] and [41] are affected to a much lesser degree and γ effects are very small. Si^C in [41], for example, has a chemical shift within 1 ppm of that for $(Me_3Si^*)_2SiMe_2$. The iron, ruthenium and rhodium π complexes of vinyl and dienylsilanes [43], [45] and [46] have silicon resonances shifted 6–14 ppm to high frequency from the uncomplexed compounds.

In the Group VI carbonyl organosilyl carbene and carbyne complexes large differences between the two types of complexes are not observed.[103] The observation that δ_{Si} goes through a minimum for M = Mo in the series Cr, Mo, W led the authors to propose that substantial π interactions between M and Si occur either by hyperconjugation or participation of the silicon d orbitals.

G. Organosilicones

For the purpose of this review the term "organosilicone" includes compounds with the general structure $(R_{3-n}SiO_n)_x$, where $n > 0$, $x > 2$ and R = alkyl, aryl or H. The spectral dispersion for organosilicones may be considerable, especially for the siloxanes $(SiRR'O)_n$. These materials comprise an important class of compounds used in resins, fluids, room temperature and heat-cured rubber industrial and consumer products. ^{29}Si NMR has played a major role in their structure determination because this large spectral dispersion allows many details of composition and microstructure to be examined. The nomenclature used to define siloxane compounds

TABLE 9

^{29}Si chemical shifts of some silicon-containing transition metal complexes.[a]

		δ_{Si}			
	Compound	A	B	C	Ref.
[36]	$Me_3SiRe(CO)_5$	-13.97^b			14
[37]	$Me_3SiMn(CO)_5$	17.95^b			14
[38]	$Me_3SiFe(CO)_2(C_5H_5)$	41.42^b			14
[39]	$Me_3SiCo(CO)_4$	44.42^b			14
[40]	$Me_3Si^BSi^AMe_2Fe(CO)_2(C_5H_5)$	16.95	-11.3		105
[41]	$Me_3Si^CSi^BMe_2Si^AMe_2Fe(CO)_2(C_5H_5)$	21.22	-36.54	-15.07	105
[42]	$Me_3SiFe(CO)(PPh_3)(C_5H_5)$	34.75			105

[43]

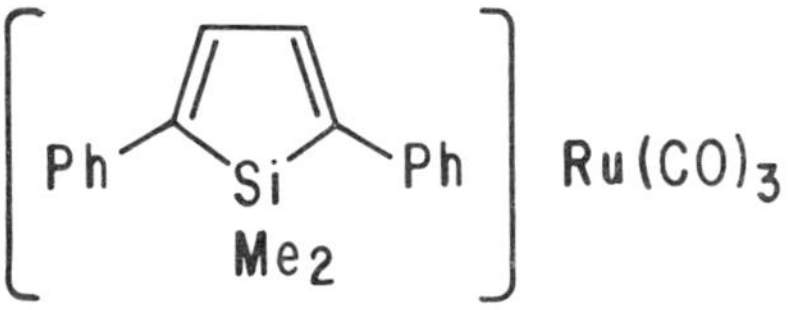

15.18 100

[44]

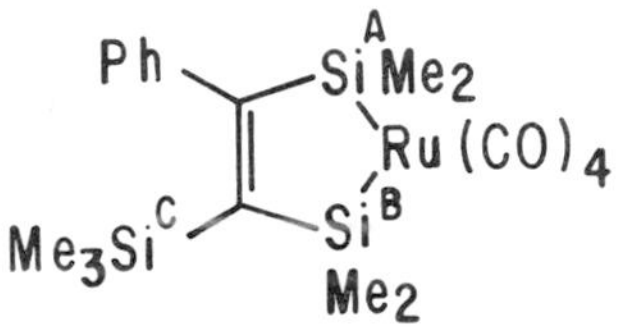

32.40 34.68 −7.92 100

| [45] | $(Me_3SiCH{=}CH_2)Fe(CO)_4$ | 4.43 | | | 100 |
| [46] | $AcacRh(CH_2{=}CHSiMe_3)_2$ | -0.62 | | | 101 |

[47]

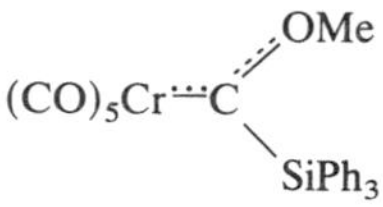

−24.2 103

[48]

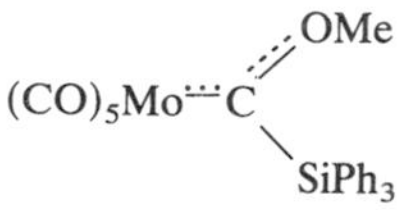

−30.3 103

[49]

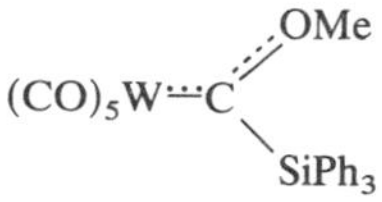

−18.3 103

| [50] | $(C_5H_5)(CO)_2W{\equiv}CSiPh_3$ | -26.7 | | | 103 |

[a] In ppm relative to Me_4Si.
[b] Data converted from reference 14 using δ_{Si} $(Me_3Si)_2O=6.97$ ppm.

incorporates the use of the letters M, D, T and Q, which represent $R_3SiO_{0.5}$, $R_2Si(O_{0.5})_2$, $RSi(O_{0.5})_3$ and $Si(O_{0.5})_4$ units respectively, where R represents aliphatic and/or aromatic substituents or H. In this discussion substituents other than methyl groups are indicated as superscripts; $D^{Ph} = PhMeSi(O_{0.5})_2$ and $M^{Ph_2} = (Ph_2MeSiO_{0.5})$, for example. Figure 6 shows a comparsion between the ^1H, ^{13}C and ^{29}Si NMR spectra of a polydimethylsiloxane. The ^{13}C and ^1H spectra of MD_3M provide much less structural information than the ^{29}Si spectrum in which not only are the M and D units separated by 28 ppm, but both D units also show unique resonances. Early ^{29}Si NMR results[5,108] reported on polydimethylsiloxanes show that individual resonance lines can be found for each unique silicon atom up to a 10-unit oligomer. The data for a series of linear and cyclic polydimethylsiloxanes are collected in Table 10. Both the cyclic trimer and tetramer, D_3 and D_4, have signals which are at higher frequency than the linear siloxanes in analogy to the ring-size effects noted for the cyclic silanes (Section III.E). A 10 ppm deshielding effect is observed on progressing from the cyclic tetramer $(Me_2SiO)_4$ to the homologous trimer $(Me_2SiO)_3$. The same occurs for S_n^H and D_n^{Ph} cyclic compounds. For example, $(MePhSiO)_4(D_4^{Ph})$ has a chemical shift centered around -30.25 ppm (multiplicity arises due to asymmetry) but the trimer is 10 ppm to high frequency at -20.66 and -20.69 ppm.[109]

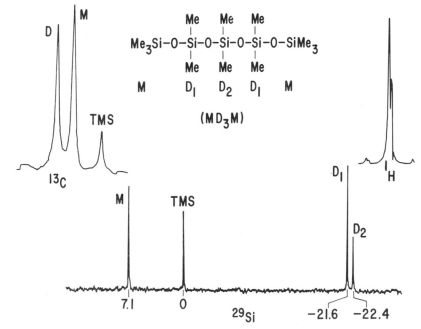

FIG. 6. Comparison of ^1H, ^{13}C and ^{29}Si NMR spectra of MD_3M.

TABLE 10

^{29}Si chemical shifts for some linear and cyclic polydimethylsiloxanes.[5]

	δ_{Si} [a]				
Compound	M	D^1	D^2	D^3	D^4
MM	6.79				
MDM	6.70	−21.5			
MD_2M	6.80	−22.0			
MD_3M	6.90	−21.8	−22.6		
MD_4M	7.0	−21.8	−23.4		
MD_5M	7.0	−21.8	−22.4	−22.3	
MD_6M	7.0	−21.8	−22.3	−22.2	
MD_7M	7.0	−21.89	−22.49	−22.33	−22.29
MD_8M	6.93	−21.86	−22.45	−22.30	−22.20
D_3 cyclic		−9.12			
D_4 cyclic		−19.51			
D_5 cyclic		−21.93			
D_6 cyclic		−22.48			

[a] In ppm relative to internal Me$_4$Si.

This appears to be a general effect for cyclic siloxanes since other studies[110,119] show the same results for a variety of mixed cyclic siloxanes in which a methyl group is replaced by several different organic substituents. Although the ^{29}Si chemical shift of $(H_2SiO)_3$ is not available, recent studies[111] show that the tetramer $(H_2SiO)_4$ (−46.78 ppm) appears at about 2 ppm below the pentamer (−48.42 ppm) and hexamer (−48.67 ppm), analogous to the magnitude and trends in the dimethylsiloxanes.

In addition to the cyclic siloxanes a number of linear siloxanes, other than the permethyl series, have been studied. Figure 7 shows the approximate chemical shift ranges for the different types of silicon resonances. The actual chemical shift is dependent upon the nature of the neighbouring groups as well as the length of the polymer chain.

Tacticity and neighbouring group effects in asymmetric and mixed siloxanes have been reported[112-116] by several workers. Harris and Kimber have examined the asymmetric siloxanes MD_n^HM. Using MD_5^HM as a model system, they assume only nearest-neighbour effects in explaining the observed spectrum. The D_1^H silicon has only one asymmetric neighbour, D_2^H, and therefore appears as a doublet (D_2^H either d or l). D_2^H and D_3^H, however, each have two asymmetric neighbours which produce the sequences $ddd = lll$, $ddl = ldd = lld = dll$ and $ldl = dld$. These sequences give rise to two 1 : 2 : 1 triplets observed at −35.52 and −35.22 ppm in the spectrum.

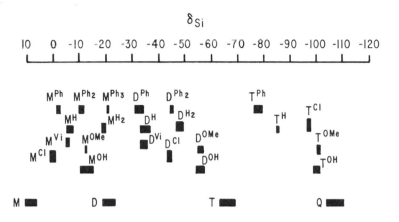

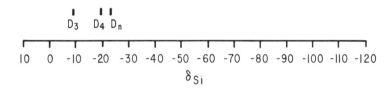

FIG. 7. ^{29}Si chemical shift ranges for some structural units in siloxanes.[7,111,116]

In a recent paper Engelhardt and Jancke have reported pentad structure in a methylphenylsilicone oil containing M, D and D^{Ph} units.[116] The spectrum shown in Fig. 8 consists of an expansion of the region for D units and shows the expected splittings for the different pentad sequences that are labeled in the structure.

A detailed study of end-group effects on oligomeric and polymeric siloxanes was undertaken by Harris and Robins.[117] They find the spectral dispersion to be greatest for the series M^{OH}D$_n$M^{OH} for which individual resonances for each unique silicon atom up to $n = 9$ are observed.

^{29}Si NMR has also been used to determine chain lengths in siloxane oligomers and block lengths in copolymers. Horn and Marsmann have demonstrated that chain terminal$^{(A)}$ and internal$^{(B)}$ silicon nuclei differ substantially in chemical shift in α,ω-dihydroxypolydimethylsiloxanes, HOSi^{A}Me$_2$(OSi^{B}Me$_2$)$_n$OSiMe$_2$OH.[120] This has been used to determine the average degree of polymerization ($\overline{DP}$) in a series of disiloxanols ($\bar{n} = 4$–80) used to study viscosity–molecular weight relationships:[121]

$$\overline{DP} = 2(B/A) + 2 \qquad (6)$$

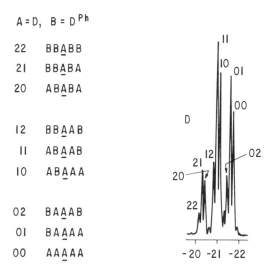

FIG. 8. D region of the ^{29}Si NMR spectrum of a mixed dimethyl and methylphenyl silicone oil.

where B and A are the integrated areas under the respective resonances. This has also been used for chain length determinations in chloroterminated siloxanes,[122] ClSi^{A}Me$_2$(OSi^{B}Me$_2$)$_n$OSiMe$_2$Cl ($\delta_{Si^A} = 3.7$ ppm, $\delta_{Si^B} = -18$ to -20 ppm), block length determinations in bisphenol-A-polycarbonate (BPAC)/polydimethylsiloxane (PDMS) block copolymers,[122] (BPAC)OSi^{A}Me$_2$(OSi^{B}Me$_2$)$_n$OSMe$_2$O(BPAC) ($\delta_{Si^A} = -13.5$ ppm, $\delta_{Si^B} = -20$ to -22 ppm), and poly(tetramethyl-p-silphenylene)siloxane (TMPS)/polydimethylsiloxane block copolymers[123] [51].

$$\left[\begin{array}{c} \text{Me} \\ | \\ \text{Si}^A \\ | \\ \text{Me} \end{array} \bigcirc \begin{array}{c} \text{Me} \\ | \\ \text{Si}^A - \text{O} \\ | \\ \text{Me} \end{array}\right]_m \left[\begin{array}{c} \text{Me} \\ | \\ \text{Si}^A \\ | \\ \text{Me} \end{array} \bigcirc \begin{array}{cc} \text{Me} & \text{Me} \\ | & | \\ \text{Si}^B - \text{O} - \text{Si}^C \\ | & | \\ \text{Me} & \text{Me} \end{array}\right] \left[\begin{array}{c} \text{Me} \\ | \\ \text{O Si}^D \\ | \\ \text{Me} \end{array}\right]_n$$

[51]

$\delta_{Si^A} = -1.11$ ppm
$\delta_{Si^B} = -2.45$ ppm
$\delta_{Si^C} = -20.65$ ppm
$\delta_{Si^D} = -21.86$ ppm

Although solvent effects in ^{29}Si NMR are usually small (<1 ppm) compared with the changes arising from structural perturbations, substantial solvent shifts are noted for the silanol silicon atoms in the hydroxy-terminated siloxanes as well as a number of other silanols.[124] The chemical shifts are found to give an excellent linear correlation with Gutmann's donor number (DN),[125] a measure of the electron pair donor ability of the solvent, and undoubtedly arises from hydrogen bonding of the type $-Si-O-H\cdots B$. The sensitivity of silicon chemical shifts to remote perturbations is evidenced by the fact that this effect is felt as far as the third silicon atom from the end of the chain in $HO(SiMe_2O)_nH$. This phenomenon provides a ready means of detecting silanol resonances in complex mixtures since a change of solvent from a poor donor ($CDCl_3$) to a good donor (DMSO) will not greatly affect other silicon atoms.[124] Solvent shifts for several silanols are collected in Table 11.

Although the discussion so far has considered only MD systems, ^{29}Si NMR has been found to be very useful in structural studies of MQ,[108,126–128,131] MT[108,129] and DT[114] systems. The chemical shifts for some MQ compounds are collected in Table 12. The greatest spectral dispersion occurs in the Q region, which spans -100 to -110 ppm.

Jancke et al. have used ^{29}Si NMR to characterize cyclic DT compounds.[114] Two possible isomers [52] and [53] exist for T_2D_3, and four structures for T_2D_4 [54]–[57]:

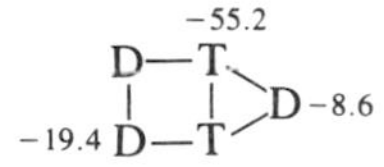

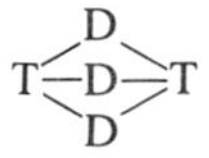

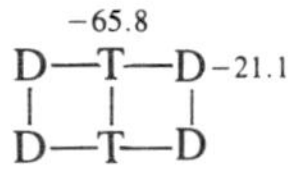

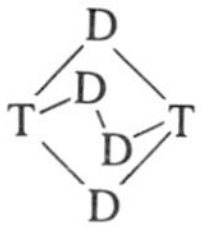

Structure [52] is identified by the existence of two D resonances, one of which is in the region expected for a cyclic trimer (-8.6 ppm). Similarly, structure [54] has been determined by the presence of only one D resonance line and the absence of any trimeric-type D unit.

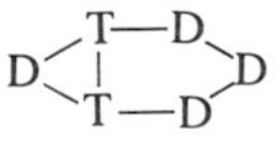

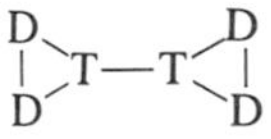

The chemical shift range for the nitrogen analog of the siloxanes, the silazanes, is much smaller.[114] The direction of the shifts with increasing nitrogen substitution is the same; if silicon has one nitrogen attached ("M" unit) the resonance is generally to high frequency of Me_4Si. Similarly, increasing nitrogen substitution effects low frequency shifts. Recently, Lavrukhin et al. have examined a series of mixed methylphenyl cyclic silazanes with units $Me_2SiNH^{(d)}$, $MePhSiNH^{(dPh)}$ and $Ph_2SiNH^{(dPh_2)}$.[130] The chemical shift ranges for the different d units are collected in Table 13. The deshielding, upon ring contraction from $n = 4$ to $n = 3$, is greatly reduced from that observed for the cyclic siloxanes in keeping with the reduced chemical shift

TABLE 11

Influence of solvent on ^{29}Si chemical shifts,[a] solvent and donor number (in parenthesis).

Compound	CHCl$_3$[b]	CH$_3$NO$_2$ (2.7)	Acetone (17.0)	DMF (26.6)	DMSO (29.8)	HMPA (38.8)	Other
(Ph)$_3$SiOH	−12.6	−14.5	−16.2	−17.51	−17.9	−19.2	−25.4 25% NaOEt/EtOH[b]
(Ph)$_2$Si(OH)$_2$	−30.7	−30.2 (50 °C)	−32.4	−34.2	−34.6	−37.5	
HO(Me$_2$SiO)$_{20}$H	−10.9	−11.5	−14.0	−14.8	−15.1	−16.5	−11.5 (25% HOAc)[b]
HO(Me$_2$SiO)$_6$H	−10.8	−11.5	−13.9	−14.7	−15.1	−16.1	
(HOSiMe$_2$)$_2$O	−10.5	−11.0	−13.6	−14.9	−15.3	−16.9	−14.8 trimethylphosphate (23.0)

[a] In ppm relative to Me$_4$Si.
[b] No donor number is available.

TABLE 12

^{29}Si chemical shifts for some silicic acid trimethylsilyl esters.[128] [a]

	Compounds	M^b	Q^b
M_4Q	M_4Q	7–6 (8.6)	−105.1 (−104.2)
M_6Q_2	M_3Q-QM_3	8.0 (9.3)	−107.5 (−106.5)
M_8Q_3	$M_3Q-M_2Q-QM_3$	8.0 (8.99) (M_3Q)	−107.7 (−106.7) (QM_3)
		8.3 (9.31) (M_2Q)	−110.1 (−109.1) (QM_2)
$M_{10}Q_4$	$M_3Q-M_2Q-M_2Q-QM_3$	7.6 (M_3Q)	−108.6 (QM_3)
		7.7 (M_2Q)	−110.04 (QM_2)
M_6Q_3	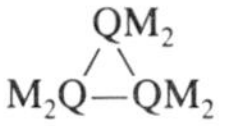 QM₂ / M₂Q—QM₂	6.4	−100.7
M_8Q_4	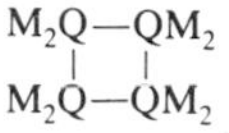 M₂Q—QM₂ / M₂Q—QM₂	9.2 (10.2)	−108.8 (−107.8)
$M_{10}Q_7$	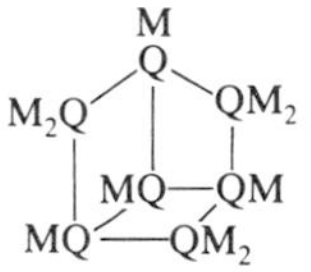 M / Q, M₂Q QM₂, MQ—QM, MQ—QM₂	9.4 (M_2Q) 10.0 (MQ)	−108.8 (QM_2) −109.8 (QM)
M_8Q_8	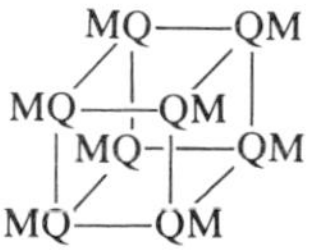 MQ—QM, MQ—QM, MQ—QM, MQ—QM	11.7	−109.3
$M_{10}Q_{10}$	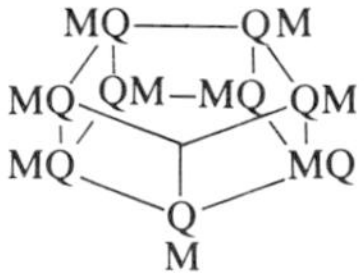 MQ—QM, MQ QM—MQ QM, MQ MQ, Q, M	12.4	−110.2

[a] In ppm relative to Me_4Si.
[b] Values in parentheses are from reference 127.

TABLE 13

^{29}Si chemical shift ranges for mixed
methylphenylcyclosilazanes[130] $(SiR_1R_2NH)_n$.

n	R_1	R_2	δ_{Si} (ppm)
3	Me	Me	−2.6 to −4.5
	Me	Ph	−12.4 to −13.8
	Ph	Ph	−21.2 to −22.4
4	Me	Me	−6.9 to −8.4
	Me	Ph	−12.5 to −17.0
	Ph	Ph	−24.6 to −25.7

range. The authors find a strong stereochemical dependence of the ^{29}Si chemical shifts in the various isomers of the cyclotetrasilazanes which is likely to arise from steric interactions of the substituents. The shifts in the cyclotrisilazanes are found to be additive and dependent on the number of positions of the phenyl groups in the ring.

H. Silicates

^{29}Si NMR has found considerable use in studies of the structures of various soluble silicates.[123–142] In this discussion, the notation of Engelhardt is used[132] to describe the silicon site. All units are Q-type silicon atoms since there are four oxygen atoms attached to each. A superscript refers to the number of bonds to other silicon atoms through an oxygen bridge: Q^1 is a monomer, Q^1Q^1 is a dimer, and $Q^1Q^2Q^1$ is a trimeric structure, for example. The degree of protonation is ignored. Early workers[133,134] report the existence of dimer, linear trimer, tricyclic and polymeric structures in sodium metasilicate solutions based on ^{29}Si NMR results. The chemical shift regions for the different building units found in a 4.1 M solution of sodium silicate with a 1:1 Na/Si ratio are shown in Fig. 9. In addition to these resonances, a broad signal from Q^4 units appears in the region −100 to −120 ppm. A large contribution to this resonance arises from the glass probe insert and sample tubes, but a signal is still present if Teflon sample tubes and insert are used.[133] The various signals are split by neighbouring group effects to produce the fine structure seen in Fig. 9.

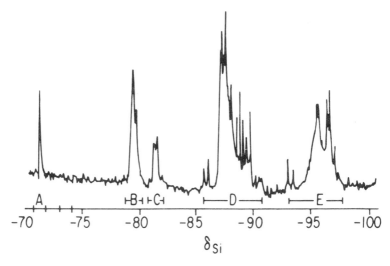

FIG. 9. ^{29}Si NMR spectrum of a 4.1 M solution of sodium silicate with a 1:1 Na/Si ratio.[132]

Changing the Na/Si ratio has a dramatic effect on the spectrum. When this ratio is small more polymeric silicate anions are present. As more Na is added, the polymer structures decrease in concentration until only the monosilicate anion remains. Recent results on isotopically enriched (^{29}Si) solutions of sodium silicate have confirmed the existence of only monomeric orthosilicate ions in solutions at 0.0096 M in silica.[140] The use of labeled materials and the resulting splitting patterns which arise from silicon–silicon coupling have facilitated the assignment of individual resonance lines for several species.[140,142,216] Harris et al. have utilized a high field spectrometer (99.36 MHz for ^{29}Si) and homonuclear decoupling to unequivocally identify several others.[142] The chemical shifts of these structures are presented in Table 14.

TABLE 14

^{29}Si chemical shifts of some silicate species.[142]

Structure	δ_{Si} [a]			
	A	B	C	D
Q^1Q^1	−78.620			
Q^2_3 [b]	−81.186			
Q^2_4	−87.096			
Q^3_6 [b]	−88.210			
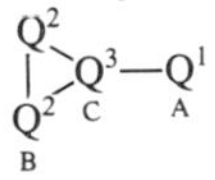	−79.072	−80.870	−99.234	
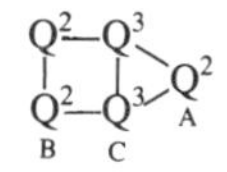	−80.863	−87.355	−88.160	
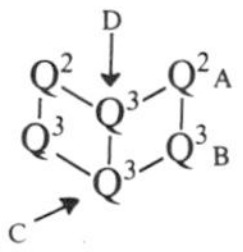	−88.048	−88.436	−87.650	−95.714
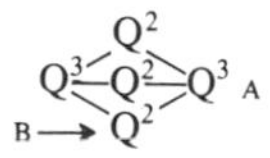	−85.222	−92.938		
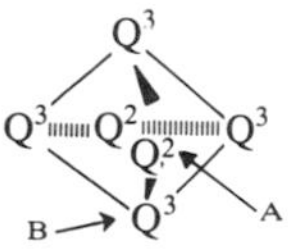	−85.511	−92.004		

[a] In ppm relative to Me$_4$Si. Data converted from reference 142 using $\delta_{Si}^{Q0} = -71$ ppm.

[b] In a three-membered ring.

I. Five- and six-coordinate silicon compounds

The possibility of valence shell expansion is one reason for the differences in chemical behaviour between silicon and carbon. Although there have not been a large number of compounds reported, sufficient information is available to indicate that coordination numbers higher than four lead to pronounced shifts to lower frequency. The silatranes [1-substituted-(2,2',2"-nitrilotriethoxy)silanes] are compounds in which a transannular bond is formed from nitrogen to silicon to form a pentacoordinate compound [58].

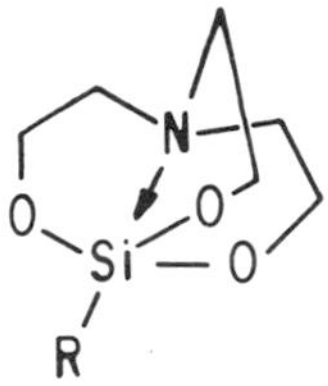

[58]

^{29}Si NMR studies of these compounds[143–147] have shown a consistent strong shielding effect associated with formation of the silatrane structure. Table 15 shows the chemical shifts for some silatranes as well as the analogous triethoxysilanes. In most cases silatrane formation effects a shift of approximately 20 ppm to lower frequency. For the electronegative substituents fluoro, ethoxy and phenoxy, the differences are only 16, 12 and 12 ppm respectively. This was assumed to indicate a longer distance between Si and

TABLE 15

^{29}Si chemical shifts for some silatranes and the analogous triethoxy silanes.[143] ^{a}

R	RSi(OCH$_2$CH$_2$)$_3$N A	RSi(OEt)$_3$ B	$\Delta\delta$ (A − B)
H	−83.6	−59.5	−24.1
Me	−64.8	−44.2	−20.6
Et	−66.5^b	−45.4^b	−21.1
Vi	−81.0	−59.5	−21.5
Ph	−81.7	−58.4	−23.3
PhC≡C	−94.7^c	−69.5^c	−25.2
F	−101.1^d	−84.5	−16.6
PhO	−98.4^e	−86.6^e	−11.8
EtO	−94.7^e	−82.4^e	−12.3

^{a} In ppm relative to Me$_4$Si.
^{b} Reference 75.
^{c} Reference 145.
^{d} For the compound FSi[OCH(CH$_3$)CH$_2$]$_3$N: reference 144.
^{e} Reference 147.

N which results in a smaller $\Delta\delta_{Si}$ for these compounds,[147] but X-ray diffraction data are required to confirm this proposal.

Studies of structural modifications to the silatrane structure have shown that carbon substituents in the silatrane skeleton lead to a strengthened transannular $N \to Si$ interaction (identified by a slight shielding effect) as a result of the retardation of conformational transitions.[146] If the oxygen in one of the bridges is replaced by carbon, or if one of the rings is enlarged, the $N \to Si$ bond is weakened.[146] The nitrogen analogue of 1-methylsilatrane, 1-methylazasilatrane, has also been examined by ^{29}Si NMR and found to contain a stronger $N \to Si$ bond.[148]

The exceptionally shielded resonance of SiF_6^{2-} ($\delta_{Si} = -191.7$ ppm) was reported as early as 1968.[3] Stable penta- and hexacoordinate complexes result only when silicon is bonded to highly electronegative atoms such as fluorine, chlorine, oxygen or nitrogen and as a result there are not a large number known.[29] NMR offers a unique method for characterizing these complexes since their chemical shifts deviate greatly from the normal range in which tetravalent silanes are usually found. Data are available for a series of cationic, neutral and anionic penta- and hexacoordinate complexes[149-150] for which chemical shifts in the range -127 to -197 ppm are found in all but one compound. Some trends are apparent. The cationic and neutral hexacoordinate complexes in which the chelate ring is derived from a 1,3-diketone display nearly identical chemical shifts regardless of charge type or substitution on the 1,3-diketone, [59]–[61], as long as silicon is bonded to six oxygen atoms. If two of the acetate ligands in [61] are replaced with CH_3 and Cl, a marked deshielding effect is observed.[62]

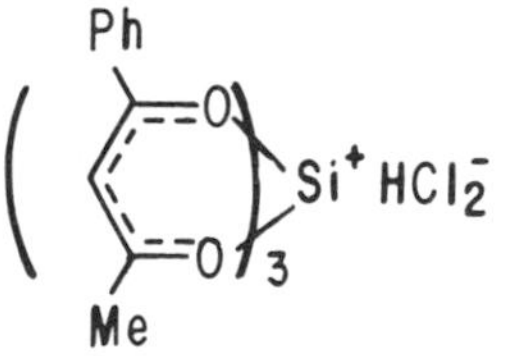

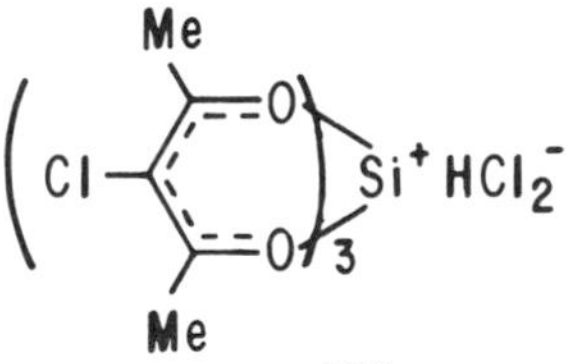

[59]   [60]

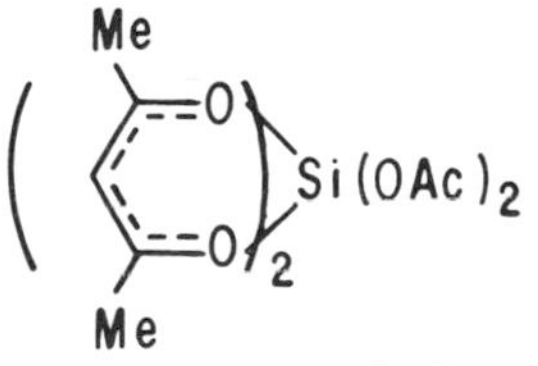

[61]

The chelate ring size also appears to be important. In the five-membered chelates such as tropolone or catchol, resonances in the region -135 to -139 ppm are observed for the cationic and anionic hexacoordinate complexes such as [63] and [64].

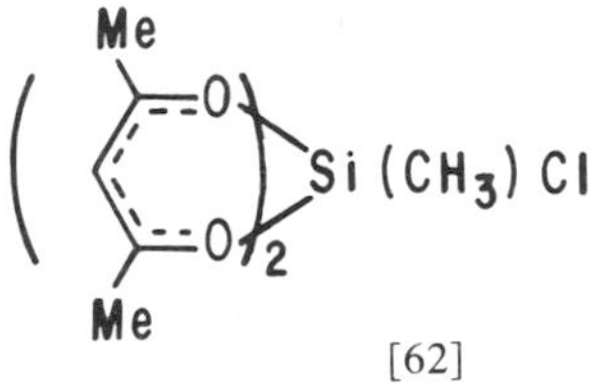

$$\left(\underset{Me}{\overset{Me}{\text{[structure]}}} \right)_2 Si(CH_3)Cl \quad \delta_{Si} = -149.5$$

[62]

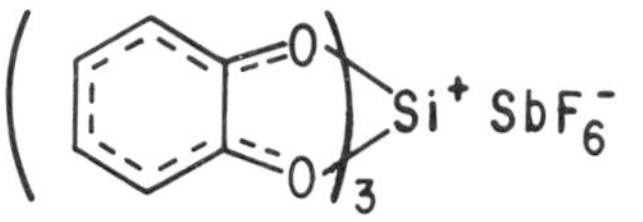

$$\left(\text{[structure]} \right)_3 Si^+ \; SbF_6^- \quad \delta_{Si} = -139.4$$

[63]

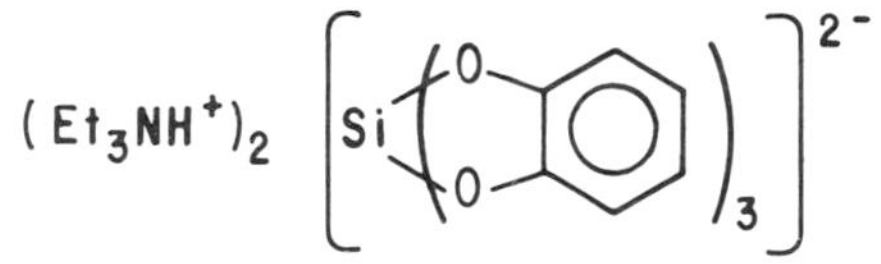

$$(Et_3NH^+)_2 \left[Si \left(\text{[structure]} \right)_3 \right]^{2-} \quad \delta_{Si} = -139.3 \; ppm$$

[64]

An anomalous shift is found for the only anionic pentacoordinate complex examined [65].

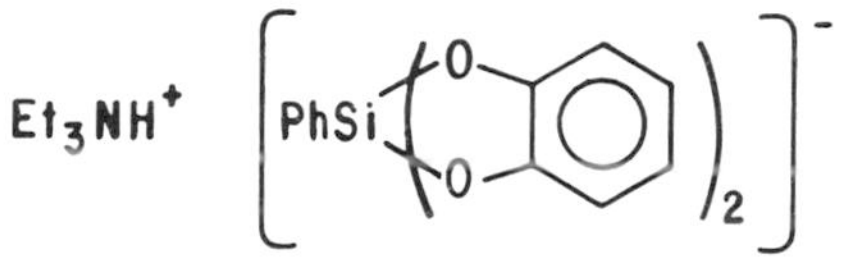

$$Et_3NH^+ \left[PhSi \left(\text{[structure]} \right)_2 \right]^- \quad \delta_{Si} = -87.1 \; ppm$$

[65]

In order to confirm this value a spectrum of solid [65] was obtained[151] and found to have a chemical shift within 1 ppm of the solution value.

It is apparent that ^{29}Si NMR is of value in determining the structure of these complexes. This has potential application in determining whether a given ligand will form a tetracoordinate or higher complex with silicon. For example, a reaction mixture of $SiCl_4$ and a bidentate ligand should show a resonance above the Q region of the spectrum (-100 ppm) if a five- or six-coordinate species has been formed.

J. Applications

The ability of ^{29}Si NMR to differentiate among minor structural changes and the use of trimethylsilylating (TMS) reagents to cap polar substituents have been combined to elucidate structures of many types of polar compounds including sugars,[155-157] steroids,[157] amines, amides and urethanes,[152,153,159] amino-, mercapto- and hydroxycarboxylic acids,[154] heterocycles,[160] and coal- and shale oil-derived fluids.[161,162] The power of this technique is illustrated in Fig. 10, which shows the variation in ^{29}Si

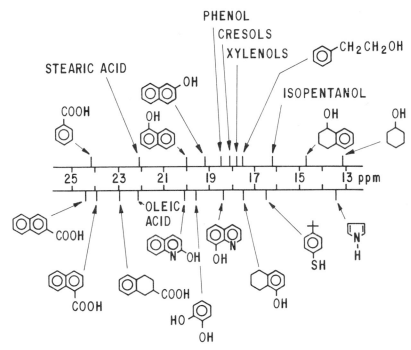

FIG. 10. ^{29}Si NMR chemical shifts of a variety of trimethylsilylated polar compounds.[161,162]

chemical shift for TMS derivatives of some polar compounds found in coal liquids and shale oil crude.[161,162]

In addition, ^{29}Si NMR has been used to study the products of reaction of SiF_2 with olefins[163] and found to lend support to the proposed involvement of monomeric SiF_2 in these reactions.[164] Several reports of ^{29}Si NMR used specifically to probe charge delocalization have appeared.[165–167] Olah *et al.* have studied α- and β-trimethylsilyl carbocations[165] in part to examine the charge-delocalizing ability of silyl substituents. The ^{29}Si chemical shift of a cation relative to the precursor alcohol, [66] and [67], is found to be dependent on the charge-delocalizing ability of R_1 and R_2. Substituents such as a cyclopropyl or phenyl group which are efficient at charge delocalization, significantly diminish the deshielding of δ_{Si} in [67] relative to [66].

$$Me_3Si-C\equiv C-C(OH)R_1R_2 \qquad Me_3Si-C\equiv C-\overset{+}{C}\overset{\displaystyle R_1}{\underset{\displaystyle R_2}{\Big\backslash}}$$

$$[66] \qquad\qquad\qquad [67]$$

The total range of this deshielding effect is 7–12 ppm. A larger effect (18 and 21 ppm) is observed upon protonation of $PhCSiMe_3$ and $MeCSiMe_3$.

The chemical shift of $Ph_2\overset{+}{C}SiMe_3$ is only 0.7 ppm to higher frequency of the precursor alcohol. This is attributed to efficient charge delocalization into the two aromatic rings leading to a smaller charge at silicon.[165]

N-substituted 2,2-diaryl-1,3-dioxa-6-aza-2-silacyclooctanes, R_1R_2Si-$(OCH_2CH_2)_2NR$, can exist in a boat–boat conformation which permits formation of a transannular bond.[167] The ^{29}Si NMR chemical shifts of these compounds are used to determine the extent of $N \rightarrow Si$ interaction. It is found that a silicon hybridization change from sp^3 to sp^3d in the compounds studied effects a 78 ppm shift to lower frequency.[167]

Both chemical shift and linewidth data for ^{27}Al and ^{29}Si are used to study the mechanism of zeolite precipitation in solutions 3 molal in SiO_2, 0–0.4 molal in Al and with a Na/Si mole ratio of 3.[168] The authors find a strong interaction of Al with Q^0 and slightly less for Q^1 units (as evidenced by linewidth increases). No change is observed in the Q^2 and cyclic trimer species, suggesting negligible interaction of Al with these systems. From this information the presence of Al^{IV}–Si^{IV} species such as $H_{6-n}SiAlO_7^{(n+1)}$ dimers and higher polymers are proposed to be present in the solutions. During nucleation and growth of the zeolite, the intensity of the Al^{IV} peak decreases with a concomitant decrease in the Q^0 linewidth in the ^{29}Si NMR spectrum. It is concluded that only initially formed Al–Si polymeric species are removed from solution by nucleation and growth of the zeolite, and that the composition and structure of the solid must be determined by the average composition of the low condensation aluminosilicate polymers formed initially.[168]

K. Solids

The early success with high resolution ^{29}Si NMR in the solid state[169] has been followed by a wealth of applications of this technique to structural studies of a number of very different silicon-containing systems. The greatest impact so far has probably been in structure determinations of silicates and aluminosilicates.[170–182] Lippmaa et al. have shown that isotropic ^{29}Si chemical shifts in solids and solutions are generally the same for silicates although the influence of cation and degree of ionization can be seen.[170] The authors find that Q_8M_8 is a very useful secondary standard ($\delta_{Si} = 11.8$ ppm for the M silicon atom) since linewidths of 0.15 ppm for the four lines of the Q resonance, which arise from distortion of the cube in the crystal, provide a means to check the magic angle setting. The data for a series of solid silicates are collected in Table 16. The main influence on the chemical shift is clearly the degree of condensation of the silicon–oxygen tetrahedra with single tetrahedra occurring at highest frequency (-66 to -74 ppm) and fully condensed Q^4 units at -107 to -109 ppm. Replacement of a silicon atom in the tetrahedron by aluminum to form aluminosilicates

TABLE 16

^{29}Si isotropic shifts in solid silicates.[170] [a]

Compound	Type of silicon–oxygen tetrahedra	Q^0 single tetrahedra	Q^1 end group	Q^2 middle group	Q^3 branching site	Q^4 cross-linking group
NaH$_3$SiO$_4$	Single	-66.4				
Na$_2$H$_2$SiO$_4$·8.5H$_2$O	Single	-67.8				
CaNaHSiO$_4$	Single	-73.5				
(CaOH)CaHSiO$_4$	Single	-72.5				
Zn$_4$(OH)$_2$[Si$_2$O$_7$]·H$_2$O (hemimorphite)	Double		-77.9			
Ca$_6$(OH)$_6$[Si$_2$O$_7$]	Double		-82.6			
K$_4$H$_4$[Si$_4$O$_{12}$]	Cyclotetra			-87.5		
Ca$_2$(OH)$_2$[SiO$_3$] (hillebrandite)	Single chain			-86.3		
Ca$_4$(OH)$_2$[Si$_3$O$_9$] (foshagite)	Single chain			-86.5		
Ca$_3$[Si$_3$O$_9$] (wollastonite)	Single chain			-88.0		
Ca$_2$NaH[Si$_3$O$_9$] (pectolite)	Single chain			-86.3		
Ca$_6$(OH)$_2$[Si$_6$O$_{17}$] (xonotlite)	Double chain			-86.8	-97.8	
Mg$_3$(OH)$_2$[Si$_4$O$_{10}$] (talc)	Layer				-98.1	
(Me$_4$N)$_8$[Si$_8$O$_{20}$]	Double four-ring				-99.3	
(Et$_4$N)$_6$[Si$_6$O$_{15}$]	Double three-ring				-90.4	
SiO$_2$ (low quartz)	Spiral					-107.4
SiO$_2$	Three-dimensional					
(Low cristobalite)	Six rings					-109.9

[a] In ppm relative to Me$_4$Si.

leads to a shift of the ^{29}Si resonance to higher frequency. As in the case of the silicates, this shift is dependent upon the number of aluminum atoms occupying the tetrahedral sites around silicon.[170] The chemical shift ranges for the various aluminosilicates are shown graphically in Fig. 11. The designation is as shown in the figure with the presence of an oxygen bridge between Si and Si or Si and Al assumed. Spectra for two aluminosilicates, thomsonite and natrolite, are shown in Fig. 12. The two different ^{29}Si resonances of natrolite correspond to Si(3Al) and Si(2Al) units whereas the single resonance for thomsonite is of a Si(4Al) type.

The ability of ^{29}Si NMR to differentiate among the various degrees of aluminum substitution for Si in the SiO_4 tetrahedra has enabled the detailed

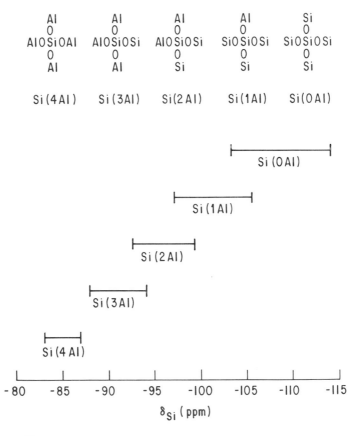

FIG. 11. ^{29}Si chemical shift ranges for different degrees of substitution by Al in aluminosilicates.[173]

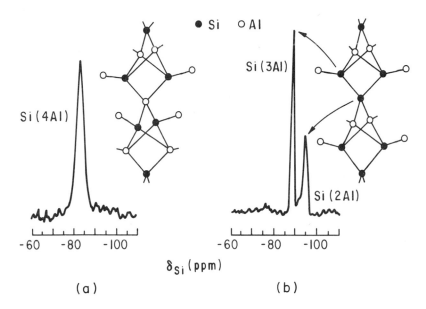

FIG. 12. High resolution solid-state ^{29}Si NMR spectra of (a) thomsonite, $Na_4Ca_8[(AlO_2)_{20}(SiO_2)_{20}].24H_2O$ and (b) natrolite, $Na_{16}[(AlO_2)_{16}(SiO_2)_{24}].16H_2O$.

distribution of Si and Al in the lattice to be determined. Using this technique it has been shown[173,174,176,179] that Loewenstein's rule,[18] which maintains that Al—O—Al linkages are to be avoided, is violated in several zeolites and should therefore be re-evaluated.

Other reports using high resolution solid state ^{29}Si NMR include studies of silica gel and silica surfaces.[188–192] Specifically, several peaks observed on the surface of silica gel could be assigned to structures: $(HO)_2\overset{*}{Si}(OSi\leqslant)_2$, $\delta_{Si} = -90.6$ ppm; $HO\overset{*}{Si}(OSi\leqslant)_3$, $\delta_{Si} = -99.8$ ppm; and $\overset{*}{Si}(OSi\leqslant)_4$, δ_{Si} -109.3 ppm. Reaction products on the surface are identified after addition of methylchlorosilanes.[189] Treatment with trimethylchlorosilane, for example, produces a resonance at 13 ppm which is undoubtedly due to silylation of surface OH groups to produce $OSiMe_3$ groups. Maciel et al. have also characterized some of the chromatographically useful organic substrates on silica gel.[190]

Several organosilicon polymers have been investigated.[193,194] These results indicate that ^{29}Si CP/MASS NMR will be extremely useful for structure elucidation in polymers, particularly since the chemical shifts are found to be very close to solution values.[194]

IV. SPIN–SPIN COUPLINGS

A. Theoretical considerations

The indirect coupling of nuclear spins through the bonding network is considered to be dependent on three factors: a pair of terms arising from the interaction of the field produced by the orbital motion of electrons with nuclear moments (the "orbital" contribution), a term arising from the dipole–dipole interaction between nuclear and electron spins ("dipolar" contribution), and a term dependent upon the properties of the electrons at the nucleus ("contact" term).[195] This third term is usually considered to be dominant in determining the coupling. Although the magnitude of the orbital and spin dipolar contributions to ^{29}Si couplings is not known, Beer and Grinter, using the finite perturbation theory (INDO level of approximation), have calculated the combined value for a series of molecules to be less than 1% of the observed $^1J(Si-C)$ coupling.[196] Furthermore, a number of correlations have been found relating various combinations of $J(Si-H)$, $J(Si-C)$, $J(Si-Si)$, $J(C-H)$ and $J(C-C)$ in the same or related systems which indicates that these couplings are influenced by the same factors. This suggests that the contact term is dominant for each and that d-orbital effects are not significant in the compounds studied. Indeed a number of calculations of $J(Si-H)$, $J(Si-C)$ and $J(Si-Si)$ using only the contact term and neglecting d orbitals have been made[210–215] with a fair amount of success. Summerhays and Deprez, however, have noted[212] that deviations for $NHSiMe_3$ and $OSiMe_3$ may be due to d-orbital participation.

Early attempts to explain silicon couplings involved the use of empirical linear additivity rules for calculating $^1J(Si-H)$.[197–200] Serious deviations occur with substituent electronegativity,[198–200] however, and the use of pairwise interaction terms is found to give a substantial improvement.[201,202] Parameters to which couplings have been related include the electronegativity of the substituents[203–205] and the s character in the bonding orbital to the coupled nucleus.[206–211] Dreeskamp and Hildenbrand have examined some vinyl- and chlorosilanes.[209] The one-bond coupling $^1J(Si-C)$ is found to be proportional to the s character of the Si-C bond. In the case of the two-bond couplings,$^2J(Si-H)$, the hybridization of both the silicon atom and the intervening carbon atom is important.

Several signs of ^{29}Si couplings have been determined. These experimental values are collected in Table 17. Theoretical attempts to calculate these signs[217,218,227] have met with some success.

B. One-bond couplings

The bulk of the one-bond couplings that have been determined are for $^1J(Si-H)$. In general the largest change occurs when an electronegative substituent is introduced directly at silicon, with the absolute magnitude of

TABLE 17

Signs of some ^{29}Si couplings.

J	Sign	References
$^1J(Si—C)$	−	217–220
$^1J(Si—F)$	+	217, 218, 221, 222
$^1J(Si—H)$	−	217–219, 221–223
$^1J(Se—Si)$	+	6
$^1J(Sn—Si)$	+	226
$^1J(Pt—Si)$	−	6
$^2J(Si—CH)$	+	219, 224
$^2J(P—Pt—Si)$	+	6
$^2J(Si—Si—H)$	−	96
$^3J(Si—C—C—H)$	−	225
$^3J(Si—Si—Si—H)$	−	96
$^3J(Si—Se—Si—H)$	−	6

J increasing with electronegativity. Multiple substitution of electronegative substituents leads to a progressively increasing coupling. Electropositive substituents, on the other hand, decrease the absolute value of $^1J(Si—H)$. The effect of β-substitution is much smaller than for α-substitution. Some representative values are shown in Table 18.

TABLE 18

Values of $^1J(Si—H)$ for some representative silanes.

Compound	$^1J(Si—H)$ (Hz)	Reference
$HSiF_3$	−381.7	228
$(EtO)_3SiH$	−287	143
H_2SiF_2	−282	228
$N(CH_2CH_2O)_3SiH$	−277	143
H_3SiBr	−240.5	234
H_3SiCN	−238	229
H_3SiF	−229	228
H_4Si	−202.5	228
Ph_2SiH_2	−198.2	230
$MeSiH_3$	−194.0	230
Me_2SiH_2	−188.6	230
Me_3SiH	−184.0	230
$(t\text{-}Bu)_2SiH_2$	−182.1	230
$(i\text{-}Pr)_2MeSiH$	−180.9	79
$(Me_3Si)Me_2SiH$	−173	231
$(Me_3Si)_3SiH$	−155	232
$(Et_3Si)_3SiH$	−147	233

Many correlations have been drawn between $^1J(Si-H)$ and parameters related to bonding and reactivity. For example, good linear relationships have been found between $^1J(Si-H)$ and Hammett σ constants in benzyldimethysilanes,[236] phenyltetramethyldisilanes,[236] and phenyl-, phenylmethyl- and phenyldimethylsilanes.[237] An excellent correlation (correlation coefficient $r = 0.994$) is found between $^1J(Si-H)$ and the sum of the Taft polar substituent constants σ^* for silanes $R^1R^2R^3SiH$ (R = H, alkyl, Ph).[238] Recently, the values of $^1J(Si-H)$ in phenoxy- and (phenylthio)dimethylsilanes were measured and found to correlate well with the Hammett σ constants for substituents on the benzene ring.[239] The values of $^1J(Si-H)$ are larger than expected for the phenylthiosilanes. This is interpreted in terms of sulfur d-orbital participation by $\sigma-\pi$ conjugation which increases the s character of the Si—H bond, and therefore increases $^1J(Si-H)$.[239]

Substituent effects on $^1J(Si-F)$ are much less straightforward than for $^1J(Si-H)$. In some cases increasing substituent electronegativity causes a decrease in J: F_3SiSiH_3, $^1J(Si-F) = 356$ Hz[240] versus F_3SiOMe, $^1J(Si-F) = 181$ Hz,[241] for example. The values for $(Me_3Si)_3SiF$ and Br_3SiF, however, show the opposite trend. These values, as well as some other representative data, are collected in Table 19.

A number of one-bond silicon–carbon couplings have been reported. The magnitude of $^1J(Si-C)$ is dependent on both the electronegativity of the substituents on silicon and the hybridization of the carbon atom. Harris and Kimber find a linear dependence on substituent electronegativity for $^1J(Si-C)$ in a series of trimethylsilyl derivatives Me_3SiX.[31] A good linear correlation is also found by the same authors for $^1J(Si-C)$ and $^1J(C-C)$ of the analogous t-butyl derivatives. This is to be expected if the same

TABLE 19

$^1J(Si-F)$ for some fluorosilanes

Compound	$^1J(Si-F)$ (Hz)	Reference
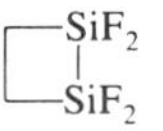 ⌐SiF₂ / ∟SiF₂	488	242
$SiFCl_2$	401	243
F_3SiSiF_3	321.8	244
$SiFH_2Me$	279.8	246
SiF_3NMe_2	201.4	247
SiF_3OMe	181	241
F_3SiSiF_2H	167	245
SiF_6^{2-}	110	248

factors affect both couplings and it is concluded that the contact term is dominant for both. They caution against over-emphasis of the relationship with electronegativity, however, since the changes in $^1J(Si{-}C)$ for the series $Me_nSi(OEt)_{4-n}$ are too large to be accounted for on the basis of electronegativity alone.[249] A collection of values of $^1J(Si{-}C)$ is presented in Table 20. From the data in the table it can be seen that the values of $^1J(Si{-}C)$ follow the order $sp > sp^2 > sp^3$ for the hybridization of the carbon atoms since there is greater s character in the Si—C bond. Since the highest values of $^1J(Si{-}C)$ occur in those compounds with electronegative substituents, it is not surprising that the value of $^1J(Si{-}C)$ for $CH_2{=}CHSiCl_3$ is very high at 113 Hz,[209] whereas the value for $[(CH_3)_3Si]_2Si^*(\overset{*}{C}H_3)_2$ is quite low (37 Hz).[72] The exceptionally low value of $^1J(Si{-}C)$ for bis(tetracarbonyliron)dimethylsilane (38.5 Hz) is taken as an indication of the decreased s-character in the Si—C(Me) bonds relative to other methylsilanes.[21]

The relationship between coupling and substituent electronegativity is extended to $^1J(Si{-}Si)$ in a series of di- and polysilanes.[72] A good linear correlation is found between $^1J(Si{-}Si)$ and the substituent electronegativities on the coupled silicon atoms. Further correlations are found for $^1J(Si{-}Si)$ and $^1J(Si^B{-}C)$ in pentamethyldisilanyl compounds

TABLE 20

$^1J(Si{-}C)$ for some representative silanes.

Compound	$^1J(Si{-}C)$ (Hz)	Reference
$CH_2{=}CHSiCl_3$	113	209
$ClCH_2SiCl_3$	97.6	209
$MeSi(OEt)_3$	96.2	250
$MeSiCl_3$	86.6	209
$Et_3Si{-}C^*{\equiv}CH$	75.0	251
$Me_2Si(OEt)_2$	73.0	250
Me_2SiCl_2	68.3	209
Me_3SiOEt	59.0	31
$Me_3SiC^*Me_3$	57.0	255
$(Me_3Si)_2NH$	56.21	31
$(Me_3^*Si)_3CMe$	51.2	255
Me_3SiH	50.8	31
$(Me_3Si)_3C^*Me$	38.0	255
$(Me_3Si)_2Si^*Me_2^*$	37.0	72
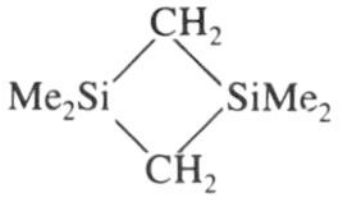 $Me_2Si\overset{\displaystyle CH_2}{\underset{\displaystyle CH_2}{\diagup\diagdown}}SiMe_2$	34.7	255

$Me_3Si^ASi^BMe_2X$, and $^1J(Si^B-C)$ and $^1J(Si-C)$ in the analogous trimethyl-silyl compounds.[72] Clearly the same factors are affecting the coupling in these compounds. It is concluded that the contact term is dominant for one-bond silicon–silicon couplings as well. Recent results[95] have shown that, for the more crowded isotetrasilanes, $(Y_3Si)_3SiX$, the values of $^1J(Si-Si)$ and $^1J(Si-C)$ for $Y = CH_3$ are dependent upon the steric effect of X as well. Data for several di- and polysilanes are collected in Table 21. Table 22 shows one-bond couplings for silicon bonded to some other elements.

C. Longer-range couplings

Longer-range couplings tend to follow the trends for one-bond couplings. For example, values of $^2J(Si-H)$ and $^3J(Si-H)$ are larger in silanes with electronegative substituents. The range of $^2J(Si-C-H)$ couplings is approximately 1–12 Hz. Much larger absolute values are obtained for $^2J(Si-F)$. Values for $^2J(Si-Si-F)$, for example, range from -14.5 Hz for $(Et_3Si)_3SiF$[233] to -90.5 Hz for Si_2F_6.[244] Some two-bond couplings $^2J(Si-C-H)$ and $^2J(Si-Si-H)$ that have recently been reported are collected in Table 23.[79,96] For more detailed discussions of couplings the reader is referred to references 6 and 9.

TABLE 21

$^1J(Si-Si)$ for some di- and polysilanes.

Compound	$^1J(Si-Si)$ (Hz)	Reference
$(Cl_3Si)_2SiCl_2$	186	72
$(MeO)_3SiSiPh_3$	160	72
$[(MeO)_3Si]_3SiOMe$	144.1	43
$[(MeO)_3Si]_3SiH$	132.4	43
$(Cl_3Si)_3SiBr$	130.2	43
$ClMe_2SiSiMe_2(OEt)$	109.8	43
$ClMe_2SiSiMe_2(OPr^n)$	107.0	43
$Me_3SiSiMe_2F$	98.7	72
$Me_3SiSiMe_2Cl$	94.0	72
$Me_3SiSiMe_2Ph$	86.1	72
$Ph_3SiSiMe_2Bu^t$	80.0	82
$(Ph_3Si)_2SiMe_2$	74.5	82
$(Me_3Si)_2SiMe_2$	73.2	72
$(Ph_3Si)_2SiMePr^n$	72.2	82
$(Ph_3Si)_2SiEt_2$	70.8	82
$(H_3Si^ASi^BH_2Si^CH_2)_2$	$J_{AB} = 69.7, J_{BC} = 61.4$	96
$(Me_3Si)_4Si$	52.5	72

TABLE 22

One-bond couplings of ^{29}Si to some other elements.

M	Compound	$^1J(\text{Si}-\text{M})$ (Hz)[a]	Ref.
^{31}P	$(\text{H}_3\text{Si})_3\text{P}$	42.2	58
^{31}P	$(\text{H}_3\text{Si})_2\text{PMe}$	39.6	58
^{31}P	$(\text{Me}_3\text{Si})_3\text{P}$	27.5	58
^{31}P	$\text{Me}_3\text{SiPMe}_2$	20.3	58
^{31}P	Me_3SiPH_2	16.2	58
^{199}Hg	$\begin{cases} \text{Me}_3\text{SiHgR} \\ \text{Et}_3\text{SiHgR} \end{cases}$	957–1367	252
^{119}Sn	$\text{Ph}_3\text{SiSnMe}_3$	650	250
^{195}Pt	$trans\text{-PtCl}(\text{SiH}_2\text{Cl})(\text{PEt}_2)$	−1600	254
^{77}Se	$(\text{H}_3\text{Si})_2\text{Se}$	+110.6	223
^{15}N	$(\text{H}_3\text{Si})_3\text{N}$	+6	253

[a] Signs given where known.

TABLE 23

Some two-bond couplings $^2J(\text{Si}-\text{C}-\text{H})^{79}$ and $^2J(\text{Si}-\text{Si}-\text{H}).^{96}$

Coupling	Compound	J (Hz)
$^2J(\text{Si}-\text{C}-\text{H})$	Ph_2MeSiH	4.7
	$\text{Me}(^n\text{Pr})_2\text{SiH}$	3.7
	$(^n\text{Bu})_3\text{SiH}$	3.2
	Et_3SiH	3.1
$^2J(\text{Si}-\text{Si}-\text{H})$	$(\text{H}_3^*\text{Si})_2\overset{*}{\text{Si}}\text{H}_2$	5.16
	$(\text{H}_3\overset{*}{\text{Si}})_2\text{SiH}_2^*$	2.76
	$(\text{H}_3\overset{*}{\text{Si}}\text{SiH}_2^*)_2$	2.9
	$(\text{H}_3^*\text{SiSi}^*\text{H}_2)_2$	5.2

V. SPIN–LATTICE RELAXATION

Four spin–lattice relaxation mechanisms are usually possible for spin $\frac{1}{2}$ nuclei: (i) dipole–dipole interaction (DD); (ii) spin–rotation interactions (SR); (iii) scalar coupling (SC); and (iv) chemical shielding anisotropy (CSA). For ^{29}Si the dominant relaxation mechanisms are usually DD and SR. SC is not generally considered, except in cases where ^{29}Si is coupled to nuclei with similar resonance frequencies such as ^{129}I. Studies conducted on Me$_3$SiI,[41] however, have not shown any unusual data to suggest that this mechanism is operative in this case. Table 24 shows a comparison of the DD, SR and CSA contributions to spin–lattice relaxation in different

TABLE 24

Contribution of ^{29}Si spin–lattice relaxation mechanisms.[5] [a]

	DD	SR	CSA	DD (intermolecular)
Very small molecules[b]				
Protonated	Minor– appreciable[c]	Appreciable– exclusive	Negligible[c]	Negligible– minor
Non-protonated	Minor– appreciable	Exclusive	Negligible	Negligible– minor
Medium and large molecules				
Protonated	Appreciable– exclusive	Negligible– exclusive	Negligible	Negligible– minor
Non-protonated	Appreciable– exclusive	Negligible– appreciable	Minor– appreciable	Minor– appreciable

[a] Relaxation near room temperature at 2.35 T; O_2 excluded.
[b] Assumes no intermolecular association.
[c] Minor = 5–10% contribution; negligible = 1–5% contribution.

types of molecules. Since detailed discussions of ^{29}Si spin–lattice relaxation are available elsewhere,[5,7,9] only general observations are presented in this section.

Spin–lattice relaxation times for organosilicon compounds generally range from 20 to 100 s. Equation (7) describes intramolecular DD relaxation for spin $\frac{1}{2}$ nuclei:

$$1/T_1(DD) = R_1(DD) = \mu_0^2 \gamma_{Si}^2 \gamma_H^2 N\hbar^2 \tau_c (16\pi^2 r_{Si-H}^6)^{-1} \qquad (7)$$

where N is the number of directly attached protons, τ_c is the molecular correlation time, γ_{Si} and γ_H are the magnetogyric ratios for ^{29}Si and ^{1}H, and r_{Si-H} is the Si—H bond length. The lower magnetogyric ratio for ^{29}Si than ^{13}C, along with the longer Si—H than C—H bond length, results in much less efficient DD relaxation for ^{29}Si than for ^{13}C in the analogous carbon compound.

As a result, the SR mechanism is more important for ^{29}Si than for ^{13}C. For ^{29}Si–^{1}H experiments the maximum NOE is -2.52.[256] This occurs when the DD mechanism dominates the relaxation and decreases when other mechanisms are present.

Both linear and cyclic polydimethylsiloxanes have been examined[5,28] (Table 25). In the linear siloxanes motional processes along the chain differ considerably. The M units at the end of the chain are free to rotate freely about their threefold axis of symmetry favouring the SR mechanism. The D units, however, are "anchored" to different degrees in the chain and show increasing contributions form the DD mechanism as evidenced by a

TABLE 25

^{29}Si relaxation in linear and cyclic polydimethylsiloxanes.[5,28]

Compound[a]	T_1 (s)[b]					NOE $(-\eta)$[b]				
	M	D^1	D^2	D^3	D^4	M	D^1	D^2	D^3	D^4
MM	39.5					0.31				
MDM	35	54				0.36	0.55			
MD$_2$M	38	78				0.53	1.2			
MD$_3$M	46	82	77			0.57	1.1	1.2		
MD$_6$M	42	64	55	59		0.7	1.6	1.9	1.9	
MD$_8$M	44	67	60	55	55	0.7	1.5	1.8	2.0	2.0
D$_3$		80					1.1			
D$_4$		100					1.83			
D$_6$		91					2.24			

[a] N$_2$ degassed.
[b] T_1 at 38 °C; D units numbered from each end, e.g. MD^{1}D^{2}D^{1}M.

larger NOE. Similar results are obtained for the polysilanes.[96] In phenyl-substituted linear and branched siloxanes, steric interactions at the ends of the chain inhibit SR relaxation and produce larger NOEs.[129]

Recent studies on some silyl transition metal complexes[105] show the effect of anchoring one end of a silane chain with a metal atom. The compound $(\eta^5\text{-}C_5H_5)Fe(CO)_2Si^AMe_2Si^BMe_2Si^CMe_3$ shows a progressive decrease in NOE from Si^A to Si^C. The percentage of DD relaxation is calculated to be as follows: Si^A, 84%; Si^B, 75%; and Si^C, 40%. These results are consistent with the SR mechanism becoming more important at the end of the chain.

The effect of molecular size can change the dominant relaxation mechanism. In the case of the cyclic siloxanes (Table 25) the DD mechanism becomes increasingly important for the larger rings, again, as shown by an increase in η. This changeover in mechanism is even more dramatically exemplified by the data for the trialkylsilanes shown in Table 26.[77]

TABLE 26

^{29}Si relaxation data for trialkylsilanes, R$_3$SiH[77]

R	T_1(s)	NOE (η)
Me	15.9	−0.21
Et	42.5	−0.67
n-Pr	45.4	−1.87
n-Bu	32.3	−2.34
n-Hexyl	11.7	−2.42

Since the DD mechanism is obviously important for ^{29}Si relaxation, it is not surprising that silicon atoms with directly bonded hydrogen atoms have shorter T_1 values. For example, the T_1 for phenylsilatrane is 70–100 s, whereas silatrane, with an Si—H bond, has a $T_1 = 23.2$ s.[143] A series of trimethylsilyl compounds has been examined.[31,46] It has been shown that for these smaller molecules relaxation is dominated by SR interactions.

In addition to molecular structure, other parameters which affect T_1 and should be considered include the medium,[31] temperature[5,77] and effects of paramagnetic additives or impurities.[5,7] Levy et al.[5] have also demonstrated the possibility of intermolecular dipole–dipole interactions by showing that Si_2Cl_6 has an NOE in benzene which is not present in benzene-d_6.

REFERENCES

1. P. C. Lauterbur, in *Determination of Organic Structure by Physical Methods*, Vol. 2 (E. C. Nachod and W. D. Phillips, eds), Academic Press, New York, 1962.
2. B. K. Hunter and L. W. Reeves, *Canad. J. Chem.*, 1968, **46**, 1399.
3. R. B. Johannesen, F. E. Brinckman and T. D. Coyle, *J. Phys. Chem.*, 1968, **72**, 660.
4. R. R. Ernst and W. A. Anderson, *Rev. Sci. Inst.*, 1966, **37**, 93.
5. G. C. Levy and J. D. Cargioli, in *Nuclear Magnetic Resonance Spectroscopy of Nuclei Other than Protons* (T. Axenrod and G. A. Webb, eds), Wiley, New York, 1974, p. 251.
6. J. Schraml and J. M. Bellama, in *Determination of Organic Structure by Physical Methods*, Vol. 6 (E. C. Nachod, J. J. Zuckerman and E. W. Randall, eds), Academic Press, New York, 1976.
7. R. K. Harris, J. D. Kennedy and W. McFarlane, in *NMR and the Periodic Table* (R. K. Harris and B. E. Mann, eds), Academic Press, London, 1978.
8. E. A. Williams and J. D. Cargioli, in *Annual Reports on NMR Spectroscopy*, Vol. 9 (G. A. Webb, ed.), Academic Press, London, 1979.
9. H. Marsmann, in *NMR: Basic Principles and Progress*, Vol. 17 (P. Diehl, E. Fluck and R. Kosfeld, eds), Springer Verlag, Berlin, 1981.
10. T. C. Farrar and E. D. Becker, *Pulse and Fourier Transform NMR*, Academic Press, New York, 1971.
11. (a) R. Freeman, K. G. R. Pachler and G. N. LaMar, *J. Chem. Phys.*, 1971, **55**, 4586; (b) O. A. Gansow, A. R. Burke and W. D. Vernon, *J. Amer. Chem. Soc.*, 1972, **94**, 2550.
12. R. Freeman, *J. Chem. Phys.*, 1970, **53**, 457.
13. S. Aa. Linde, H. J. Jakobsen and B. J. Kimber, *J. Amer. Chem. Soc.*, 1975, **97**, 3219.
14. S. Li, D. L. Johnson, J. A. Gladysz and K. L. Servis, *J. Organomet. Chem.*, 1979, **166**, 317.
15. R. D. Bertrand, W. B. Moniz, A. N. Garroway and G. C. Chingas, *J. Amer. Chem. Soc.*, 1978, **100**, 5227.
16. P. D. Murphy, T. Taki, T. Sogabe, R. Metzler, T. G. Squires and B. C. Gerstein, *J. Amer. Chem. Soc.*, 1979, **101**, 4055.
17. D. M. Doddrell, D. T. Pegg, W. Brooks and M. R. Bendall, *J. Amer. Chem. Soc.*, 1981, **103**, 727.
18. G. A. Morris and R. Freeman, *J. Amer. Chem. Soc.*, 1979, **101**, 760.
19. B. J. Helmer and R. West, *Organometallics*, 1982, **1**, 877.
20. R. Radeglia, *Z. Phys. Chem. (Leipzig)*, 1975, **256**, 453.

21. A. L. Bikovetz, O. V. Kuzmin, V. M. Vdovin and A. M. Krapivin, *J. Organomet. Chem.*, 1980, **194**, C33.
22. E. D. Becker, in *Nuclear Magnetic Resonance Spectroscopy of Nuclei Other than Protons* (T. Axenrod and G. A. Webb, eds), Wiley, New York, 1974, p. 1.
23. G. C. Levy, J. D. Cargioli, G. E. Maciel, J. J. Natterstad, E. B. Whipple and M. Ruta, *J. Magn. Reson.*, 1973, **11**, 352.
24. G. C. Levy, *J. Amer. Chem. Soc.*, 1972, **94**, 4893.
25. R. G. Scholl, G. E. Maciel and W. K. Musker, *J. Amer. Chem. Soc.* 1972, **94**, 6376.
26. R. B. Johannesen, *J. Chem. Phys.*, 1967, **47**, 955.
27. M. Vongehr and H. C. Marsmann, *Z. Naturforsch*, 1976, **31b**, 1423.
28. G. C. Levy, J. D. Cargioli, P. C. Juliano and T. D. Mitchell, *J. Amer. Chem. Soc.*, 1973, **95**, 3445.
29. N. N. Zemlyanskii and O. K. Sokolikova, *Zhur. Anal. Khim.*, 1981, **36**, 1990.
30. R. K. Harris and R. H. Newman, *J. Chem. Soc., Faraday Trans. II*, 1977, **1977**, 1204.
31. R. K. Harris and B. J. Kimber, *J. Magn. Reson.*, 1975, **17**, 174.
32. A. Saika and C. P. Slichter, *J. Chem. Phys.*, 1954, **22**, 26.
33. W. E. Lamb, Jr, *Phys. Rev.*, 1941, **60**, 817.
34. C. J. Jameson and H. S. Gutowsky, *J. Chem. Phys.*, 1964, **40**, 1714.
35. G. Engelhardt, R. Radeglia, H. Jancke, E. Lippmaa and M. Mägi, *Org. Magn. Reson.*, 1973, **5**, 561.
36. F. F. Roelandt, D. F. van de Vondel and E. A. van den Berghe, *J. Organomet. Chem.*, 1975, **94**, 377.
37. R. Wolff and R. Radeglia, *Z. Phys. Chem.* (*Leipzig*), 1976, **257**, 181.
38. R. Wolff and R. Radeglia, *Z. Phys. Chem.* (*Leipzig*), 1977, **258**, 145.
39. R. Wolff and R. Radeglia, *Org. Magn. Reson.*, 1977, **9**, 64.
40. R. Radeglia, *Z. Phys. Chem.* (*Leipzig*), 1975, **256**, 453.
41. R. Radeglia, *Z. Naturforsch.*, 1977, **32b**, 1091.
42. R. Wolff and R. Radeglia, *Z. Phys. Chem.* (*Leipzig*), 1980, **261**, 726.
43. G. Engelhardt, R. Radeglia, H. Kelling and R. Stendel, *J. Organomet. Chem.*, 1981, **212**, 51.
44. V. S. Lyubimov and S. P. Ionov, *Russ. J. Phys. Chem.*, 1972, **46**, 486.
45. J. A. Cella, J. D. Cargioli and E. A. Williams, *J. Organomet. Chem.*, 1980, **186**, 13.
46. R. K. Harris and B. Lemarie, *J. Magn. Reson.*, 1976, **23**, 371.
47. W. McFarlane and J. M. Seaby, *J. Chem. Soc. Perkin Trans. II*, 1972, **1972**, 1561.
48. C. R. Ernst, L. Spialter, G. R. Buell and D. R. Whilhite, *J. Amer. Chem. Soc.*, 1974, **96**, 5375.
49. J. Schraml, P. Koehler, K. Licht and G. Engelhardt, *J. Organometal. Chem.*, 1976, **121**, C1.
50. V. A. Pestunovich, M. E. Larin, E. I. Dubinskaya and M. G. Voronkov, *Dokl. Akad. Nauk SSSR*, 1977, **233**, 378.
51. J. Schraml, J. Pola, M. Černý, H. Jancke, G. Engelhardt and V. Chvalovský, *Coll. Czech. Chem. Comm.*, 1976, **41**, 360.
52. J. Schraml, R. Ponec, V. Chvalovský, G. Engelhardt, H. Jancke, H. Kriegsmann, M. Г. Larin, V. A. Pestunovich and M. G. Voronkov, *J. Organomet. Chem.*, 1979, **178**, 55.
53. L. Delmulle and G. P. van der Kelen, *J. Mol. Struct.*, 1980, **66**, 315.
54. L. Delmulle and G. P. van der Kelen, *J. Mol. Struct.*, 1980, **66**, 309.
55. C. G. Pitt, *J. Organomet. Chem.*, 1973, **61**, 49.
56. T. N. Mitchell and H. C. Marsmann, *Org. Magn. Reson.*, 1981, **15**, 263.
57. E. Lippmaa, M. Mägi, G. Engelhardt, H. Jancke, V. Chvalovský and J. Schraml, *Coll. Czech. Chem. Comm.*, 1974, **39**, 1041.
58. J. Schraml, J. Pola, V. Chvalovský, M. Mägi and E. Lippmaa, *J. Organomet. Chem.*, 1973, **49**, C19.

59. J. Schraml, J. Včelák and V. Chvalovský, *Coll. Czech. Chem. Comm.*, 1974, **39**, 267.
60. N. D. Chuy, V. Chvalovský, J. Schraml, M. Mägi and E. Lippmaa, *Coll. Czech. Chem. Comm.*, 1975, **50**, 875.
61. J. Schraml, J. Včelák, G. Engelhardt and V. Chvalovský, *Coll. Czech. Chem. Comm.*, 1976, **41**, 3758.
62. J. Schraml, V. Chvalovský, M. Mägi and E. Lippmaa, *Coll. Czech. Chem. Comm.*, 1977, **42**, 306.
63. J. Schraml, N. D. Chuy, V. Chvalovský, M. Mägi and E. Lippmaa, *Org. Magn. Reson.*, 1975, **7**, 379.
64. J. Schraml, V. Chvalovský, M. Mägi and E. Lippmaa, *Coll. Czech. Chem. Comm.*, 1975, **40**, 897.
65. M. Čapka, J. Schraml and H. Jancke, *Coll. Czech. Chem. Comm.*, 1978, **43**, 3347.
66. J. Schraml, V. Chvalovský, M. Mägi and E. Lippmaa, *Coll. Czech. Chem. Comm.*, 1978, **43**, 3365.
67. J. Schraml, V. Chvalovský, M. Mägi and E. Lippmaa, *Coll. Czech. Chem. Comm.*, 1979, **44**, 854.
68. E. Lukevits, E. Liepin'sh, E. P. Popova, V. D. Shatts and V. A. Belikov, *Zh. Obshch. Khim.*, 1980, **50**, 388.
69. E. Lukevits, O. A. Pudova, Yu. Popelis and N. P. Erchak, *Zh. Obshch. Khim.*, 1981, **51**, 369.
70. U. Niemann and H. C. Marsmann, *Z. Naturforsch.*, 1975, **30b**, 202.
71. G. E. Maciel, in *Topics in Carbon-13 NMR Spectroscopy*, Vol. 1 (G. C. Levy, ed.), Wiley, New York, 1974.
72. K. G. Sharp, P. A. Sutor, E. A. Williams, J. D. Cargioli, T. C. Farrar and K. Ishibitsu, *J. Amer. Chem. Soc.*, 1976, **98**, 1977.
73. H. C. Marsmann, W. Raml and E. Hengge, *Z. Naturforsch*, 1980, **35b**, 35.
74. R. Löwer, M. Vongehr and H. C. Marsmann, *Chem. Z.*, 1975, **99**, 33.
75. E. A. Williams and J. D. Cargioli, unpublished results.
76. J. Schraml, N. D. Chuy, V. Chvalovský, M. Mägi and E. Lippmaa, *J. Organomet. Chem.*, 1973, **51**, C5.
77. R. K. Harris and B. J. Kimber, *Advanc. Mol. Relax. Process.*, 1976, **8**, 15.
78. D. M. Grant and E. G. Paul, *J. Amer. Chem. Soc.*, 1964, **86**, 2984.
79. E. Liepins and E. Lukevits, *Izv. Akad. Nauk SSSR Ser. Khim.*, 1979, **3**, 351.
80. A. G. Brook, S. C. Nyburg, W. F. Reynolds, Y. C. Poon, Y.-M. Chang, J.-S. Lee and J.-P. Picard, *J. Amer. Chem. Soc.*, 1979, **101**, 6750.
81. G. Fritz and N. Braunagel, *Z. Anorg. Allg. Chem.*, 1973, **399**, 280.
82. E. A. Williams, J. D. Cargioli and P. E. Donahue, *J. Organomet. Chem.*, 1980, **192**, 319.
83. D. Kovar, K. Utvary and E. Hengge, *Monats. Chem.*, 1979, **110**, 1295.
84. A. Marchand, P. Gerval, M. Joanny and P. Mazerolles, *J. Organomet. Chem.*, 1981, **217**, 19.
85. M.-L. Filleux-Blanchard, N. D. An and G. Manuel, *Org. Magn. Reson.*, 1978, **11**, 150.
86. R. H. Cragg and R. D. Lane, *J. Organomet. Chem.*, 1981, **212**, 301.
87. R. L. Lambert, Jr and D. Seyferth, *J. Amer. Chem. Soc.*, 1972, **94**, 9246.
88. D. Seyferth, D. C. Annarelli and S. C. Vick, *J. Amer. Chem. Soc.*, 1976, **98**, 6382.
89. D. Seyferth and D. C. Annarelli, *J. Amer. Chem. Soc.*, 1975, **97**, 2273.
90. D. Seyferth and D. C. Annarelli, *J. Organometal. Chem.*, 1976, **117**, C51.
91. D. A. Stanislawski and R. West, *J. Organomet. Chem.*, 1981, **204**, 295.
92. J. J. Burke and P. C. Lauterbur, *J. Amer. Chem. Soc.*, 1964, **86**, 1870.
93. M. Ishikawa, J. Iyoda, H. Ikeda, K. Kotake, T. Hashimoto and M. Kumada, *J. Amer. Chem. Soc.*, 1981, **103**, 4845.
94. D. A. Stanislawski and R. West, *J. Organomet. Chem.*, 1981, **204**, 307.

95. H. C. Marsmann, W. Raml and E. Hengge, *Z. Naturforsch.*, 1980, **35b**, 1541.
96. J. Hahn, *Z. Naturforsch.*, 1980, **35b**, 282.
97. G. Engelhardt, H. Jancke, R. Radeglia, H. Kriegsmann, M. E. Larin, V. A. Pestunovich, E. I. Dubinskaja and M. Voronkov, *Z. Chem.*, 1977, **17**, 376.
98. S. Cradcock, E. A. V. Ebsworth, N. S. Hosmane and K. M. Mackay, *Angew. Chem.*, 1975, **87**, 207.
99. T. N. Mitchell and H. C. Marsmann, *J. Organomet. Chem.*, 1978, **150**, 171.
100. H. Sakurai, Y. Kamiyama, A. Mikoda, R. Kobayashi, K. Sasaki and Y. Nakadaira, *J. Organomet. Chem.*, 1980, **20–1**, C14.
101. K. H. Pannell, A. R. Bassindale and J. W. Fitch, *J. Organomet. Chem.*, 1981, **209**, C65.
102. W. Malisch and W. Ries, *Chem. Ber.*, 1979, **112**, 1304.
103. F. H. Kohler, H. Hollfelder and E. O. Fischer, *J. Organomet. Chem.*, 1979, **168**, 53.
104. K. M. Abraham and G. Urry, *Inorg. Chem.*, 1973, **12**, 2850.
105. K. H. Pannell and A. R. Bassindale, *J. Organomet. Chem.*, 1982, **229**, 1.
106. E. Colomer, R. J. P. Corriu, C. Marzin and A. Vioux, *Inorg. Chem.*, 1982, **21**, 368.
107. O. V. Kuzmin, A. L. Bikovetz, V. M. Vdovin and A. M. Krapivin, *Izv. Akad Nauk SSSR Ser. Khim.*, 1979, **1979**, 2815.
108. G. Engelhardt, H. Jancke, M. Mägi, T. J. Pehk and E. Lippmaa, *J. Organomet. Chem.*, 1971, **28**, 293.
109. R. K. Harris, B. J. Kimber, M. D. Wood and A. Holt. *J. Organomet.. Chem.*, 1976, **116**, 291.
110. V. A. Pestunovich, M. F. Larin, M. G. Voronkov, G. Engelhardt, H. Jancke, V. P. Mileshkevich and Y. A. Yuzhelevski, *Zhur. Strukt. Khim.*, 1977, **18**, 578.
111. D. Seyferth, private communication, 1982.
112. R. K. Harris and B. J. Kimber, *Appl. Spectrosc. Rev.*, 1975, **10**, 117.
113. R. K. Harris and B. J. Kimber, *J. Organomet. Chem.*, 1974, **70**, 43.
114. H. Jancke, G. Engelhardt, M. Mägi and E. Lippmaa, *Z. Chem.*, 1973, **13**, 435.
115. H. Jancke, G. Engelhardt, H. Kriegsmann and F. Keller, *Plaste Kautsch*, 1979, **26**, 612.
116. G. Engelhardt and H. Jancke, *Polymer Bull.*, 1981, **5**, 577.
117. R. K. Harris and M. L. Robins, *Polymer*, 1978, **19**, 1123.
118. R. K. Harris and B. J. Kimber, *Chem. Comm.*, 1974, **1974**, 559.
119. R. K. Harris, B. J. Kimber and M. D. Wood, *J. Organomet. Chem.*, 1976, **116**, 291.
120. H. G. Horn and H. C. Marsmann, *Makromol. Chem.*, 1972, **162**, 255.
121. R. W. LaRochelle, J. D. Cargioli and E. A. Williams, *Macromolecules*, 1976, **9**, 85.
122. E. A. Williams, J. D. Cargioli and S. Y. Hobbs, *Macromolecules*, 1977, **10**, 782.
123. R. W. LaRochelle, J. D. Cargioli and E. A. Williams, unpublished results.
124. E. A. Williams, J. D. Cargioli and R. W. LaRochelle, *J. Organomet. Chem.*, 1976, **108**, 153.
125. V. Gutmann, *Chem. Brit.*, 1971, **7**, 102.
126. E. Lippmaa, M. A. Alla, T. J. Pehk and G. Engelhardt, *J. Amer. Chem. Soc.*, 1978, **100**, 1929.
127. R. K. Harris and R. H. Newman, *Org. Magn. Reson.*, 1977, **9**, 426.
128. D. Hoebbel, G. Garzó, G. Engelhardt, H. Hancke, P. Franke and W. Wieker, *Z. Anorg. Allg. Chem.*, 1976, **424**, 115.
129. G. Engelhardt and H. Jancke, *Z. Chem.*, 1974, **14**, 206.
130. B. D. Lavrukhin, B. A. Astapov, A. A. Zhadanov, G. Engelhardt and H. Jancke, *Org. Magn. Reson.*, 1982, **18**, 71.
131. D. Hoebbel, G. Garzo, G. Engelhardt and A. Till, *Z. Anorg. Allg. Chem.*, 1979, **450**, 5.
132. G. Engelhardt, D. Ziegan, H. Jancke, D. Hoebbel and W. Wieker, *Z. Anorg. Allg. Chem.*, 1975, **418**, 17.
133. H. C. Marsmann, *Z. Naturforsch*, 1974, **29b**, 495.

134. R. O. Gould, B. M. Lowe and N. A. MacGilp, *Chem. Comm.*, 1974, **1974**, 720.
135. G. Engelhardt, W. Altenburg, D. Hoebbel and W. Wieker, *Z. Anorg. Allg. Chem.*, 1977, **428**, 43.
136. G. Engelhardt, W. Altenburg, D. Hoebbel and W. Wieker, *Z. Anorg. Allg. Chem.*, 1977, **437**, 249.
137. B. D. Mosel, W. Muller-Warmuth and H. Dutz, *Phys. Chem. Glasses*, 1974, **15**, 154.
138. R. K. Harris and R. H. Newman, *J. Chem. Soc. Faraday Trans.*, 1977, **73**, 1204.
139. G. Engelhardt, H. Jancke, D. Hoebbel and W. Wieker, *Z. Chem.*, 1974, **14**, 109.
140. R. K. Harris, J. Jones, C. T. G. Knight and D. Pawson, *J. Mol. Struct.*, 1980, **69**, 95.
141. R. K. Harris, C. T. G. Knight and D. N. Smith, *Chem. Comm.*, 1980, **1980**, 726.
142. R. K. Harris, C. T. G. Knight and W. E. Hull, *J. Amer. Chem. Soc.*, 1981, **103**, 1577.
143. R. K. Harris, J. Jones and S. Ng, *J. Magn. Reson.*, 1978, **30**, 521.
144. V. A. Pestunovich, S. N. Tandura, M. G. Voronkov, V. P. Baryshok, C. I. Zelchan, V. I. Glukhikh, G. Englehardt and M. Witanowski, *Spectrosc. Lett.*, 1978, **11**, 339.
145. V. A. Pestunovich, S. N. Tandura, M. G. Voronkov, G. Engelhardt, E. Lippmaa, T. Pehk, V. F. Sidorkin, G. I. Zelchan, V. P. Baryshok, *Dokl. Akad. Nauk SSSR*, 1978, **240**, 914.
146. E. E. Liepinsh, I. S. Birgele, I. I. Solomennikova, A. F. Lapsinya, G. I. Zelchan and E. Lukevits, *Zhur. Obshch. Khim.*, 1980, **50**, 2462.
147. P. Henscei and H. C. Marsmann, *Acta Chem. Acad. Sci. Hung.*, 1980, **105**, 79.
148. S. N. Tandura, V. A. Pestunovich, M. G. Voronkov, G. Zelchan, I. I. Solomennikova and E. Lukevits, *Khim. Geterotsikl. Soedin.*, 1977, **1977**, 1063.
149. J. A. Cella, J. D. Cargioli and E. A. Williams, *J. Organomet. Chem.*, 1980, **186**, 13.
150. H. C. Marsmann and R. Löwer, *Chem. Z.*, 1973, **97**, 660.
151. E. A. Williams, and J. D. Cargioli, unpublished results (courtesy of T. Zens, Varian Assoc.).
152. H. Jancke, G. Engelhardt, S. Wagner, W. Dirnens, G. Herzog, E. Thieme and K. Ruhlmann, *J. Organometal. Chem.*, 1977, **134**, 21.
153. B. Heinz, H. C. Marsmann and U. Niemann, *Z. Naturforsch.*, 1977, **32b**, 163.
154. J. Schraml, J. Pola, V. Chvalovský, H. C. Marsmann and K. Bláha, *Coll. Czech. Chem. Comm.*, 1977, **42**, 1165, and references therein.
155. A. H. Haines, R. K. Harris and R. C. Rao, *Org. Magn. Reson.*, 1977, **9**, 432.
156. D. J. Gale, A. H. Haines and R. K. Harris, *Org. Magn. Reson.*, 1975, **7**, 635.
157. J. Schraml, J. Pola, H. Jancke, G. Engelhardt, M. Černý and V. Chvalovský, *Coll. Czech. Chem. Comm.*, 1976, **41**, 360.
158. L. Riesel, A. Claussnitzer and C. Ruby, *Z. Anorg. Allg. Chem.*, 1977, **433**, 200.
159. A. R. Bassindale and T. B. Posner, *J. Organomet. Chem.*, 1979, **175**, 273.
160. A. Steigel, *Chem. Ber.*, 1980, **113**, 3915.
161. W. M. Colemann, III and A. R. Boyd, *Anal. Chem.*, 1982, **54**, 153.
162. K. D. Rose and C. G. Scouten, in *Chemistry and Physics of Coal Utilization*, 1980, American Physical Society Proceedings, Morgantown, American Institute of Physics, New York, 1981, p. 82.
163. J. C. Thompson, A. P. G. Wright and W. E. Reynolds, *J. Amer. Chem. Soc.*, 1979, **101**, 2236.
164. D. Seyferth and D. P. Duncan, *J. Amer. Chem. Soc.*, 1978, **100**, 7734.
165. G. A. Olah, A. L. Berrier, L. D. Field and G. K. S. Prakash, *J. Amer. Chem. Soc.*, 1980, **104**, 1349.
166. G. A. Olah and R. J. Hunadi, *J. Amer. Chem. Soc.*, 1980, **102**, 6989.
167. R. W. Hoffmann, M. Lotze, M. Reiffen and K. Steinbach, *Liebigs Ann. Chem.*, 1981, **1981**, 581.
168. W. E. Dibble, Jr, B. H. W. S. DeJong and L. W. Cary, *Proc. 3rd Int. Symp. Water-Rock Interaction*, 1980, 47.

169. E. T. Lippmaa, M. A. Alla, T. J. Pehk and G. Engelhardt, *J. Amer. Chem. Soc.*, 1978, **100**, 1929.
170. E. Lippmaa, M. Mägi, A. Samosan, G. Engelhardt and A. R. Grimmer, *J. Amer. Chem. Soc.*, 1980, **102**, 4889.
171. G. Engelhardt, D. Ziegan, E. Lippmaa and M. Mägi, *Z. Anorg. Allg. Chem.*, 1980, **468**, 35.
172. M. Mägi, A. Samosan, M. Tarmak, G. Engelhardt and E. Lippmaa, *Dokl. Akad. Nauk SSSR*, 1981, **261**, 1169.
173. E. Lippmaa, M. Mägi, A. Samosan, M. Tarmak and G. Engelhardt, *J. Amer. Chem. Soc.*, 1981, **103**, 4992.
174. J. Klinowski, J. M. Thomas, C. A. Fyfe and J. S. Hartman, *J. Phys. Chem.*, 1981, **85**, 2590.
175. G. Engelhardt, E. Lippmaa and M. Mägi, *Chem. Comm.*, 1981, **1981**, 712.
176. J. M. Thomas, L. A. Bussill, E. A. Lodge, A. Cheetham and C. A. Fyfe, *J. Chem. Soc. Chem. Comm.*, 1981, **1981**, 276.
177. A.-R. Grimmer, R. Peter, E. Fechner and G. Molgedey, *Chem. Phys. Lett.*, 1981, **77**, 331.
178. J. Klinowski, J. M. Thomas, M. Audier, S. Vasudevan, C. A. Fyfe and J. S. Hartman, *Chem. Comm.*, 1981, **1981**, 570.
179. S. Ramdas, J. M. Thomas, J. Klinowski, C. A. Fyfe and J. S. Hartman, *Nature (Lond.)*, 1981, **292**, 228.
180. G. Engelhardt, E. Lippmaa and M. Mägi, *Chem. Comm.*, 1981, **1981**, 712.
181. C. A. Fyfe, G. C. Gobbi, J. S. Hartman, R. F. Lenkinski, J. H. O'Brien, E. R. Beange and M. A. R. Smith, *J. Magn. Reson.*, 1982, **47**, 1681.
182. G. Engelhardt, D. Hoebbel, M. Tarmak, A. Samosan and E. Lippmaa, *Z. Anorg. Allg. Chem.*, 1982, **484**, 22.
183. W. Loewenstein, *Amer. Mineral*, 1954, **39**, 92.
184. A. Pines, M. G. Gibby and J. S. Waugh, *J. Chem. Phys.*, 1973, **59**, 569.
185. E. R. Andrew, A. Bradbury and R. G. Eades, *Nature (Lond.)*, 1958, **182**, 1659.
186. E. R. Andrew, *Prog. Nucl. Reson. Spectrosc.*, 1971, **8**, 1.
187. J. Schaefer and E. O. Stejskal, in *Topics in Carbon-13 NMR Spectroscopy* (G. C. Levy, ed.), Wiley, New York, 1979, p. 283.
188. G. E. Maciel and D. W. Sindorf, *J. Amer. Chem. Soc.*, 1980, **102**, 7607.
189. D. W. Sindorf and G. E. Maciel, *J. Amer. Chem. Soc.*, 1981, **103**, 4263.
190. G. E. Maciel, D. W. Sindorf and V. J. Bartuska, *J. Chromatog.*, 1981, **205**, 438.
191. E. T. Lippmaa, A. V. Samosan, V. V. Brei and Yu. I. Gorlov, *Dokl. Akad. Nauk SSSR*, 1981, **259**, 403.
192. A. R. Siedle and R. A. Newmark, *J. Amer. Chem. Soc.*, 1981, **103**, 1240.
193. G. E. Maciel, M. J. Sullivan and D. W. Sindorf, *Macromolecules*, 1981, **14**, 1607.
194. G. Engelhardt. H. Jancke, E. Lippmaa and A. Samosan, *J. Organomet. Chem.*, 1981, **210**, 295.
195. N. F. Ramsey and E. M. Purcell, *Phys. Rev.*, 1952, **85**, 143.
196. M. D. Beer and R. Grinter, *J. Magn. Reson.*, 1978, **31**, 187.
197. E. R. Malinowski, *J. Amer. Chem. Soc.*, 1961, **83**, 4479.
198. E. A. V. Ebsworth and J. J. Turner, *J. Chem. Phys.*, 1962, **36**, 2628.
199. C. Juan and H. S. Gutowsky, *J. Chem. Phys.*, 1962, **37**, 2198.
200. H. J. Cambell-Ferguson, E. A. V. Ebsworth, A. G. MacDiarmid and T. Yoshioka, *J. Phys. Chem.*, 1967, **71**, 723.
201. F. O. Bishop and M. A. Jensen, *Chem. Comm.*, 1966, **1966**, 922.
202. E. P. Malinowski and T. Vladimiroff, *J. Amer. Chem. Soc.*, 1969, **86**, 3575.
201. A. G. Brook, F. Abdesaken, B. Gutekunst, G. Gutekunst and R. K. Kallury, *Chem. Comm.*, 1981, **1981**, 191.
202. R. West, M. J. Fink and J. Michl, *Science*, 1981, **214**, 1343.
203. M. A. Jensen, *J. Organomet. Chem.*, 1968, **11**, 423.

204. J. E. Drake and N. Goddard, *J. Chem. Soc. A*, 1970, 2587.
205. E. A. V. Ebsworth, A. G. Lee and G. M. Sheldrik, *J. Chem. Soc. A*, 1968, **1968**, 2294.
206. C. Juan and H. S. Gutowsky, *J. Chem. Phys.*, 1962, **37**, 2198.
207. A. Rastelli and S. A. Pozzoli, *J. Mol. Struct.*, 1973, **18**, 463.
208. G. C. Levy, D. M. White and J. D. Cargioli, *J. Magn. Reson.*, 1972, **8**, 280.
209. H. Dreeskamp and K. Hildenbrand, *Annalen*, 1975, **1975**, 712.
210. K. Kovacević and Z. B. Maksić, *J. Mol. Struct.*, 1973, **17**, 203.
211. K. Kovacević, K. Krmpotic and Z. B. Maksić, *Inorg. Chem.*, 1977, **16**, 1421.
212. K. D. Summerhays and D. A. Deprez, *J. Organomet. Chem.*, 1976, **118**, 19.
213. M. D. Beer and R. Grinter, *J. Magn. Reson.*, 1977, **26**, 421.
214. M. D. Beer and R. Grinter, *J. Magn. Reson.*, 1978, **31**, 187.
215. H. B. Jansen, A. Meeuwis and P. Pyykkö, *Chem. Phys.*, 1979, **38**, 173.
216. R. K. Harris and C. T. G. Knight, *J. Mol. Struct.*, 1982, **78**, 273.
217. A. H. Cowley and W. D. White, *J. Amer. Chem. Soc.*, 1969, **91**, 1913, 1917.
218. C. J. Jameson and H. S. Gutowsky, *J. Chem. Phys.*, 1969, **51**, 2790.
219. W. McFarlane, *J. Chem. Soc. A*, 1967, **1967**, 1275.
220. R. R. Dean and W. McFarlane, *Mol. Phys.*, 1967, **12**, 289, 364.
221. R. B. Johannesen, *J. Chem. Phys.*, 1967, **47**, 955.
222. S. S. Danyluk, *J. Amer. Chem. Soc.*, 1964, **86**, 4504.
223. G. Pfisterer and H. Dreeskamp. *Ber. Bunsenges. Phys. Chem.*, 1969, **73**, 654.
224. S. A. Linde, H. J. Jakobsen and B. J. Kimber, *J. Amer. Chem. Soc.*, 1975, **97**, 3219.
225. S. Danyluk, *J. Amer. Chem. Soc.*, 1965, **87**, 2300.
226. J. D. Kennedy, W. McFarlane, G. S. Pyne and B. Wrackmeyer, *J. Chem. Soc. Dalton Trans.*, 1975, **1975**, 386.
227. A. H. Cowley, W. D. White and S. L. Manatt, *J. Amer. Chem. Soc.*, 1967, **89**, 6433.
228. E. A. V. Ebsworth and J. J. Turner, *J. Chem. Phys.*, 1962, **36**, 2628.
229. D. E. I. Arnold, S. Cradcock, E. A. V. Ebsworth, J. D. Murdoch, D. W. H. Rankin, D. C. J. Skea, R. K. Harris and B. J. Kimber, *J. Chem. Soc. Dalton Trans.*, 1981 **1981**, 1349.
230. C. Shumann and H. Dreeskamp. *J. Magn. Reson.*, 1970, **3**, 204.
231. J. V. Urenovitch and R. West, *J. Organomet. Chem.*, 1965, **3**, 138.
232. H. Bürger and W. K. Kilian, *J. Organomet. Chem.*, 1969, **18**, 299.
233. H. Bürger and W. Kilian, *J. Organomet. Chem.*, 1971, **26**, 47.
234. E. Hengge and F. Holler, *Z. Naturforsch*, 1971, **26a**, 768.
235. V. A. Pestunovich, S. N. Tandura, M. G. Voronkov, V. P. Baryshok, G. I. Zelchan, V. I. Glukhikh, G. Engelhardt and M. Witanowski, *Spectrosc. Lett.*, 1978, **11**, 339.
236. F. K. Cartledge and K. H. Riedel, *J. Organomet. Chem.*, 1972, **34**, 11.
237. Y. Nagai, M. Ohtsuki, T. Nakano and H. Watanabe, *J. Organomet. Chem.*, 1972, **37**, 81.
238. Y. Nagai, H. Matsumoto, T. Nakano and H. Watanabe, *Bull. Chem. Soc. Japan*, 1972, **45**, 2560.
239. H. Watanabe, R. Akaba, T. Iezumi and Y. Nagai, *Bull. Chem. Soc. Japan*, 1980, **53**, 2981.
240. J. E. Drake and N. P. C. Westwood, *J. Chem. Soc. A*, 1971, **1971**, 3300.
241. W. Airey and G. M. Sheldrick, *J. Inorg. Nucl. Chem.*, 1970, **32**, 1827.
242. J. C. Thompson and J. L. Margrave, *Chem. Comm.*, 1966, **1966**, 566.
243. F. Hofler and W. Veigl, *Angew. Chem.*, 1971, **83**, 977.
244. R. B. Johannesen, T. C. Farrar, F. E. Brinckman and T. D. Coyle, *J. Phys. Chem.*, 1966, **44**, 962.
245. J. F. Bald, Jr, K. G. Sharp and A. G. MacDiarmid, *J. Fluorine Chem.*, 1973, **3**, 433.
246. E. A. V. Ebsworth and S. G. Frankiss, *Trans. Faraday Soc.*, 1963, **59**, 1518.
247. J. J. Moscony and A. G. MacDiarmid, *Chem. Comm.*, 1965, **1965**, 307.
248. E. L. Muetterties and W. D. Phillips, *J. Amer. Chem. Soc.*, 1959, **81**, 1084.
249. R. K. Harris and B. J. Kimber, *Org. Magn. Reson.*, 1975, **7**, 460.

250. H. Elser and H. Dreeskamp, *Ber. Bunsenges. Phys. Chem.*, 1969, **73**, 619.
251. K. Kamienska-Trela, *J. Organomet. Chem.*, 1978, **159**, 15.
252. T. N. Mitchell and H. C. Marsmann, *J. Organomet. Chem.*, 1978, **150**, 171.
253. R. Gruning, P. Krommes and J. Lorberth, *J. Organomet. Chem.*, 1977, **127**, 167.
254. D. W. W. Anderson, E. A. V. Ebsworth and D. W. H. Rankin, *J. Chem. Soc. Dalton Trans.*, 1973, **1973**, 2370.
255. B. Wrackmeyer and W. Biffar, *Z. Naturforsch.*, 1979, **34**, 1270.
256. J. H. Noggle and R. E. Schirmer, *The Nuclear Overhauser Effect*, Academic Press, New York, 1971.

NMR of Organic Compounds Adsorbed on Porous Solids

HARRY PFEIFER, WOLFGANG MEILER AND
DETLEF DEININGER

Sektion Physik, Karl-Marx-Universität, DDR-7010 Leipzig, Linnéstrasse 5, German Democratic Republic

I. INTRODUCTION

Porous solids such as zeolites (molecular sieves), silica and alumina play an important role in many industrial processes, especially in the chemical industry, where they are used mainly as catalysts or as adsorbents. The state of adsorbed molecules, which is in several respects intermediate between the liquid and solid state, has been the subject of numerous investigations. Most of them make use of classical thermodynamic methods like the measurement of adsorption isotherms and adsorption heats or of well established spectroscopic techniques, mainly with infrared spectroscopy. The advent of special NMR techniques to enhance sensitivity, e.g. FT NMR spectroscopy and cross-polarization (CP), to measure translational

ANNUAL REPORTS ON NMR SPECTROSCOPY
VOLUME 15   ISBN 0-12-505315-0

motion directly (pulsed field gradient technique), and to reduce linewidths through special pulse sequences and magic angle spinning, has produced an increasing number of papers dealing with the study of surface phenomena using NMR.[1-5] Characteristic examples which demonstrate the possibilities of NMR methods in this field of research are the measurement of proton spin relaxation of adsorbed molecules,[2] proton resonance studies of surface OH groups,[6,7] and the direct measurement of translational molecular motion in zeolites;[8,9] in the last-mentioned case it is shown that the values for the intracrystalline self-diffusion coefficients in zeolites, as determined by the classical sorption technique, are erroneous by up to five orders of magnitude.

In 1972 the first highly resolved ^{13}C NMR spectra of adsorbed molecules were measured.[10,11] With these experiments it became possible to study the change of the electronic state of the various carbon atoms in a molecule due to its interaction with the surface. Moreover, from the ^{13}C relaxation times and the nuclear Overhauser enhancement (NOE) factors, additional information about the geometry and the microdynamic behaviour of the adsorbed molecule can be derived.

Recent developments include the measurement of highly resolved ^{15}N resonances which seem to be more sensitive to specific interactions between the adsorbed molecule and the surface[12,13] than ^{13}C resonances, and the study of the nuclei of the adsorbent using magic angle spinning of the sample (MASS) in addition to CP.[14] In the present account it is not possible to give a review of the whole field without neglecting important details of the methods, of the different problems and of the theoretical background as, for instance, in the case of relaxation and line shape analysis. Instead, we shall review papers which are connected with a study of highly resolved NMR spectra of organic compounds adsorbed in porous solids and their theoretical interpretation in terms of a quantum mechanical treatment of the surface complexes formed.

II. HIGHLY RESOLVED NMR SPECTRA OF MOLECULES ADSORBED ON POROUS SOLIDS

A. Porous solids, their structure and properties

Because of their relatively low sensitivity, NMR signals can only be observed in the case of adsorbents having a high specific surface area:[2]

$$S \geqslant 10 \, m^2 \, g^{-1} \tag{1}$$

On the other hand, just the same condition must be fulfilled to ensure high productivity of the adsorbents if they are used as catalysts or molecular sieves in industry. Typical examples are alumina, silica, porous carbons and especially zeolites. To understand the problems and the interpretation of the results to be described below, we choose examples where zeolites and silica are used, and present some details of the structure and properties of these adsorbents.

1. Zeolites[15-22]

The surface properties of zeolites have received much attention since the pioneering work of Barrer.[18] Because of their well defined porous structure and their high specific surface area (up to $1000 \, m^2 \, g^{-1}$), they are attractive solids on which to make experimental and theoretical studies of adsorption. Moreover, they are widely used in industry as molecular sieves and selective adsorbents, as catalysts of high activity, selectivity and thermal stability, and as ion exchangers. At present, for instance, more than 70% of all paraffins are obtained by selective adsorption in zeolites, and the introduction of cracking catalysts based on zeolite Y improved gasolene yields in industry by as much as 25%. The common formula of zeolites is

$$(AlO_2)^- M^+ (SiO_2)_n$$

where M^+ denotes an exchangeable cation and n is a number greater than or equal to 1.

The examples given here are concerned mainly with molecules in the pores of a special "family" of zeolites (the faujasite group), to which belong the so-called zeolites NaX and NaY. These synthetic zeolites are available as small crystallites with a mean diameter between 1 and 100 μm, and have the same structure as the mineral faujasite; they differ only in their value of n.

The faujasite structure is built up of SiO_4 and AlO_4 tetrahedra, stacked in such a way that two interconnected pore systems, one large and one small, result (see Fig. 1). The only one accessible to hydrocarbon molecules is the large pore system, which is a three-dimensional network of nearly spherical cavities commonly referred to as supercages. A supercage has a mean free diameter of 1.16 nm which is sufficient to adsorb up to about 30 water molecules or five benzene molecules, the diameters of the entrances to it are about 0.8–0.9 nm. The small pore system is inaccessible to hydrocarbons, but is accessible to cations and small molecules such as water. The free diameter of the small cavities, called cubo-octahedra, is 0.66 nm, corresponding to a volume of about four water molecules. The diameter of the entrances from the supercages to the cubo-octahedra (six-membered rings in Fig. 1) is 0.25 nm. There are eight supercages and eight cubo-octahedra in each crystallographic unit cell. The crystalline dimensions depend but little on the degree of filling of the intracrystalline space. Because each silicon or aluminium ion is surrounded tetrahedrally by four larger oxygens, the internal surfaces consist of oxygen, the only other elements

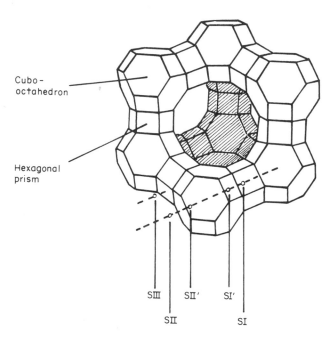

FIG. 1. Model of zeolite X or Y. T-atoms (T = Al or Si) are located at the corners of the geometrical figures shown, and oxygen atoms on the lines connecting neighbouring corners, so that each T-atom is tetrahedrally surrounded by four oxygen atoms.

exposed to adsorbates being the exchangeable cations, which are sodium ions in the case of NaX and NaY zeolites. By X-ray measurements[23] it was possible to localize only a part of these cations. In Table 1 results are summarized for the number of exchangeable cations per cubo-octahedron (denoted as sites SI′) and per supercage (non-localizable cations and cations on SII, which are sites in front of the windows of the cubo-octahedra at the wall of the supercage; see Fig. 1) in dehydrated (hydrated) zeolites NaX and NaY. For completeness it should be mentioned that cations situated in the hexagonal prisms (SI), which connect neighbouring cubo-octahedra (see Fig. 1), have been omitted in Table 1 since these cations are not accessible at all to adsorbed molecules.

The model of another important member of the family of zeolites, the so-called zeolite A, is shown in Fig. 2. Its silicon-to-aluminium ratio is 1, and the cubo-octahedra are arranged in such a way that the large cavity has about the same mean free diameter (1.14 nm) as for zeolite X or Y, but that of the entrances is much smaller (0.41–0.45 nm). One large cavity and one cubo-octahedron form the crystallographic unit cell, and the exchangeable cations are located only in the large cavity. For a NaA zeolite, eight sodium ions are located in front of the six-membered rings (sites 1

TABLE 1

**Number of exchangeable cations (Na^+) per cubo-octahedron (SI')
and supercage (SII, non-localized) in dehydrated (hydrated) NaX and
NaY zeolites.**

	NaX ($n = 1.37$)	NaY ($n = 2.26$)
SI'	1.7 (0.9)	2.2 (1.6)
SII	3.2 (2.9)	3.4 (2.6)
Non-localized	4.0 (5.2)	1.1 (3.0)

in Fig. 2) and the remaining four are distributed over positions near the centres of the entrances (sites 2 in Fig. 2) and of the cavity, depending on the degree of hydration.

The adsorption and catalytic properties of zeolites can be modified through an exchange of the sodium ions with other cations (e.g. Li^+, K^+, Rb^+, Cs^+, Tl^+, Ag^+, Ca^{2+}, Co^{2+}), denoted in general as M^+, or through formation of hydroxyl groups (Broensted centres) or Lewis centres, depending on the heat treatment after an exchange with NH_4^+, or finally through the silicon-to-aluminium ratio of the zeolite lattice. In the following, 60 $AgNaX_{1.8}$ denotes a zeolite X with a silicon-to-aluminium ratio n of 1.8 and where 60% of the sodium ions were exchanged with Ag^+.

2. Silica[24-27]

In contrast to zeolites, the various types of silica which are used as adsorbents have no well defined structure. Their surface area, pore volume,

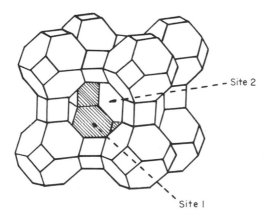

FIG. 2. Model of zeolite A. Key as for Fig. 1.

pore size and particle size are to some extent independently controllable, and it is this fact which makes the amorphous silicas important commercially. They can be subdivided into aqueous solution-based silica gels and into pyrogenic silicas. The primary particles of silica gel have diameters from 5 to 500 nm and contain small pores with diameters less than 1 nm. Through aggregation, secondary particles with diameters up to 50 000 nm[25] are formed and a whole spectrum of pores results. The values for the diameters of these pores range from 1 nm to more than 100 nm.

The primary particles of pyrogenic silicas (e.g. Aerosil) are non-porous spheres with diameters between about 1 and 50 nm, depending on the method of production used.

The atomic structure of silica is determined by SiO_4 tetrahedra interconnected via the oxygen atoms. At the surface the bonds are saturated through OH groups which are classified as single (isolated), geminal and vicinal hydroxyl groups, respectively, as shown in Fig. 3.

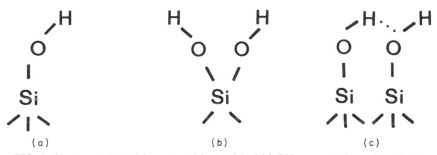

FIG. 3. Single or isolated (a), geminal (b) and vicinal (c) OH groups at the surface of silica.

At a pretreatment temperature of 200 °C the surface of pyrogenic silica contains between four and six OH groups per square nanometre, which is lower than for silica gel, where apparent values of up to 20 OH groups per square nanometre can be measured.[28] A certain number of the latter hydroxyl groups, however, are not accessible even for small molecules. These OH groups are located in the interior or at the contact area between neighbouring primary particles, so that the number of accessible OH groups is of the same order as for pyrogenic silicas.

There are several methods available to modify the surface of silica. Replacing the OH groups with methyl groups makes the surface hydrophobic, the introduction of phosphorus changes the acidity of the hydroxyl groups, and of course metal atoms or ions may be deposited on the surface.

B. Applicability of NMR

To study highly resolved NMR spectra of adsorbed organic compounds, only nuclei with a spin of $\frac{1}{2}$ are suitable. Among the constituents of organic compounds, such nuclei comprise mainly protons, and the isotopes ^{13}C and ^{15}N. In Table 2 quantities are given which characterize the NMR properties of these nuclei and their applicability to a study of adsorbed molecules.

In accordance with these values, ^{1}H NMR spectra of adsorbed molecules are successfully resolved in only a few cases.[29-35] More general attempts are based on an exchange of the adsorbed molecules with the bulk liquid[36,37] or on the various NMR line narrowing techniques.[38] The first experiment published in 1970 used MASS.[39] It seems, however, that residual molecular mobility of the adsorbed molecules prohibits a general application of line narrowing techniques. The reason is that line narrowing due to thermal motion and line narrowing techniques are not additive, and this holds for homonuclear[38] and heteronuclear[40] line broadening.

For the ^{13}C and ^{15}N NMR of adsorbed molecules, resolution is no problem at temperatures which are not too low. The term "low" depends on the strength of interaction with the surface. Most of the experiments can be performed at room temperature, but sometimes it is necessary to raise the temperature of measurement above 100 °C. A more difficult problem is sensitivity. Accumulation times from about 10 min up to 10 h are sufficient to study the ^{13}C NMR of adsorbed molecules having a natural abundance of ^{13}C. In contrast, measurement of ^{15}N spectra requires molecules enriched in ^{15}N, as can be seen from Table 2 by comparing the corresponding values for sensitivity and natural abundance.

TABLE 2

Characteristic NMR properties of the spin $\frac{1}{2}$ nuclei which are the main constituents of organic compounds.

	^{1}H	^{13}C	^{15}N
Magnetogyric ratio (s^{-1} T^{-1})	26.75×10^7	6.73×10^7	-2.71×10^{-7}
Resonance frequency (MHz) for $B_0 = 2.348\,69$ T	100	25.14	10.13
Relative intensity of the NMR signal for a given value of B_0	1	1.59×10^{-2}	1.04×10^{-3}
Natural abundance of the isotope (%)	99.985	1.108	0.365
Product of relative intensity and natural abundance (rounded values)	100	1.8×10^{-2}	3.8×10^{-4}
Interval of chemical shift (isotropic values) (ppm)	10	300	500
Typical NMR linewidths of adsorbed molecules (Hz) at about room temperature	500	100	100

In Fig. 4 some typical spectra are shown for acetonitrile and but-1-ene adsorbed in a zeolite NaX with about two and four molecules per supercage, respectively. The resonance frequencies are 90, 22.6 and 9.12 MHz, and the accumulation times are 150, 2000 and 8000 s for the ^{1}H, ^{13}C and ^{15}N NMR spectra, respectively. The ^{1}H and ^{13}C NMR spectra shown correspond to nuclei with natural abundance, whereas in the ^{15}N spectra the ^{15}N nuclei are enriched to 95%. All measurements are performed at room temperature using a commercial FT spectrometer with cylindrical samples of 10 mm diameter.

It can be seen from Fig 4 that all resonance lines are bell-shaped. This means that thermal motion of the molecules averages out the anisotropy of the nuclear shielding and valuable information is lost in these spectra. To measure the three principal values of the shielding tensor and not simply its average value (one-third of its trace), it is necessary to slow down thermal motion by cooling the sample and to apply techniques which reduce the homonuclear and/or heteronuclear magnetic dipolar line broadening without affecting the shielding tensor.[38] Compared with the liquid-like spectra shown for ^{13}C in Fig. 4, the resulting powder spectra may be drastically broadened (due to the anisotropy of the shielding tensor) and accompanied by an increasing overlap of the lines and a decreasing signal-to-noise ratio.[40,41] To circumvent this disadvantage, a partial reduction of the influence of the anisotropy should be performed through magic angle or near-magic angle spinning of the sample;[42-44] a first example can be found in a short note published recently.[182]

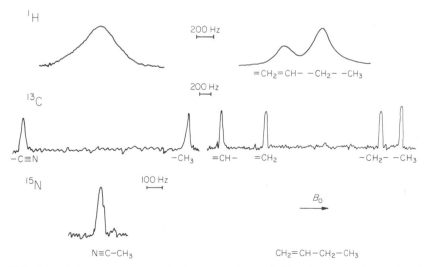

FIG. 4. Typical NMR spectra of hydrocarbons (acetonitrile and but-1-ene) adsorbed in porous solids (zeolite NaX). For details see text.

In reference 45, MASS is used to narrow the ^{13}C NMR line of CO_2, which is drastically broadened by the magnetic susceptibility heterogeneity of the adsorbent studied in this work (Na^+–mordenite).

C. Measurement of induced shifts

In most cases the observed shifts in the ^{13}C NMR spectra of molecules adsorbed on diamagnetic surfaces are less than about 10 ppm. On the other hand, the influence of the bulk susceptibility (cf. Section II.C.1) and of the van der Waals interactions (cf. Section II.C.2) may be of the same order of magnitude. Hence, it is necessary to eliminate these effects in order to determine that contribution to the observed shifts which is due to adsorption. If the surface is heterogeneous, i.e. if molecules are adsorbed both on adsorption sites forming adsorption complexes and on the surface between these sites (physically adsorbed molecules), the observed NMR spectrum may depend not only on the respective shifts but also on the rate of molecular exchange between the adsorption complexes and physically adsorbed molecules (cf. Section II.C.3).

1. Influence of susceptibility

Using the Faraday method, the diamagnetic, the paramagnetic and the ferromagnetic contributions to the susceptibility of various zeolite specimens of type X, Y and A were measured.[46] Some results are shown in Table 3,

TABLE 3

Diamagnetic, paramagnetic and ferromagnetic contributions to the magnetic susceptibility of various zeolite specimens having different content of paramagnetic impurities.[46]

Zeolite	Si/Al	Fe^{3+} (ppm)	$\chi_d \times 10^6$ (theoretical) $(cm^3\,g^{-1})$	$\chi_d \times 10^6$ $(cm^3\,g^{-1})$	$\chi_p \times 10^6$ ($T = 295$ K) $(cm^3\,g^{-1})$	$\chi_f \times 10^6$ $T = 295$ K, $H_0 = 10$ kOe) $(cm^3\,g^{-1})$
NaX_S	1.37	< 5	-0.401	-0.395	0	0
NaY_S	1.76	40		-0.390	0.008	0
NaY_S	2.26	110		-0.385	0.02	0
NaY_S	2.69	280	-0.409	-0.420	0.05	0.025
NaX_T	1.2	550		-0.350	0.100	0.090
NaY_T	2.6	500		-0.340	0.080	0.080
$NaCaY_T$	2.6	700		-0.360	0.080	0.165
NaA_T	1		-0.398	-0.365	0.040	0.065

To derive values for the volume susceptibility, the values given in this table must be multiplied by the apparent density of zeolites, $\rho \approx 0.5$ g cm^{-3}.

where theoretical values for the diamagnetic contribution to the susceptibility and numerical values for the amount of paramagnetic impurities (Fe^{3+}) are also included.

According to these results, the total susceptibility χ of the pure zeolite (NaX_S) is about $-0.4 \times 10^{-6}\,cm^3\,g^{-1}$ and does not depend on the temperature T or on the applied magnetic field H_0. In contrast, for the commercial zeolite NaX_T, having an impurity of about 550 ppm Fe^{3+} ions, the total susceptibility χ depends on both T and H_0. For $H_0 = 10$ kOe (which corresponds to a magnetic induction of $B_0 = 1$ T) the total susceptibility χ increases from $-0.3 \times 10^{-6}\,cm^3\,g^{-1}$ to $-0.2 \times 10^{-6}\,cm^3\,g^{-1}$ if T decreases from about 550 K to 170 K, and at room temperature χ increases by a factor of about 1.3 if H_0 decreases to half of the above value. Denoting the total volume susceptibilities (cf. legend to Table 3) as χ_V, the magnetic field acting on a nucleus is given by

$$H_i = \left[1 + \left(\frac{4\pi}{3} - \alpha_i + k_i\right)\chi_{Vi} - \sigma_i\right]H_0 \tag{2}$$

where the index $i = a$ characterizes the adsorbate–adsorbent system and $i = r$ characterizes the reference. α_i is the demagnetization factor, $k_i\chi_{Vi}H_0$ the local field inside the hypothetical Lorentzian sphere, and σ_i the shielding constant. The experimentally determined quantity

$$\delta_{obs} = \frac{(H_r - H_a)}{H_0} \tag{3}$$

and the real shift

$$\delta_{real} = \sigma_r - \sigma_a \tag{4}$$

are therefore related by

$$\delta_{real} = \delta_{obs} + \left[\frac{4\pi}{3} - \alpha_r + k_r\right]\chi_{Vr} - \left[\frac{4\pi}{3} - \alpha_a + k_a\right]\chi_{Va} \tag{5}$$

Because of the relatively fast isotropic reorientational and/or translational motion, even for molecules adsorbed in zeolites (with mean lifetimes of $\tau_c \lesssim 10^{-9}$ s in their equilibrium positions at room temperature), the influence of the local field is averaged out, and the factors k_r and k_a are zero. According to Wiedemann's rule we assume that the total volume susceptibility of the adsorbate–adsorbent system can be written as the weighted average

$$\chi_{V\,a} = (1 - p_A)\chi_{V\,solid} + p_A\chi_{V\,ads} \tag{6}$$

of the adsorbent ($\chi_{V\,solid}$) and of the adsorbate ($\chi_{V\,ads}$).

If the experimental values δ_{obs} are extrapolated to zero pore-filling factor ($p_A \rightarrow 0$), it follows from equation (5) (with $k_r = k_a = 0$ and $\alpha_a = \alpha_r = \alpha$) that

$$\delta_{real} = \delta_{obs} + \left(\frac{4\pi}{3} - \alpha\right) (\chi_{V\ r} - \chi_{V\ solid}) \qquad (7)$$

Using the gaseous state of the adorbate molecules ($\chi_{V\ r} = 0$) as a reference, and assuming that $\delta_{real} \approx 0$, which is justified (cf. below) for the proton resonance of such molecules as methane, cyclohexane and tetramethyl-silane,[47] it follows that

$$\delta_{obs} = \left(\frac{4\pi}{3} - \alpha\right) \chi_{V\ solid} \qquad (8)$$

from which the volume susceptibility of the adsorbent can be deetermined.[47] The critical assumption of this method is $\delta_{real} \approx 0$, since in general, for non-specifically adsorbed molecules, like those mentioned above, there may be a contribution of non-specific (van der Waals) interactions to the observed shift. The order of magnitude of this contribution can be estimated from the resonance shift between the gaseous and the liquid states.[48] Since this shift is equal to or smaller than 0.5 ppm for the proton resonance, the above assumption is approximately fulfilled. In contrast, it is found,[46] for the same molecules as studied in reference 47, that the ^{13}C shifts due to the van der Waals interactions may be significantly greater than those due to the bulk susceptibility. The same holds for ^{15}N shifts, so that this method[47] is limited to ^1H NMR.

Taking into account the usual form of the samples (cylindrical tubes perpendicular to the external magnetic field H_0 with a value of about 3 for the ratio between the filling height and the diameter of the sample), the quantity α can be put equal to 1.8π.[46] Using liquid benzene as a reference ($\chi_{V\ r} = -0.6 \times 10^{-6}$), the influence of the bulk susceptibility of zeolites (cf. Table 3) leads to an apparent shift of the resonance line (second term on the right-hand side of equation (7)) of about +0.7 ppm, where the positive sign denotes as usual a shift to lower magnetic fields or higher frequencies.[46,49]

2. Influence of the van der Waals interactions

The question of a suitable choice for the reference to measure shifts caused by adsorption is closely connected with the difference of resonance frequencies between the gaseous and liquid states. While for proton resonances these differences are negligible (<0.5 ppm), they may be much greater[48] than the influence of bulk susceptibility for ^{13}C and ^{15}N resonances, which is especially true if the nuclei under study are found on the periphery of the molecule (cf. Table 4).

These differences are due to a change in the electron density distribution of the molecule caused by intermolecular interactions which apparently depend on pressure. In the adsorbed state it is both the (non-specific) molecule–surface and molecule–molecule interactions which contribute to the van der Waals shift of the NMR frequency. The ^{13}C shifts of several saturated hydrocarbons which are known to be non-specifically adsorbed (only van der Waals interactions) are shown in Fig. 5 as a function of coverage.

Through an extrapolation to zero coverage the contribution of the molecule–surface term of the van der Waals interactions to the shift can be determined. Together with results which are published in reference 46, we arrive at the following conclusion: the contribution of the van der Waals part of the molecule–surface interaction to the ^{13}C shift is, in general (except for CH_4, cf. reference 46), very small, in contrast to the contribution of molecule–molecule interactions, which can be much greater than the influence of the bulk susceptibility.

TABLE 4

Differences between NMR frequencies for the liquid and gaseous states of ^{13}C and ^{15}N nuclei.[a]

Molecule	Shift difference					Pressure ($\times 10^5$ Pa)	Reference
	C_1	C_2	C_3	C_4	N		
CH_3-CH_3	2.23	—	—	—	—	0^b	50
$CH_2=CH_2$	5.30	—	—	—	—	0^b	50
$CH_3-CH_2-CH_3$	3.39	2.24	—	—	—	0^b	50
$(CH_3)_4C$ ⎫ $(CH_3)_3CH$ ⎭	3.95	3.10	—	—	—	0^b	50
$(CH_3)_3CH$	2.4	0.6	—	—	—	2.53	51
$CH_3-CH_2-CH_2-CH_3$	2.1	0.5	—	—	—	2.03	51
$CH_2=CH-CH_2-CH_3$	2.4	0.5	0.7	2.1	—	2.53	51
$CH_3-CH=CH-CH_3$							
cis	3.0	1.4	—	—	—	1.82	51
trans	2.8	1.2	—	—	—	2.03	51
$CH_2=C(CH_3)_2$	2.5	0.3	2.5	—	—	2.53	51
$O=C(CH_3)_2$ ⎧	10.1	—	—	—	—	0.13	Unpublished
⎨	9.3	—	—	—	—	0.31	Unpublished
⎩	8.8	4.2	—	—	—		52
C_5H_5N	—	—	—	—	-8.9		53
NH_3	—	—	—	—	18		54
$N≡C-CH_3$	—	—	—	—	-9.9		55

[a] Values are in ppm and are positive if the resonance of the liquid is to high frequency.
[b] Extrapolated.

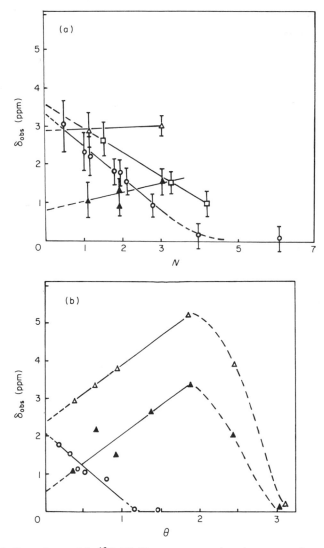

FIG. 5. Dependence of the ^{13}C shift ($\delta_{obs} = \sigma_{liqu} - \sigma_{ads}$) on the coverage for various saturated hydrocarbons adsorbed in a zeolite and on silica gel.[46] (a) NaX_T (Si/Al = 1.2); N = number of molecules per supercage: TMS (O—O); cyclohexane (□—□); butane CH_3 (△—△); butane CH_2 (▲—▲). (b) Silica gel; θ = number of (statistical) monolayers: TMS (O—O); butane CH_3 (△—△), butane CH_2 (▲—▲).

Hence it is proposed[46] to use as a reference for very low and high coverages the NMR frequency of the adsorbate molecule in the gaseous and liquid states, respectively, if one wants to study the specific influence of the surface on the NMR spectrum of an adsorbate, e.g. via adsorption sites.

3. *Exchange effects*

A typical property of adsorbate–adsorbent systems is the existence of regions where the molecules have a different mobility and, in general, a modified electronic structure which gives rise to a change in linewidth and resonance frequency, respectively. In most cases, because of exchange effects, the observed spectrum is not simply a superposition of the spectra which characterize the molecules of the different regions (e.g. molecules adsorbed at adsorption sites of different strength, liquid-like molecular clusters, gaseous state of molecules between the crystallites) but may depend on the rate of molecular exchange between these regions.[56–60] If the adsorption sites are paramagnetic, which is the case for important systems such as transition metals on diamagnetic surfaces (a group of catalysts), the linewidths of the molecules adsorbed at these sites are so big, and their signals so weak, that a determination of the linewidth and line shift is only possible via the resulting signal, which is caused by exchange with a sufficiently large number of diamagnetically adsorbed molecules. Such a measurement is described in reference 61, where the more general case is treated in which the paramagnetic sites are not equal but give rise to different NMR signals for the molecules adsorbed on each of them.

In what follows, a method is described which has been proven to be useful in an analysis of highly resolved NMR spectra of adsorbate–adsorbent systems with diamagnetic adsorption sites. At such a site, a molecule is bound as a complex, the strength of which depends on the kind of adsorption site. The most important problem to be solved in the interpretation of the shifts, measured as a function of the number of adsorbed molecules, is the determination of the shifts for the molecules bound in complexes. In most cases this information cannot be drawn directly from the experiments because the experimental shifts extrapolated to zero coverage are not necessarily identical with the shift for the molecules bound in complexes. One way to do it is to consider the equilibrium between the physisorbed molecules (M), the free adsorption sites (A) and the complexes formed (MA) and to derive the number of sites (N_A) and the equilibrium constant from a fitting of the experimental shifts as a function of the total number N of adsorbed molecules.[13,62,63,181]

We consider here only one sort of adsorption site. In this case it is possible to describe the equilibrium

$$M + A \underset{}{\overset{K}{\rightleftharpoons}} MA$$

from which the relation

$$K = \frac{N_C}{(N - N_C)(N_A - N_C)} \tag{9}$$

can be derived. N_C denotes the number of molecules bound in complexes and N_A the number of adsorption sites which is equal to the number of adsorption centres multiplied by the coordination number. Hence, $N - N_C$ is the number of physisorbed molecules which are not bound in complexes and $N_A - N_C$ is the number of free sites. Furthermore, we denote by δ the observed shift with respect to the physisorbed state (i.e. shift $\delta_M = 0$ for physisorbed molecules) and by δ_C the respective shift for a molecule bound in a complex which has to be derived from the experimental data. If the condition for rapid exchange is fulfilled, both shifts are related by the equation

$$\delta = \delta_C \frac{N_C}{N} \tag{10}$$

By combination of equations (9) and (10), a quadratic equation in δ/δ_C results. A suitable form of this equation, with respect to an analysis of experimental data, is given by

$$\frac{\delta_C}{\delta} = \frac{1 + KN_A}{KN_A} + \frac{N}{N_A}\left(1 - \frac{\delta}{\delta_C}\right) \tag{11}$$

The solution of this equation can be written as

$$\frac{\delta}{\delta_C} = \frac{1}{2}\left(1 + \frac{1}{KN} + \frac{N_A}{N}\right) \pm \sqrt{\frac{1}{4}\left(1 + \frac{1}{KN} + \frac{N_A}{N}\right)^2 - \frac{N_A}{N}} \tag{12}$$

If we denote the value of δ extrapolated to $N = 0$ by δ_0, it follows from equation (12) that

$$\delta_0 = \frac{KN_A}{1 + KN_A} \delta_C \tag{13}$$

Introducing this quantity into equation (12), we find

$$N\left(\frac{\delta}{\delta_0}\right)^2 = \left(\frac{\delta_C}{\delta_0}\right)N\left(\frac{\delta}{\delta_0}\right) + \left(\frac{\delta_C}{\delta_0}\right)^2 N_A\left(\frac{\delta}{\delta_0}\right) - N_A\left(\frac{\delta_C}{\delta_0}\right)^2 \tag{14}$$

which can be written as

$$y = x\frac{\delta_C}{\delta_0} - N_A\left(\frac{\delta_C}{\delta_0}\right)^2 \tag{15}$$

where $y = N(\delta/\delta_0)^2/1 - (\delta/\delta_0)$ and $x = N(\delta/\delta_0)/1 - (\delta/\delta_0)$. The quantities x and y are given by the experiment and, from the plot of y versus x, the shift δ_C and the number of adsorption sites N_A can be determined.[181]

It is noteworthy to consider also the two special cases of a strong and of a weak complex. A complex is said to be strong if the condition

$$KN \gg 1$$

is fulfilled. Then it follows from equation (12) that

$$\delta = \begin{cases} (N_A/N)\delta_C & \text{if } N_A \lesssim N \\ \delta_C & \text{if } N_A \gtrsim N \end{cases} \tag{16}$$

This is the well known result for fast strong interaction where all molecules are adsorbed at adsorption sites as long as their number N_A is still higher than the total number N of molecules involved (cf. Fig. 6). For a weak complex, i.e. if the condition

$$KN \ll 1$$

is fulfilled, we obtain

$$\delta = \left(\frac{KN_A}{1 + KN_A} \right) \delta_C \tag{17}$$

Since the number of sites N_A is reasonably less than the total number of molecules N, or at least of a comparable magnitude, we also have $KN_A \ll 1$. Thus, the observed shift δ is less than the shift δ_C for the complex. A further characteristic feature is that in this case the quantity δ does not depend on the total number N of molecules adsorbed (see Fig. 6).

III. INTERACTION OF MOLECULES WITH DIAMAGNETIC ADSORPTION SITES

A. ^{13}C NMR

1. Adsorption in zeolites

After the first successful ^{13}C measurements of molecules adsorbed in zeolites,[10,11,64] an increasing number of adsorbate–adsorbent systems have been studied using this method. These investigations show that, to a certain degree, the exchangeable cations determine the observed shifts in zeolites. Since the shifts are strongly correlated with the electronic state of the molecules, they can be used to characterize the molecule–ion interaction or more generally the adsorption complexes. Of course it is necessary to select molecules (e.g. unsaturated hydrocarbons) which are able to interact specifically with adsorption sites.

A direct proof for the predominant interaction between unsaturated hydrocarbons and the exchangeable cations in zeolites can only be given in the case of silver cations because silver salt solutions exist where the unsaturated hydrocarbons can be dissolved in order to compare both spectra.[65–67] The formation of silver–olefin complexes in solution is a well known fact and is characterized by significant ^{13}C signal shifts. A similar

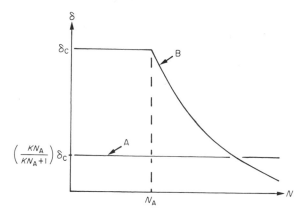

FIG. 6. Schematic plot of shifts δ observed as a function of the total number N of molecules in the case of weak (curve A) and strong (curve B) adsorption complexes.

behaviour is shown by aromatics such as methylbenzene.[49] If the sodium ions of a zeolite are completely or only partially exchanged by silver ions, the same characteristic line shifts occur as in silver salt solutions. This can be seen from the spectra shown in Fig. 7 for but-1-ene and the values given in Table 5 for various unsaturated hydrocarbons.

Similar results are found[49] for toluene, o-xylene, and p-xylene adsorbed in a 50 AgNaX zeolite at 390 K. All these shifts are qualitatively different from those found for the same hydrocarbons adsorbed in Na zeolites, a fact which is discussed in detail later. A characteristic feature of the molecules adsorbed in Ag-exchanged zeolites is the experimentally observed fact that between about 290 K and 360 K the electronic state of the Ag^+–olefin complex does not change (the shifts are constant) while the molecular mobility increases significantly, as can be seen from the linewidths of the signals shown in Fig. 8.

The agreement between the ^{13}C spectra for the olefin molecules dissolved in a silver salt solution and adsorbed in silver-exchanged zeolites leads to the conclusion that in both cases the same olefin–silver complexes are formed. Attention should be paid to the fact that, as a consequence of rapid exchange, the shifts decrease with an increasing ratio of the number of molecules per silver ion. It can be seen from Table 5 that at least for the X zeolite the silver ions are located in the supercages (where they are accessible), even at a very low degree of exchange (20%). From the interpretation of the ^{13}C spectra for the olefin–silver salt solutions it can be stated that olefins with a symmetrical double bond, such as but-2-ene, form adsorption complexes of the π type, where the ion is located symmetrically below the double bond,[68] while different values for the shifts of the olefinic carbon atoms for olefins such as but-1-ene and isobutene (cf. Table

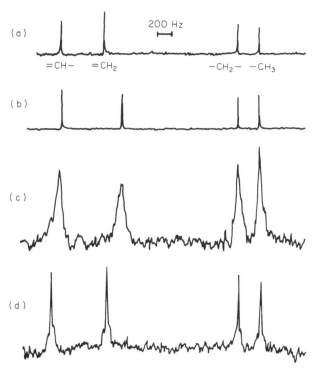

FIG. 7. Proton-decoupled ^{13}C NMR spectra of but-1-ene molecules.[66] (a) In the liquid state (293 K, pulse width 6 μs, pulse distance 0.2 s, 512 scans, 2k data points (1k = 1024) at 22.6 MHz); (b) dissolved in a AgBF$_4$ × H$_2$O/CHCl$_3$ solution (ratio of moles Ag$^+$: olefin = 1 : 2, 4k scans, other data as for (a)); (c) adsorbed in a 60 AgNaX$_{1.8}$ zeolite (four molecules per supercage, 355 K, pulse width 25 μs, pulse distance 0.2s, 4k scans, other data as for (a)); (d) adsorbed in a NaX$_{1.8}$ zeolite (333 K, other data as for (c)).

5) indicate an asymmetric arrangement of the ion with respect to the double bond.[69]

Besides silver, other metal cations may act as adsorption centres for hydrocarbons if they are able to undergo a specific interaction with the metal cations. This follows both from thermodynamic data[17,70] and from the results of infrared measurements.[71,72] Moreover, an analysis of proton relaxation data for such systems gives direct evidence for the existence of adsorption complexes formed by metal cations and hydrocarbons.[1,2,73,74]

In the case of monovalent cations such as Li$^+$, Na$^+$, K$^+$, Rb$^+$, Cs$^+$ and Tl$^+$, it is necessary to check systematically whether other sites may be responsible for the effects observed in zeolites.[175,179] In order to treat the various steps of this proof systematically, we start with experiments using but-1-ene as adsorbate and compare this discussion with the results for isobutene:[75-77]

TABLE 5

^{13}C shifts of butene molecules adsorbed in various zeolites or solved in silver salt solutions with respect to the liquid state.[a]

	Molecules per supercage	^{13}C shifts					Temperature (K)
		=CH$_2$	=CH–	=C<	–CH$_2$–	–CH$_3$	
But-1-ene/NaX$_{1.8}$	4.1	–2.8	5.4		–0.5	–0.9	330
But-1-ene/20 AgNaX$_{1.8}$	5	–4.8	1.9		–0.5	–0.9	333
But-1-ene/40 AgNaX$_{1.8}$	4.6	–9.6	1.3		–0.9	–1.6	333
But-1-ene/60 AgNaX$_{1.8}$	4.7	–12.2	0.4		–0.1	–0.1	333
Solved in AgBF$_4$×H$_2$O/CHCl$_3$	Ag$^+$:olefin, 1:2	–12.9	–1.6		–1.2	–0.9	293
Isobutene/NaX$_{1.8}$	4.4	–2.9		9.3		0.4	298
Isobutene/60 AgNaX$_{1.8}$	2.5	–20.4		10.0		–0.3	308
	5.0	–16.2		9.4		–0.3	308
cis-But-2-ene/NaX$_{1.8}$	4.4		1.5			–1.1	313
cis-But-2-ene/60 AgNaX$_{1.8}$	3.0		–5.9			0.0	353
Solved in AgBF$_4$×H$_2$O/CHCl$_3$	Ag$^+$:olefin, 3.5:1		–5.6			1.5	293

[a] Values are in ppm and are positive for high frequency shifts. The maximum pore filling is about 5.5 butene molecules per supercage. Corrections of 0.7 ppm, for a shielding increase, are applied for susceptibility effects. The experimental error is ±0.5 ppm.

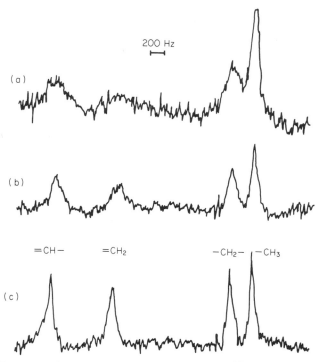

FIG. 8. Temperature dependence of the proton-decoupled ^{13}C NMR spectra of but-1-ene molecules adsorbed in a 60 AgNaX$_{1.8}$ zeolite[66] (4.7 molecules per supercage, pulse width 25 μs, pulse distance 0.2s, 2k data points (1k = 1024) at 22.6 MHz). (a) 300 K, 32k scans; (b) 323 K, 16k scans; (c) 355 K, 16k scans.

(1) If the contributions from the van der Waals interactions and bulk susceptibility to the ^{13}C shifts are eliminated, the characteristic effect observed for the pure sodium forms of X, Y zeolites (Si/Al < 2.6) is the high frequency shift of the groups =CH— for but-1-ene and =C< for isobutene (Table 6). A striking fact is that these resonances are much less shifted if the same molecules are adsorbed in Y zeolites with a very low content of aluminium and sodium (Si/Al = 47). This result clearly rules out any role involving the Si and O atoms of the zeolitic lattice.

(2) In the next step the influence of the structural OH groups is examined. Since their number is rather small for the NaX and NaY zeolites used (less than about 0.5 OH groups per supercage), the interaction of but-1-ene with them should give rise to a pronounced decrease in the typical shifts for the group =CH— with increasing pore filling. This effect should arise because of exchange between the small number of molecules complexed with OH groups and the other molecules. To enable precise measurements over a wide range of coverage (from 0.2 to 5 molecules per large cavity), but-1-ene

TABLE 6

^{13}C shifts for the olefinic carbon atoms of olefins adsorbed in NaX and NaY zeolites.^{a}

	Molecules per supercage	^{13}C shifts			Temperature (K)
		$=CH_2$	$=CH-$	$=C<$	
Ethene/NaX$_{1.35}$	2.0	−0.8			305
Propene/NaX$_{1.35}$	2.1	−2.1	5.9		305
Isobutene/NaX$_{1.35}$	1.6	−7.9		12.6	330
Isobutene/NaY$_{2.6}$	3.3	−4.0		9.6	330
Isobutene/NaY$_{47}$	3.5	−2.8		0.0	220
But-1-ene/NaX$_{1.35}$	1.1	0.8	7.2		305
But-1-ene/NaY$_{2.6}$	1.1	0.6	6.5		305
But-1-ene/NaY$_{47}$	3.5	0.5	1.6		220
trans-But-2-ene/NaY$_{2.6}$	3.3		1.5		330
Pent-1-ene/NaX$_{1.35}$	2.2	−2.1	7.3		305
2-Methylbut-1-ene/NaX$_{1.35}$	3.0	−4.9		9.2	310
2-Methylbut-2-ene/NaX$_{1.35}$	3.0		−0.7	5.0	308
Hex-1-ene/NaX$_{1.35}$	1.3	−2.8	7.5		305
Oct-1-ene/NaX$_{1.35}$	1.0	−1.7	5.0		305
Dodec-1-ene/NaY$_{2.6}$	1.0	−2.3	5.2		323

^{a} Values are in ppm with respect to the liquid state and are positive for high frequency shifts. Corrections of 0.7 ppm are applied for susceptibility effects. The experimental error is ±0.5 ppm.

molecules are used, enriched with ^{13}C in the $=CH-$position (enrichment of 50%).

The slight dependence of the shifts on the pore filling in NaX and NaY zeolites (Fig. 9) cannot be explained by an exchange between a few molecules complexed with OH groups and the other molecules. Also, if one takes into account sites of low adsorption energy,[51] this effect can only be explained if the number of sites responsible for the shifts is so large that even at the highest pore filling each molecule interacts with one adsorption site. Hence, the action of a small number of efficient sites such as OH groups can be ruled out.

(3) Finally, it has to be decided whether the shift in NaX and NaY zeolites is due to interactions with aluminium atoms of the lattice or is in fact caused by interactions with sodium ions. This problem is not difficult to solve on the basis of investigations which were performed with zeolites modified by various monovalent diamagnetic cations (Table 7). In particular, the drastic reduction of the shifts found for the larger cations Tl^+ and Cs^+ can only be understood if the Na^+−olefin interaction in the pure sodium forms is predominant. In a similar way, the experiments using isobutene have to be explained. In addition to the effects mentioned, we observe a

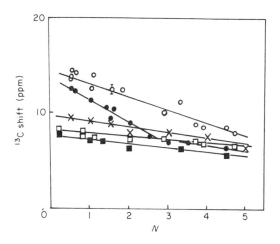

FIG. 9. ^{13}C shifts for the groups $=CH-$ of but-1-ene and $=C<$ of isobutene as a function of the number N of molecules per supercage:[75] isobutene in $NaX_{1.35}$ (○—○), $KK_{1.35}$ (●—●) and $NaY_{2.6}$ (×—×); but-1-ene in $NaX_{1.35}$ (□—□), $KX_{1.35}$ (■—■). Values are in ppm with respect to the gaseous state and are positive for high frequency shifts. Susceptibility effects are taken into account by a shielding increase correction of 0.7 ppm.

TABLE 7

^{13}C shifts of but-1-ene and isobutene adsorbed in modified X, Y zeolites with respect to the gaseous state.[a]

	Molecules per supercage	^{13}C shifts		
		$=CH_2$	$=CH-$	$=C<$
But-1-ene adsorbed in				
73 $LiNaX_{1.35}$	1.0	−1.5	6.7	
$NaX_{1.8}$	1.0	−2.0	7.9	
$KX_{1.8}$	1.0	−1.9	6.4	
72 $CsNaX_{1.8}$	1.0	−1.3	5.1	
	4.0	−1.3	4.8	
$TlX_{1.2}$	3.8	2.8	1.4	
Isobutene adsorbed in				
73 $LiNaX_{1.35}$	1.0	−2.7		12.1
$NaX_{1.35}$	0.5	−6.3		14.7
$KX_{1.35}$	0.5	−5.6		12.7
82 $TlNaX_{1.35}$	1.1	1.2		7.0

[a] Values are in ppm and are positive for high frequency shifts. Corrections of 0.7 ppm are applied for susceptibility effects. Measuring temperatures between 308 and 330 K.

stronger dependence of the shifts on the pore filling in X zeolites compared with Y. As already mentioned, X and Y zeolites have the same structure but differ in the number of exchangeable cations, which are sodium ions in the case of NaX and NaY zeolites. Some of these Na^+ ions are not accessible to hydrocarbons since they are localized at sites SI and SI' outside the supercages (cf. Fig. 1), some are localized at the supercages opposite to the windows (SII) and some are assumed to be loosely bound (SIII) to the walls of the supercages between sites SII. For the NaY zeolite used, the number of exchangeable cations per 1/8 unit cell (corresponding to one supercage) is 3.3 (SII) and 0.4 (SIII). For NaX the respective values are 3.2 (SII) and 4.0 (SIII).[23] Since the stronger dependence of the shift on the pore filling observed for isobutene in NaX correlates with a higher population of the SIII sites, compared with NaY, it may be argued that there is a predominant interaction with sodium ions on SIII sites.

More directly, the stronger interaction with sodium ions at SIII sites can be studied by means of the proton spin relaxation, even in those cases where the change of the ^{13}C shift is not sensitive enough to detect whether the molecules interact with sodium ions at SII or SIII sites, respectively. This situation arises if, for example, the adsorption of benzene and the n-butenes is studied. From the proton spin relaxation it has been reported[78] that in NaY all benzene molecules are strongly bound to the localized sodium ions (sites SII) which are fixed to the wall of the supercage. This explains the low mobility of benzene molecules in the NaY zeolites. In NaX zeolites, however, the molecules are bound to the non-localizable ions (sites SIII) since these ions are more exposed than Na^+ at sites SII and act as primary adsorption centres. So the higher mobility in NaX zeolites (factor of 5 at 300 K) can be explained by a lower height of the potential wall between neighbouring non-localized sodium ions. The proton relaxation measurements of n-butenes in NaX and NaY zeolites are similarly explained.[74]

In the case of *acetone* the observed ^{13}C shifts[93] are much larger than those of the olefins adsorbed in the same zeolites (Fig. 10). This can be explained by a strong interaction between the lone pair electrons of the acetone oxygen and the adsorption centres. Of course this holds only for the carbon attached to the oxygen while the shift of the methyl group carbon is small and within the limits given by the difference between the gaseous and liquid states (cf. Fig. 10). To measure the shifts, down to a pore filling of about 0.2 molecules per supercage, it is necessary to use adsorbates enriched by up to 95% with ^{13}C.

As can be seen from Table 7, for Li-exchanged zeolites there is a deviation from the correlation between the ^{13}C shifts and the radii of the metal cations. A similar anomaly is observed for the heat of adsorption of water and explained by a reduced accessibility of the lithium ion since, due to its small radius, it may be screened more strongly by the oxygen atoms of the

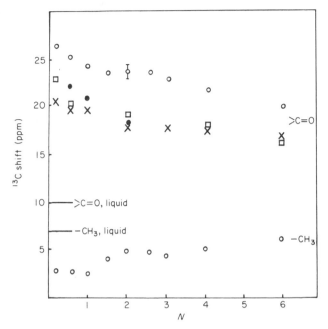

FIG. 10. ^{13}C shifts for the groups $>$C=O and $-$CH$_3$ of acetone as a function of the number of molecules per supercage: NaX$_{1.35}$ (O—O), KX$_{1.35}$ (●—●), 70 RbNaX$_{1.35}$ (□—□), and NaY$_{2.6}$ (×—×). Values are in ppm with respect to the gaseous state and are positive for high frequency shifts. Shielding increase corrections of 0.7 ppm are applied for susceptibility effects.

zeolite lattice in which it is embedded.[79] It was also suggested[80] that Li$^+$ may be localized preferentially at SI' sites so that the total number of metal cations in the supercage is reduced. Both effects should lead to a reduced shift for the adsorbed molecules. In the case of acetonitrile, however, no such deviation for the Li-exchanged zeolite is observed: the ^{13}C shift of the N$\equiv$C$-$ group shows a monotonous change on going from Li- to Na- and K-exchanged zeolites.[172] This result is discussed later, together with the similar behaviour found for ^{15}N shifts.

The adsorption of benzene and its derivatives in X and Y zeolites leads to such a strong restriction of their molecular mobility that, in general, at room temperature no highly resolved ^{13}C spectra are observed.[49,174] The shift for the ^{13}C resonance of benzene adsorbed in sodium-exchanged zeolites is similar to that of ethene (Table 6) and of the order of the experimental error. For the methylbenzenes one can observe a shift of 2–4 ppm to higher frequency for ring carbon atoms to which the methyl group is attached.[49] According to a quantum mechanical treatment (cf. Section VI), these shifts can be explained by the polarizing effect of the sodium cation, which gives rise to a change in the electronic charge distribution within the molecules. As in the case of olefins and acetone, adsorbed benzene, and its derivatives, in Li-exchanged zeolites produce a ^{13}C shift which is smaller than that expected on the basis of the relatively large electric field of this cation. On the other hand, if sodium is exchanged by alkali cations with a larger diameter, the mobility of the adsorbed aromatics is drastically reduced. In the case of toluene adsorbed in a CsX zeolite, an anisotropy of about 130 ppm is observed for the resonance of the ring carbon to which the methyl group is attached.[81] This is explained by restricted rotation of the toluene molecule about its C_2-axis (correlation time $\leqslant 10^{-3}$ s). In contrast, for a NaX zeolite, no anisotropy occurs and therefore an isotropic reorientation of the toluene molecule must be assumed.[81]

The ^{13}C NMR spectra of CO_2 and related compounds adsorbed in NaX, NaA zeolites, and in mordenite, have been observed with and without MASS.[45] From the temperature dependence of the observed anisotropy of the ^{13}C signal, a model for the thermal reorientation of CO, CO_2, COS and CS_2 molecules adsorbed on various cation-exchanged mordenites is developed.[82] The isotropic part of the ^{13}C shift in these experiments is only small. On the other hand, it is well known that certain metal–carbonyl compounds formed in the liquid state show rather large high frequency ^{13}C shifts.[83] Moreover, for CO molecules which interact with transition metal ions, shifts to low frequency have been reported[84] (cf. Section V). Therefore it seems reasonable to study systematically the influence of adsorption sites in zeolites on the electronic state of adsorbed CO molecules in comparison with adsorbed CO_2.[85] The adsorbate molecules are enriched with ^{13}C to about 50%. Values for the ^{13}C shifts of CO and CO_2 adsorbed in NaX, NaY and NaA zeolites are not very different from the values for the gaseous state. The same result holds for these adsorbate–adsorbent systems if Na^+ is exchanged by other alkali cations or Tl^+. Significant effects, namely extremely large shifts to high frequency, are observed for CO adsorbed on decationated zeolites (Fig. 11). The values are largest for deep-bed treated specimens, but even for shallow-bed zeolites they are much greater than for zeolites containing alkali cations.[85] No such effects occur for CO_2. The strong temperature dependence of the ^{13}C shift of CO adsorbed in decationated zeolites, as shown in Fig. 11, is a direct hint that the observed shifts are controlled by an exchange process. Below a temperature of 210 K the same values result as in the case of sodium zeolites. The linewidth of the ^{13}C signal is about 150 Hz at 210 K and increases with increasing temperature (up to about 800 Hz). A study of the ^{13}C shifts for CO adsorbed on silica gel and on aluminosilicates yields values which are independent of temperature and of coverage and differ only slightly from the value for CO in the gaseous state. These results, and the fact that the ^{13}C shifts of CO adsorbed in decationated zeolites are not correlated with the number

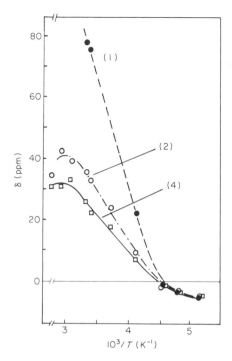

FIG. 11. ^{13}C shifts δ for CO adsorbed in 72 DeNaY$_{2.6}$-DB as a function of $10^3/T$: one (●—●), two (○—○) and four (□—□) molecules per supercage.[85] Values are in ppm with respect to the gaseous state and are positive for high frequency shifts. Susceptibility effects are taken into account by a shielding increase correction of 0.7 ppm.

of acidic OH groups (a comparison between 50 DeNaY and 72 DeNaY), lead to the suggestion that the observed strong effects are not due to an interaction with adsorption sites of the Broensted type. Instead, the extremely strong effects which occur in the case of deep-bed treated specimens lead to the conclusion that an interaction with Lewis sites, e.g. with the extra-lattice aluminium ions,[86] is responsible for the large values of the ^{13}C shifts of CO. Further work along these lines is in progress.

2. *Adsorption on silica*

As shown in Section III.A.1 for zeolites, the exchangeable cations may act as efficient adsorption centres. In silica gel there are no such cations and instead of them the surface OH groups play an important role for the interaction with adsorbed molecules. Saturated hydrocarbons show only a weak interaction with surface hydroxyl groups, as can be inferred from the relatively small shift $\Delta\nu_{OH} = 30$–$40 \, cm^{-1}$ of the OH stretching vibration band. As is well known,[71] OH band shifts of more than $100 \, cm^{-1}$ occur for adsorbed alkenes and for molecules like acetone and pyridine which characterize the stronger interaction between their π-electron system and lone pair electrons, respectively, and the silanol groups.

^{13}C shifts of hydrocarbons adsorbed on silica gel have also been measured.[77,87–89] Compared with zeolites, however, the values for the shifts are relatively small, so that an interpretation and an unambiguous correlation of the effects observed with certain types of adsorption centres such as OH groups, oxygen atoms of the silica gel, and lattice defects is rather difficult. One way to overcome these problems, at least partially, is by a combined interpretation of the results of different spectroscopic methods, such as ^{13}C and/or ^{15}N NMR and IR spectroscopy.[63,90,91] Surface OH groups act as adsorption sites and are responsible for the strong interaction between isobutene molecules and silica gels. This interaction gives rise to a strong decrease in wavenumber of the OH stretching vibration ($\Delta \nu_{OH} = 220 \text{ cm}^{-1}$), a shift of the quaternary carbon to higher frequency and a decrease in frequency of the $\nu_{C=C}$ vibration of the isobutene molecule.[90] In Fig. 12 the ^{13}C shift for the quaternary carbon atom of isobutene is shown as a function of coverage for adsorption on a partially dehydroxylated and a methylated silica gel.

From these results[90] it must be concluded that, in contrast to reference 88, a possible interaction of the isobutene molecules with the oxygens in siloxane bridges has no influence on the ^{13}C shifts. Moreover, the combination with IR data clearly points out the dominant role of the surface silanol groups for the adsorption of isobutene.[90] The shifts for the carbons in the $CH_2=$ and CH_3- groups are relatively small and do not depend significantly

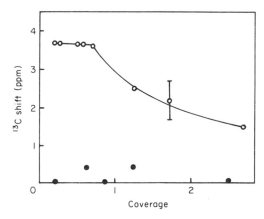

FIG. 12. ^{13}C shifts for the quaternary carbon $>C=$ of isobutene as a function of coverage given in statistical monolayers (area of the isobutene molecule, 0.4 nm^2). Values are in ppm with respect to the gaseous state and are positive for high frequency shifts.[90] Isobutene on silica gel (O) and on methylated silica gel (●).

on the coverage. For acetone adsorbed on silica gel there is no dependence of the ^{13}C shifts on the coverage for the carbons of the methyl groups, but a significant effect for the carbon of the keto group, as can be seen from Fig. 13. The large, high frequency shift in the case of the adsorption on silica gel and the striking difference in the values obtained for the methylated silica gel indicate clearly the existence of a strong interaction between the keto group of acetone and the surface hydroxyl groups.[63] The residual high frequency shift for the carbon of the keto group on methylated samples arises from an interaction with residual OH groups and a small number of non-hydroxylic centres.[63]

The results shown in Fig. 13 may be analysed according to the theoretical treatment of exchange effects, as described in Section II.C.3. Applying the method of least squares, it follows that for the shift at an adsorption centre $\delta_{AM} = 10–15$ ppm and 7–10 ppm for the silica gel and the methylated silica gel, respectively. The corresponding mean numbers of adsorption centres are 1.4 ± 0.2 nm^{-2} and 0.3 ± 0.1 nm^{-2} respectively. The values are in reasonable agreement with the concentrations of surface hydroxyl groups for these samples (about 5 and 0.1 nm^{-2}, respectively). Comparing these values, one has to bear in mind that not all hydroxyl groups can interact with the molecules, as a result of steric effects. On the other hand, for the methylated specimen with its small number of surface OH groups, some non-hydroxylic adsorption sites may contribute to the observed effect.

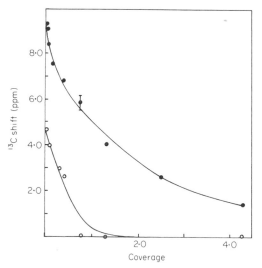

FIG. 13. ^{13}C shifts for the carbon of the keto group of acetone as a function of coverage given in statistical monolayers (area of the acetone molecule 0.4 nm^2). Values are in ppm with respect to the gaseous state and are positive for high frequency shifts.[63] Acetone on silica gel (●—●) and on methylated silica gel (○—○).

In the IR spectra of acetone adsorbed on silica gel,[63] four bands for the C=O stretching vibration are found which are ascribed to interactions with geminal surface hydroxyl groups (1690 cm^{-1}), with isolated surface hydroxyl groups (1710 cm^{-1}) and with non-hydroxylic adsorption sites (1720–1722 cm^{-1}), and to the gaseous state (1735–1737 cm^{-1}). Another experimental result is that the shift of the stretching vibration band of the hydroxyl groups involved in hydrogen bonding interactions increases with growing acetone adsorption. This fact is explained in terms of different spatial arrangements of the hydrogen bond. The resulting ^{13}C shift is therefore an average value due to fast exchange of the molecules between several states.

As a last example, the adsorption of pyridine is mentioned.[91] In this case ^{13}C and ^{15}N shifts can be measured; together with IR data a more detailed interpretation of the experimental results is possible.

The study of adsorption on phosphorus-modified silica gel, where formation of pyridinium ions has been suggested, is discussed in Section V.

There is only a small influence on the ^{13}C shielding of pyridine resulting from adsorption on SiO_2. The values do not change significantly as a function of the coverage, and are of the same order of magnitude as in the case of non-specific interactions. On the other hand, the large and coverage-dependent ^{15}N signal shifts to lower frequency, which indicates an appreciable influence of adsorption sites on the electronic density at the nitrogen atom.[91] Moreover, the shifts of approximately -25 ppm with respect to the liquid state are constant for coverages lower than ~ 0.5 statistical monolayers. This behaviour indicates the formation of a strong complex: the ^{15}N shielding remains constant as long as the number, N, of adsorbed molecules is smaller than the number, N_A, of adsorption sites. For $N > N_A$ a fast exchange between adsorbed and free molecules occurs, giving rise to a decreasing shift. The specific interaction between pyridine molecules and adsorption sites on partially dehydroxylated silica gel occurs via the formation of hydrogen bonds. These bonds are formed between a proton of the hydroxyl group and the lone pair electrons of the nitrogen atom and cause a significant increase in the nitrogen shielding. The small changes observed in the ^{13}C NMR spectra indicate that no specific interaction of the π-electron system with surface sites occurs. Characteristic bands for hydrogen bonding appear in the IR spectrum.

B. ^{15}N NMR

The first detailed ^{15}N NMR study[13] has revealed that this is a powerful tool for the study of surface phenomena: the shifts are very sensitive to the nature of the adsorption sites and the changes in the spectra are often much larger than those found in analogous ^{13}C NMR measurements.

A study of ^{15}N NMR spectra of adsorbed molecules seems to be of special interest because the nitrogen atoms of a variety of molecules possess lone pair electrons which may participate in the formation of hydrogen bonds, giving rise to large ^{15}N shielding changes. Moreover, molecules like pyridine and acetonitrile may be directly protonated at the nitrogen atom. Consequently, the difference between the ^{15}N shifts of protonated and non-protonated species may be considerably larger than the shifts for the adjacent ^{13}C nuclei and protons in the molecule. Hence, ^{15}N NMR measurements should be more favourable for a study of acidic sites on surfaces (cf. Section V). In addition, through a ^{15}N NMR study, the electronic state of the ammonia molecule can be investigated in more detail than with the ^{1}H NMR because ^{1}H shifts are small and often less than the linewidths, which prevents their accurate measurement.

As is shown in Fig. 14, the shifts remain constant for *acetonitrile* adsorbed in NaY zeolites as long as the number, N, of adsorbed molecules is less than about five per supercage. This behaviour suggests that acetonitrile molecules interact with a relatively large number of adsorption sites. From the plot of δ as a function of N by means of equation (13), a number of 6 ± 0.5 active sites per supercage and a value $\delta_{C} = -19.5$ ppm for the shift

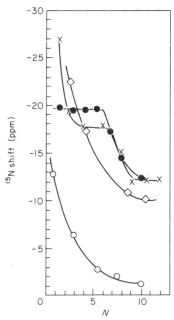

FIG. 14. ^{15}N shifts of acetonitrile as a function of the number N of molecules per supercage: NaX$_{1.35}$ ($\times$—$\times$), NaY$_{2.6}$ ($\bullet$—$\bullet$), NaY$_{47}$ (Us-Ex) ($\bigcirc$—$\bigcirc$), 88 DeNaY$_{2.6}$ ($\diamond$—$\diamond$). Values are in ppm with respect to the liquid state and are positive for high frequency shifts.[13] All measurements were performed at 300 K.

of the complexed molecules can be calculated, in agreement with the simple interpretation of the plateau (strong adsorption complex). The number of active sites is equal to about twice the number of accessible sodium ions (about 3.3 Na^+ at SII sites) so that the coordination number of the sodium ions for CH_3CN molecules in NaY zeolites is about 2. For the physisorbed molecules the shift is assumed to be the same as in the liquid state. This assumption is supported by the following results. In order to check whether sodium ions can act as adsorption sites, similar measurements are performed using a strongly dealuminated zeolite NaY_{47} (denoted as Us-Ex). The different behaviour of acetonitrile molecules due to the appreciable reduction of the number of sodium ions can be clearly inferred from Fig. 14. At nearly complete pore filling of the dealuminated zeolite, the resonance lines appear in an interval which is typical for the liquid state of acetronitrile.

Thus, in contrast to the sodium forms, the influence of molecule–molecule interactions is directly reflected by the measured shifts at higher coverages. The influence of an interaction with the remaining adsorption sites (Na^+ and/or OH groups) appears at lower coverage, where the resonance lines are more and more shifted towards that value which is typical for the $NaY_{2.6}$ zeolite. Obviously, the number of sites is very small because a plateau cannot be observed within the limits of experimental error. This is reasonable, since the number of sodium ions in the dealuminated zeolite is very small. Measurements were also performed on a NaX zeolite (cf. Fig. 14). For higher coverages the ^{15}N shifts are nearly the same as those measured for NaY: the values for the shifts decrease monotonously with increasing coverage as long as the total number of molecules is larger than about 6.5 ± 0.5 per supercage; and the shift values are independent of the number, N, of molecules for $4.0 \leqslant N \leqslant 6.5$. In contrast to the NaY zeolite, for $N < 4.0$ an additional shift to lower frequency occurs. Presumably, the CH_3CN molecules adsorbed at low coverage interact preferentially with sodium ions at SIII sites (about four per large cavity), which practically do not occur in NaY zeolites. It is noteworthy that the number of complexed acetonitrile molecules derived from the whole curve for NaX shown in Fig. 14 is again about six to seven per supercage. This value, however, is less than the total number of sodium ions per supercage for the NaX zeolite. Obviously, the fraction of acetonitrile molecules which can interact simultaneously with sodium ions located in the supercages is limited here because of steric reasons.

For the highest coverages ($N > 10$) the shift of the molecules adsorbed in the supercages remains constant. In addition, a separate resonance line appears at a frequency which is typical for liquid acetonitrile (see Fig. 15). Obviously, under these conditions, a slow exchange occurs between molecules adsorbed in the zeolite cavities and as a liquid-like phase at the outer surface of the crystallites.

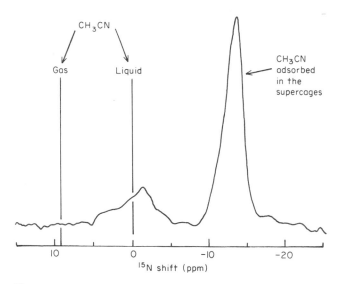

FIG. 15. ^{15}N NMR spectrum of the acetonitrile/NaX$_{1.35}$ system for an apparent coverage of 12 molecules per supercage (a complete pore filling corresponds to about 10 molecules per supercage).[13]

The influence of the type of cation on the ^{15}N shift is much greater than for ^{13}C shifts, even of the same molecule (see Fig. 16). These results demonstrate unambiguously the strong interaction between the nitrogen atom of the acetonitrile molecule and the exchangeable cation of the zeolites.[172] Together with the small effect which occurs for the ^{13}C shift of the methyl group, these results yield valuable information about the charge redistribution within the molecule during the formation of the adsorption complex and can be used to check quantum mechanical calculations concerning the structure of this complex (see Section VI). On the other hand, the strong interaction between the nitrogen atom and the exchangeable cations may explain why, for the adsorption of acetonitrile, no anomaly with respect to the lithium ion occurs: it is this strong interaction which makes the lithium ion accessible in contrast to the adsorption of other molecules (see Table 7) which leaves the location of Li$^+$ unchanged.

Values for the ^{15}N shifts of acetonitrile molecules adsorbed in a decationated zeolite (88 DeNaY$_{2.6}$) are also plotted in Fig. 14.[13] In contrast to the adsorption of ammonia and pyridine molecules in such decationated zeolites, where the formation of the protonated forms can be followed in the spectra (see Section V), the observed shifts are of the same order of magnitude as for unprotonated acetonitrile molecules. From the plot of the shift, δ, as a function of N, by means of equation (13) a number of 3.0 ± 0.5 active sites per supercage and a value $\delta_C = -43.6$ ppm for the shift of the complexed

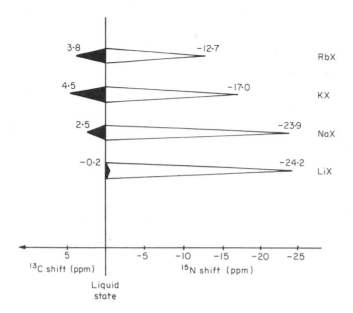

FIG. 16. Plot of the ^{13}C and ^{15}N shifts of the $-C\equiv N$ group of acetonitrile molecules adsorbed in modified X zeolites with respect to the liquid state. The accuracy of the values given is approximately ± 0.5 ppm.

molecules can be determined. This value differs appreciably from that for protonated acetonitrile species in superacid solutions (-102.6 ppm referred to liquid CH_3CN^{98}). The number of sites, however, is of the same magnitude as obtained from the ^{15}N NMR data on pyridine molecules adsorbed in the same zeolites. This result reveals that even in decationated zeolites the CH_3CN molecules are not protonated but interact with the structural OH groups via hydrogen bonds. The ^{15}N signal of acetonitrile molecules adsorbed in *stabilized* decationated zeolites is shifted to high frequency, with respect to the liquid state, by about 27 ppm (for $N = 8$; see reference 13). This result is of special interest because it is substantially different from that for systems either containing Broensted acid sites of various strength (SiO_2, 88 $DeNa_{2.6}$) or sodium ions. Clearly, a protonation of acetonitrile cannot account for this shift. Other weak interactions, already described, cannot be responsible for the shift because all of them produce low frequency ^{15}N shifts. Since a shift to high frequency (relative to the gaseous state) occurs only if acetonitrile molecules are adsorbed in stabilized zeolites, it is attributed to an interaction with Lewis acid sites created during the stabilization process. This interpretation is supported by the analogous experiments using ammonia, as described below.

The ^{15}N shift for *ammonia* molecules adsorbed in NaY, NaX, NaA and Na–mordenite depends strongly on the coverage. The plot (Fig. 17) is characterized by the approach of the shift to the values of liquid (0 ppm) and gaseous (-18 ppm[54]) ammonia for nearly complete pore filling and for zero coverage, respectively.[12,13] For the strongly dealuminated Y zeolite (Us-Ex), the ^{15}N shift does not change as a function of coverage. Its value is about the same as for liquid ammonia (see Fig. 17).

The very surprising result is that the ^{15}N shifts for ammonia molecules adsorbed in the sodium forms of various zeolites (A, X, Y and mordenite) are mainly due to molecule–molecule interactions, although a strong influence of the sodium ions can be clearly inferred from results of adsorption heat measurements.[94,95] The ^{15}N shifts reveal, however, that even in the limit of zero coverage, where the greatest adsorption effects are measured, only a small deviation of the shift from the value for the gas phase (-18 ppm) occurs. Moreover, for the various zeolites (NaY, NaX, NaA, Na–mordenite) with a different number of sodium ions, approximately the same dependence

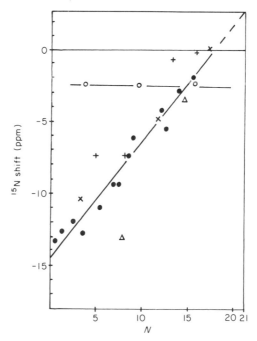

FIG. 17. ^{15}N shifts of ammonia adsorbed in $NaX_{1.35}$ ($\times$), $NaY_{2.6}$ ($\bullet$), NaA ($+$), Na–mordenite ($\triangle$) and NaY_{47} (Us-Ex) ($\circ$) as a function of the number N of molecules per supercage.[12,13] Values are in ppm with respect to the liquid state and are positive for high frequency shifts. All measurements were performed at 300 K. A complete pore filling at this temperature corresponds to about 21 molecules per supercage.

of the shift on the coverage is observed. Obviously the variation in shift with decreasing coverage, between the values measured for the liquid and the gaseous states, can simply be explained by the fact that association of the ammonia molecules is more and more prevented due to interaction with the sodium ions. This interaction, however, must lead to an almost negligible influence on the electron density at the nitrogen atom.

Furthermore, the same value for the indirect spin–spin coupling, 62 ± 5 Hz, is measured at low coverages as for ammonia in the gaseous or liquid states (61.6 Hz). Since it is well known[96] that the indirect spin–spin coupling of the ammonia molecule depends on the bonding angle, it follows from this fact that the geometrical structure of the ammonia molecule also remains unchanged. At higher coverages ($N > 7$) the multiplet splitting disappears in NaY zeolites, which is presumably due to an increasing proton exchange rate. For the NaX samples a multiplet splitting is not observed, while in the case of NaA zeolites the splitting appears up to the highest coverages ($N \approx 17$).[97] ^{15}N spectra of ammonia molecules adsorbed in stabilized decationated zeolites can only be measured at temperatures of 340 K and higher because of a strong line-broadening. This is the reason why the systems discussed here were investigated at 380 K. In spite of the high enrichment with ^{15}N nuclei, the signal-to-noise ratio is not sufficient to permit the study of samples with less than about two molecules per super-cage. But the small change of the shifts below a coverage of six molecules per supercage justifies an extrapolation of the values to zero coverage. The shift thus obtained (116 ppm to high frequency with respect to the gaseous state) is considerably larger than values which are reported in the literature[98] for ammonium ions in different media and has not been interpreted in terms of a dominant interaction of ammonia molecules with residual hydroxyl groups acting as Broensted acid sites. On the other hand, *stabilized* decationated zeolites (see reference 70) contain, besides a more or less relatively small number of residual OH groups, adsorption sites of the Lewis type which are generated during the stabilization process. In agreement with the results for adsorbed acetonitrile molecules, we attribute the observed shifts to an interaction with these sites. In order to derive the number of Lewis acid sites which interact with adsorbed ammonia molecules, the coverage dependence of the shift[13] was analysed by means of equation (11). The fitting procedure yields $\delta_C = 116$ ppm, which can also be derived directly by means of a simple extrapolation of the experimental shifts to zero coverage. This finding is an experimental proof for the existence of a strong adsorption complex. The corresponding number of sites for the ammonium molecules obtained from the slope is $N_A = 12.5$. Assuming a coordination number of 2–3, this value corresponds to about four to six Lewis sites per supercage, which is greater than the value determined indirectly from sorption measurements. Details can be found in the literature.[13,99]

IV. INTERACTION OF MOLECULES WITH PARAMAGNETIC ADSORPTION SITES

A. Analysis of paramagnetic NMR spectra

NMR is a powerful tool for the investigation of complexes of organic molecules with paramagnetic metal ions in solutions. The method gives information about the structure of complexes, the nature of metal–ligand bonds, the reorientation time of complexes, the residence time of molecules in the solvation sphere, etc.[100,101] NMR spectra of paramagnetic systems are characterized by (a) very large shifts caused by an interaction with the magnetic moment of the unpaired electrons – the measured ^{1}H shifts are, for example, of the order of some 100 ppm,[101] in comparison with about 10 ppm in diamagnetic systems (see also Section IV.B.2); (b) a strong line-broadening due to dipolar and/or scalar electron–nucleus interaction; (c) a strong temperature dependence of the position and width of the resonance lines.

Favourable conditions for observing NMR spectra of paramagnetic surface complexes are given for paramagnetic ions with short electron relaxation times[101–103] and for small concentrations of these ions on the surface, so that the exchange between the coordinated molecules and the physisorbed molecules on the remaining surface leads to sufficiently narrow NMR lines. To obtain detailed information on the geometrical and the electronic structure of the ion–molecule complex, the analysis includes two steps: (a) from the measured spectra the "absolute" parameters, i.e. the paramagnetic shift, the linewidth and the exchange time for the co-ordinated molecules, must be determined; (b) these "absolute" values should be used, together with other information available, to develop a model for the geometrical and electronic structure of the surface complex.

In reference 104 a method is described to determine the number of active ions and the paramagnetic shift of coordinated molecules on the ions from the observed shift dependence on the simultaneously measured amount of adsorption. For the benzene–Co^{2+} complex on Aerosil, for example, the paramagnetic ^{1}H shift has been determined to be $\delta_P^H = 150$ ppm with respect to the physisorbed benzene on pure Aerosil.

In general the NMR parameters of coordinated molecules may be derived from the measured temperature dependence of spectra which are controlled by exchange processes. A quantitative theory for exchange between sites characterized by two or more lines of Lorentzian shape can be found in reference 105. For adsorbent–adsorbate systems, however, an inhomogeneous broadening of lines, resulting from the heterogeneity of the adsorption sites, must be taken into account.[61,106] In reference 61, the influence of a distribution of Lorentzian lines (inhomogeneous broadening) on the observed NMR spectrum has been investigated by numerical calculations. It is shown that the influence of the distribution of lines in the paramagnetic region on the observed spectrum depends strongly on the ratio of the width of the distribution, $\Delta\nu_P$, to the difference of the resonance positions of the paramagnetic and the diamagnetic lines, i.e. the paramagnetic line shift $\delta\nu_P$. For exchange rates with $2\pi\delta\nu_P\tau \leqslant 10^{-2}$ (fast exchange) and $2\pi\delta\nu_P\tau \geqslant 1$ (slow exchange) the derived values for $\delta\nu_P$, assuming a Lorentzian line of the same width as the distribution, coincide within 10% with the values taking into account the distribution; i.e. an approximate analysis by the usual methods is justified. Between these limits, however, i.e. for $10^{-2} < 2\pi\delta\nu_P\tau < 1$, the distribution must be taken into account, otherwise the $\delta\nu_P$ values can be too large, with deviations up to 50% (see reference 61).

In some special cases, e.g. for slow exchange ($2\pi\delta\nu_P\tau \gg 1$), fast exchange ($2\pi\delta\nu_P \ll 1$) or if the relative number of molecules in the paramagnetic region is very small, approximate formulae may be applied.[61,105–107]

The isotropic paramagnetic shift, $\delta\nu_P$, is caused by a contact interaction and/or a dipolar interaction. The methods that may be used for qualitative and quantitative separations of dipolar and contact shifts are extensively discussed in reference 101. In the case of adsorbent–adsorbate systems, the analysis is in general confined to a more qualitative discussion because of the lack of reliable data for the surface state of the paramagnetic ions. Therefore the following procedures may be applied:[49] (a) NMR measurements for different nuclei in the same molecule, e.g. ^{1}H and ^{13}C NMR (see Section IV.B); (b) comparison of the relative shifts of various nuclei with the calculated dipolar shifts according to a simple geometrical dependence (see reference 101); (c) neglect of the dipolar shift for ions with an isotropic **g** tensor (see reference 101); (d) comparison of the shift of a proton and of a methyl group which replaces that proton (for dominating contact interaction the sign of the ^{1}H shift usually changes; see reference 101); (e) quantum chemical calculation of the spin density distribution and the corresponding contact shifts for model complexes.

B. NMR studies of paramagnetic surface complexes

1. Adsorption in zeolites

In the literature only a few NMR studies can be found on the interaction of molecules with transition metal ions in zeolites, though these complexes are of interest both from the theoretical (ion–molecule interaction) and practical (heterogeneous catalysis) points of view. This is largely the result of experimental problems encountered in recording highly resolved NMR spectra of adsorbed organic molecules and the necessity to combine the results of NMR measurements with those of other experimental methods in order to derive detailed information. ^{1}H shifts of hydrogen adsorbed in CoNaY and NiNaY zeolites were studied over a wide temperature range at a constant hydrogen pressure.[103] From the temperature dependence of the ^{1}H shift, evidence is obtained for the formation of weak complexes between H_2 and Co^{2+} and Ni^{2+} at temperatures below 270 K, with an enthalpy of complexation of about 16 kJ mol^{-1}.[103]

^{13}C shifts of 60–70 ppm to lower frequency have been reported[84] at room temperature for CO molecules adsorbed in transition metal zeolites relative to gaseous CO. These shifts should be typical for cationic carbonyls with transition metal ions and are explained by an electron transfer from the lone pair orbital of the carbon into orbitals of the transition metal ion.

Both ^{1}H and ^{13}C NMR studies have been employed for benzene and cyclohexane adsorbed in CoNaX and CoNaY zeolites with varying amounts of Co^{2+}.[49] Up to a cation exchange of 35% and 45% for CoNaX and CoNaY zeolites, respectively, the measured shifts of the broadened resonance lines (some four to six times wider in comparison with the resonance lines in NaX and NaY zeolites) correspond to the change in the susceptibility of the specimen, i.e. there is no complex formation of the molecules with the Co^{2+} ions because they are not located in the supercages. For higher values of the cation exchange, paramagnetic Co^{2+}–molecule complexes should be formed within the supercages, but the corresponding signals are not observed by the conventional FT technique because of the extremely broadened resonance lines produced by the strong electron–nucleus interaction.

2. Adsorption on silica

To study transition metal ion–molecule complexes on surfaces by high resolution NMR methods, SiO_2 adsorbents such as silica gel and Aerosil are preferable to zeolites because the ion–molecule ratio may be varied to a larger extent and, in particular, fast chemical exchange may be realized for high coverages to yield narrower NMR lines. Surface complexes of olefins and saturated hydrocarbons with Ni^{2+} and Co^{2+} ions on Aerosil have been studied by ^{1}H NMR.[103] In addition to line-broadening, considerable paramagnetic shifts occurred due to the formation of surface complexes involving these ions. The observed increase in shift with decreasing coverage is explained by a fast exchange between coordinated and physically adsorbed molecules. To interpret the spectra it is assumed that the shifts observed are mainly caused by contact interactions.

The ^{1}H NMR spectra of *benzene* adsorbed on Aerosil containing Co^{2+} and Ni^{2+} have been studied.[102,104,108,109] For Co^{2+}, large ^{1}H shifts to low frequency, in comparison with the line position of benzene on pure Aerosil, occur, whereas for the bivalent nickel much smaller shifts result.[109] The differences in these ^{1}H NMR spectra of benzene are explained by the different symmetry of the wave-functions of the tetrahedral transition metal ions Co^{2+} and Ni^{2+} on the Aerosil surface for the interaction with the π orbitals of benzene.[103] ^{13}C NMR yields additional information about the formation of these complexes, as shown in reference 108 in the case of Co^{2+}–benzene complexes and for Ni^{2+}–benzene in reference 109. In contrast to the ^{1}H spectra, the ^{13}C spectra both for Co^{2+} and Ni^{2+} show a large shift to high frequency with respect to the corresponding line of physisorbed benzene. The large ^{13}C shift for the Ni^{2+}-containing surface cannot be explained by the model given above,[103] which considers only the contact interaction. Taking into account the dipolar interaction, the observed shifts are the sum of both contributions, which should nearly cancel at the protons of the Ni^{2+}–benzene complex, so that only small ^{1}H shifts result. At the carbons of the coordinated benzene both contributions have the same sign and add to the observed large shift. In the case of Co^{2+} the dipolar contribution is less than that for Ni^{2+}, so that both the ^{1}H and the ^{13}C shifts are mainly caused by contact interactions.[49,103] Additionally, in the case of Co^{2+}–benzene complexes the dipolar interaction exerts an influence on the shifts, as shown[108] from the temperature dependence of the ^{1}H and ^{13}C shifts.

The ^{1}H NMR data of *propene* adsorbed at 283 K on Co^{2+} and Ni^{2+} containing Aerosil has been reported.[103] The ^{1}H and ^{13}C spectra of propene adsorbed on Aerosil and silica gel containing small amounts of Ni^{2+} and Co^{2+} ions (varying the content between 0.25 and 1 wt%) have been discussed.[49,101] To improve the signal-to-noise ratio, and in particular to measure small coverages, propenes enriched with ^{13}C (propene-1-^{13}C and propene-2-^{13}C) have been used. Below a critical temperature, where chemical reactions start (this temperature depends on the chemical composition of the catalyst), only a single broad ^{13}C line due to the enrichment of the olefinic line is observed, which is strongly shifted to high frequency in comparison with the corresponding line position of physisorbed propene on pure Aerosil (see Fig. 23). In contrast to these shifts to high frequency, the methyl carbon signal is shifted to low frequency. From the dependence on coverage and temperature, the "absolute" NMR parameters are derived[49] as $\delta_P^C = 2000$ ppm and $\delta_P^C = 1600$ ppm for the shift of the olefinic carbons of coordinated propene at the Ni^{2+} and Co^{2+} ions on the SiO_2 catalyst, respectively, with linewidths of 10–20 kHz. The corresponding mean residence time at 273 K for propene at a Co^{2+} ion is 2×10^{-6} s and about 10 times smaller at the Ni^{2+} ion. The shifts can be interpreted in terms of the formation of a π complex involving the π orbitals of propene and the corresponding d orbitals of the transition metal ion. In accordance with the π complex, the ^{1}H spectra show low frequency shifts for the protons of the double bond and a high frequency shift for the protons in the methyl group; the corresponding "absolute" ^{1}H shifts are in the range 100–

300 ppm. Differences in the values between the propene–Ni^{2+} and propene–Co^{2+} complexes can be explained by a larger dipolar contribution to the shifts in the case of Ni^{2+} complexes than for the Co^{2+} complexes, as is also found for complexes of benzene with these ions. To derive more detailed information on the structure of the surface complexes, additional information about the electronic states of the ions on the surface is necessary.

As another example, the ^{13}C NMR data of *acetone* adsorbed on Aerosil with Co^{2+} ions (0.5 wt%) is mentioned.[111] From an analysis of the measured shifts and linewidths and their dependence on temperature (measured between 200 and 340 K), "absolute" shifts of $\delta_P^{CO} = 300$ ppm and $\delta_P^{CH_3} = 200$ ppm are obtained for the CO and the CH_3 groups, respectively at 300 K, with a non-Curie temperature dependence (increasing shift with increasing temperature). The dipole contribution to the resulting shift is calculated using the measured axially symmetric **g**-tensor of the Co^{2+} ions in analogous H_2O–Co^{2+} complexes on Aerosil[112] and assuming a reasonable structure for the surface complex. The calculation shows a large low frequency shift ($\delta_P^{dip} = -800$ to -1000 ppm for the CO group and $\delta_P^{dip} = -200$ to -300 ppm for the CH_3 groups) so that the observed ^{13}C shift to high frequency appears to be a result of the contact contribution ($\delta_P^{con} = 1100$–1300 ppm for the CO group and $\delta_P^{con} = 400$–500 ppm for the CH_3 groups). This contact shift can be explained by spin delocalization involving the σ orbitals of the surface complex, which is supported by preliminary quantum chemical calculations.[111]

In summary it can be stated that the observed shifts of paramagnetic surface complexes are best interpreted by taking into account both the contact and the dipolar interactions.

V. MOLECULAR REACTIONS IN THE ADSORBED STATE

The study of molecular reactions in adsorbate–adsorbent systems has become an important field of NMR application.[176–178] This is mainly the result of the fact that the occurrence of molecular species formed in such processes is accompanied by corresponding lines in the ^{13}C and/or ^{15}N NMR spectra. The most extensively studied reactions are the protonation of molecules and the isomerisation of butene molecules, discussed below. The dehydration of formic acid on titanium oxides has been studied both experimentally[113,114] and theoretically.[115] In a few cases ^1H NMR spectra have also been investigated.[116]

A. Protonation of adsorbed molecules

The formation of carbocations in superacid solution and their ^{13}C NMR spectra are well known.[117–119] They are characterized by a significant reduction of the shielding of the carbon nuclei. It is also possible to observe the formation of protonated organic compounds on surfaces by means of ^{13}C and/or ^{15}N NMR. In these cases the proton transfer is realized by acidic OH groups (Broensted acid sites) of the surface, occurring especially in zeolites which are decationated or exchanged with bi- and/or trivalent ions. Difficulties arise mainly because of the rather short lifetime and the small concentration of the protonated species.

By means of highly resolved ^{13}C NMR spectra, Broensted acid sites in zeolites were studied via the formation of *pyridinium ions.*[120] This method allows a direct determination of the number of accessible acidic OH groups from the ^{13}C shifts in those systems where a fast exchange occurs between pyridine molecules and pyridinium ions. As a basis for this analysis, samples with pyridinium ions were produced by HCl addition to normally non-acidic zeolites (see Fig. 18 and Table 8). The number of acidic OH groups obtained for DeNaY and CaNaY zeolites are in good agreement with the results from the application of IR spectroscopy.

Additionally, dynamic properties of the pyridine protonation were studied. If the number of pyridine molecules is less than the number of acidic OH groups, the thermal mobility of the pyridinium ions is strongly reduced. As is already well known from IR data, it follows that there is a

TABLE 8

^{13}C shifts of pyridine molecules and pyridinium ions.[a]

Species	System or adsorbent	^{13}C shifts			
		C2	C3	C4	C2–C3
Pyridine	Liquid	21.4	−4.5	7.5	25.9
	$NaY_{2.6}$	21.7	−4.2	9.4	25.9
	NaY_{47}	21.2	−5.5	6.5	26.7
	Silica gel	—	—	—	25.3
Pyridinium ion	Solution[b]	13.5	0.2	19.7	13.3
	NaY[c]	13.0	−1.0	18.5	14.0
	NaY_{47} [c]	12.1	−1.2	17.4	13.3
	Silica gel[d]	—	—	—	12.9

[a] Values are in ppm with respect to liquid benzene and are positive for high frequency shifts.[120] C2, C3 and C4 denote the carbon nuclei in the positions *ortho, meta* and *para* with respect to the nitrogen nucleus, respectively.

[b] Produced by mixing liquid pyridine with concentrated H_2SO_4; molar ratio of pyridine to H_2SO_4 is 1:3.

[c] Produced by co-adsorption of HCl with pyridine at a molar ratio of pyridine:HCl less than 1.

[d] Values from reference 121.

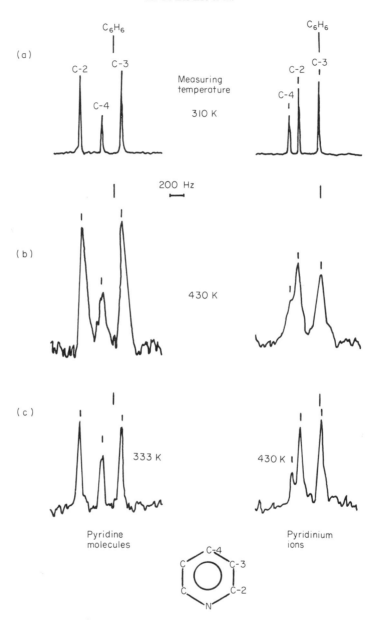

FIG. 18. ^{13}C NMR spectra of pyridine molecules and pyridinium ions in the liquid and in the adsorbed state. Pyridine molecules: (a) pure liquid; (b) adsorbed in NaY$_{2.6}$, three molecules per supercage, 16k scans; (c) adsorbed in NaY$_{47}$ (Us-Ex), three molecules per supercage, 4k scans. Pyridinium ions: (a) in H$_2$SO$_4$ (molar ratio 1:3); (b) produced by co-adsorption of HCl to pyridine molecules in NaY$_{2.6}$ (n_{Py}:$n_{HCl} \approx 3$:6), 16k scans; (c) produced in NaY$_{47}$ (Us-Ex) by co-adsorption of HCl with pyridine molecules (n_{Py}:$n_{HCl} \approx 3$:6), 16k scans.

strong interaction between Broensted acid sites and the pyridine molecules. This conclusion is in accordance with results of ^{1}H NMR measurements concerning the mobility of hydroxyl protons.[6] Their increased mobility under the influence of adsorbed pyridine has been interpreted by the action of pyridine as a "transport vehicle" for the hydroxyl protons. In contrast to ^{13}C NMR measurements relating to the formation of pyridinium ions mainly on silica with added HCl,[87,121] ^{13}C and ^{15}N spectra have been used to study the formation of pyridinium ions resulting from an interaction with the OH groups of the adsorbent.[13,120] This interaction is accompanied by a dramatic broadening of the NMR lines due to a lower mobility of the protonated species.

The formation of *pyridinium ions* in decationated zeolites can be clearly followed in the ^{15}N NMR spectra,[13] which reveal relatively large shifts and which are therefore better suited for such an analysis than analogous ^{13}C NMR measurements.[120]

Since the lifetime of pyridinium ions ($\tau = 5 \times 10^{-7}$ s at 313 K[122]) in 88 DeNaY$_{2.6}$ zeolites is much smaller than the inverse value of the difference of Larmor frequencies of adsorbed pyridine molecules and pyridinium ions ($\omega^{-1} = 2 \times 10^{-4}$ s according to a shift difference of 89 ppm), the condition of fast exchange is fulfilled. Moreover, because of the observed coverage dependence, the pyridinium ion can be considered as a strong adsorption complex: as long as the total number of molecules adsorbed is less than the number of Broensted acid sites, the shift, δ, observed is equal to the shift, δ_I, of the pyridinium ion, while for higher coverages a mean shift

$$\delta = (1 - p_I)\delta_M + p_I\delta_I \tag{18}$$

occurs, where δ_M denotes the shift for the adsorbed pyridine molecules and p_I is the relative amount of pyridinium ions (see equation (11), where $\delta_M = 0$ is chosen). Using for δ_M the shift observed in NaY zeolites, where only non-protonated molecules occur, and for δ_I the value found for pyridinium ions in aqueous solutions ($\delta_I \approx -115$ ppm, see reference 53), the relative amount (p_I) and hence the total number of pyridinium ions per supercage ($p_I N$) can be determined. The results agree quite well with values derived from analogous ^{13}C NMR measurements (see Table 9).

In accordance with this interpretation, the observed ^{15}N shift is about the same as that measured for the pyridinium ions in solution, if the coverage is only slightly in excess ($N = 2.1$) of the number of accessible OH groups (about two per supercage). At a lower coverage ($N = 1.7$), no ^{15}N NMR signal is observed. An analogous situation occurs in the case of ^{13}C measurements. The disappearance of the signal has been attributed to a much lower mobility of the pyridinium ions compared with pyridine molecules adsorbed in zeolites which have been studied independently by means of proton spin relaxation.[122]

TABLE 9

^{15}N shifts of pyridine adsorbed in NaY$_{2.6}$ and 88 DeNaY$_{2.6}$ zeolites.[13] [a]

	Coverage (molecules and pyridinium ions per supercage), N	^{15}N shifts (ppm)	p_1	Number of pyridinium ions per supercage, Np_1
NaY$_{2.6}$	2.4	−26.8	—	—
	5.6	−25.3	—	—
88 DeNaY$_{2.6}$	2.1 (2.6)	−107.3	0.912	1.9 (1.9)
	3.0 (3.0)	−101.5	0.847	2.5 (2.2)
	4.6 (3.6)	−84.0	0.650	3.0 (2.5)

[a] Values are in ppm with respect to the liquid state and are positive for high frequency shifts. p_1 denotes the relative amount of pyridinium ions, as determined by equation (18). Values from analogous ^{13}C measurements[120] are given in parentheses.

Formation of pyridinium ions has been observed by the adsorption of pyridine molecules on phosphorus-modified silica gel.[91] From the observed shifts (Table 10), a concentration of ~0.05 protonated molecules per square nanometre is derived, and it is shown by IR that protonation proceeds on the POH groups.[91]

The adsorption of ammonia in decationated zeolites is characterized by a strong interaction with the acidic sites giving rise to the formation of ammonium ions. Typical ^{15}N NMR spectra are shown in Fig. 19.[13]

TABLE 10

^{13}C and ^{15}N shifts for pyridine adsorbed on SiO$_2$-P.[91] [a]

Statistical monolayers	^{13}C shifts			^{15}N shift
	C2	C3	C4	
5.08	−0.9	−0.2	0.1	−2.7
1.68	−2.2	0.4	0.5	−18.8
1.19	−2.2	0.5	0.4	−26.6
0.81	−3.7	−0.6	0.5	−26.7
0.61	−3.3	−0.2	0.4	−27.7
0.31	−3.6	−0.2	2.1	−31.0
Pyridinium ion (in solution)[b]	−7.7	5.0	12.4	−115.1

[a] Values are in ppm with respect to the liquid state and are positive for high frequency shifts. The notation of carbon atoms is as in Table 8.
[b] References 53 and 123.

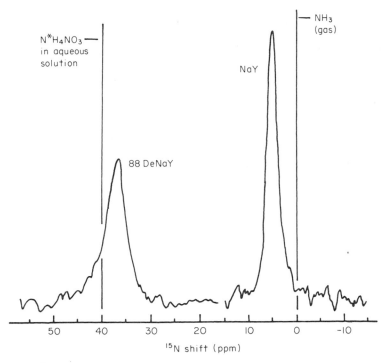

FIG. 19. ^{15}N NMR spectra of ammonia adsorbed in a decationated zeolite (88 DeNaY$_{2.6}$, $N = 9$ molecules and ions per supercage) and in a sodium zeolite (NaY$_{2.6}$, $N = 3.4$). Measurements were performed at 300 K.[13]

With respect to the Na forms, the ^{15}N lines for ammonia molecules adsorbed in 88 DeNaY$_{2.6}$ zeolites are appreciably shifted to high frequency. At low coverages (less than four molecules per supercage) the shift remains constant at a value which is typical for NH$_4^+$ solutions (not containing Cl$^-$ ions). From this result we conclude that all molecules are converted into ammonium ions as a consequence of their interaction with structural hydroxyl groups in the decationated zeolite. On the other hand, the interaction with structural hydroxyl groups leads to a strong reduction in the molecular mobility,[122] so that the ^{15}N lines are broadened and disappear if the number of molecules is equal to, or less than, the number of accessible hydroxyl groups; this is quite analogous to the behaviour found for pyridine.

B. Reactions of simple olefins

The conversion of *isobutene* molecules adsorbed in CaNaY$_{2.6}$ zeolites was studied using isobutene molecules which are partially ^{13}C enriched in the positions =CH$_2$ and —CH$_3$.[124] Therefore it is possible to follow the transformation of definite groups during the reaction. The time dependence of the ^{13}C spectra, for which an example is shown in Fig. 20, is explained by simple dimerization processes with final products which are more or less branched hydrocarbons.[124]

In contrast, the spectra of isobutene adsorbed in NaX, KX and NaY zeolites do not show a time dependence at room temperature, which means that no reaction takes place. The isomerization of *n-butene* is studied in the adsorbed state in zeolites[125-129] and on alumina.[130]

For pure NaY zeolites, a catalytic reaction can be observed in the spectra only when the samples are heated up to 430 K for some hours. For but-1-ene

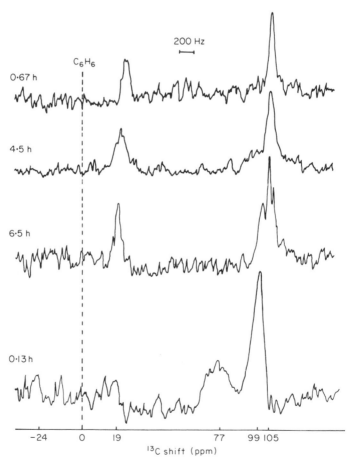

FIG. 20. Time dependence of ^{13}C NMR spectra of, initially, two isobutene molecules $^*CH_2{=}C(^*CH_3)_2$ adsorbed in a 40 $CaNaY_{2.6}$ zeolite after 0.67, 4.5, 6.5 and 13 h at 312 K. The asterisk denotes a ^{13}C enrichment of $\sim 10\%$, and the spectrum at the beginning is about the same as for isobutene molecules adsorbed in a NaX zeolite.[124]

molecules adsorbed in an 88 CaNaY$_{2.6}$ zeolite, the conversion rate at room temperature is so fast that the process cannot be studied by observing the line intensities unless the temperature is lowered. Thus, the investigations are carried out with 65–70 CaNaY$_{2.6}$ zeolites, allowing a convenient measurement to be made at room temperature or above. In the interval from 330 to 350 K the intensities of the =CH$_2$ and =CH— signals of but-1-ene decrease while an olefinic line occurs with a shift of $\delta \approx 1.3$ ppm (to low frequency of benzene). The doublet splitting of this line in the spectra without proton decoupling indicates the presence of the =CH— groups of cis- and trans-but-2-ene (which overlap mutually at the beginning of the reaction). More detailed information about the reactions which take place in these systems can be derived by using but-1-ene which is partially ^{13}C enriched, i.e. *CH$_2$=CHCH$_2$CH$_3$ and CH$_2$=*CHCH$_2$CH$_3$.[125,126] At lower temperatures (290–350 K) only the double bond isomerization of butenes proceeds in 65–70 CaNaY$_{2.6}$ zeolites. Within the limits of experimental error, an enrichment of ^{13}C nuclei is found in the —CH$_3$ and =CH— groups of the but-2-ene molecules if the but-1-enes are enriched in the positions =CH$_2$ and =CH—, respectively. As an example, some ^{13}C NMR spectra, taken during double bond isomerization of but-1-ene ^{13}C enriched in the =CH$_2$ position for 3.3 molecules per supercage, are shown in Fig. 21.

Relative equilibrium concentrations, as determined from the line intensities, are in fair agreement with known data for the thermal equilibrium of the n-butene isomers in the gas phase. Thus, possible differences between NOE factors of the different ^{13}C lines are so small that they do not give rise to significant errors in determining concentrations via the enhanced resonance line intensities. At temperatures above 350 K polymerization reactions occur which can be inferred from the overlap of numerous lines in that region of ^{13}C signals where aliphatic CH$_n$ ($n = 0, \ldots, 3$) groups appear (see Fig. 22).

For an initial coverage of 3.3 but-1-ene molecules per supercage, partly branched octanes are formed, while for a low coverage (0.4 molecules per supercage) octenes are found predominantly. These experimental results are explained by a different availability of hydrogen as an important controlling factor in these reactions.

Finally, some results of ^{13}C NMR studies of *propene* adsorbed on NiO/SiO$_2$ catalysts are mentioned.[110]

Figure 23 shows the temperature dependence of the ^{13}C NMR spectra of propene-2-^{13}C adsorbed on a NiO/silica gel catalyst with a coverage of one monolayer. Before reaction one observes a broad ^{13}C line, which is strongly shifted to high frequency in comparison with the position of physisorbed propene on pure Aerosil ($\delta \approx 50$ ppm at 213 K). The shift of this line increases with increasing temperature because of hindered exchange between the chemisorbed propene molecules (at the coordinatively unsatur-

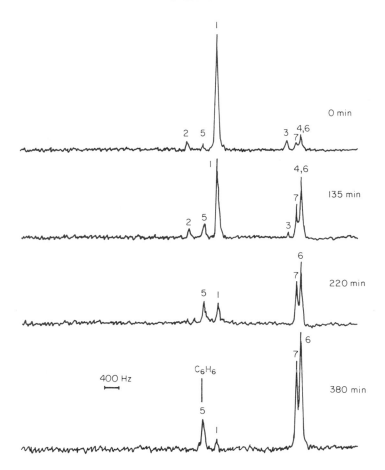

FIG. 21. ^{13}C NMR spectra taken during the double bond isomerization of but-1-ene enriched with ^{13}C nuclei in the position $=CH_2$ for 3.3 molecules per supercage in a 67 CaNaY$_{2.6}$ zeolite at 333 K. Assignment of carbons: but-1-ene, 1 ($=CH_2$, 15% enriched with ^{13}C); but-1-ene, 2, 3, 4 ($=CH-$, $-CH_2-$, $-CH_3$, respectively, with natural abundance of ^{13}C nuclei); *cis*-but-2-ene, 5 ($=CH-$); *cis*-but-2-ene 6 ($-CH_3$, ^{13}C-enriched after reaction); *trans*-but-2-ene, 5 ($=CH-$); *trans*-but-2-ene, 7 ($-CH_3$, ^{13}C-enriched after reaction).

ated Ni^{2+} ions) and the physisorbed molecules on the surface.[104] From an analysis of the temperature dependence, the following information is extracted for the chemisorbed propene molecules:[131] (a) mean residence time at Ni^{2+}, $\tau \approx 0.5 \times 10^{-6}$ s at 230 K with an activation energy of $\Delta E \approx$ 13 kJ mol^{-1}; (b) ^{13}C shift $\delta_C \approx 2000$ ppm. Furthermore, the same value of the ^{13}C shift is observed for propene-1-^{13}C as for propene-2-^{13}C. This equal influence on the two carbon atoms of the double bond can be explained by the formation of a π complex between chemisorbed propene and coordina-

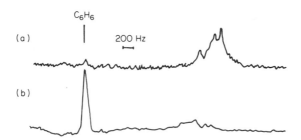

FIG. 22. ^{13}C NMR spectra after polymerization reactions at temperatures above 370 K (initial but-1-ene in 67 CaNaY$_{2.6}$ zeolite, 3.3 molecules per supercage). (a) ^{13}C enrichment (15%) initially in the $=CH_2$ position of but-1-ene. Measurements (20 000 scans, 330 K) after reaching equilibrium. (b) ^{13}C enrichment (17%) initially in the $=CH-$ position of but-1-ene. Measurements (70 000 scans, 300 K) after reaching equilibrium.

tively unsaturated Ni^{2+}.[103] In the course of the reaction, additional ^{13}C lines occur as a result of dimerization products. From the line splitting, without proton decoupling, it follows that the lines belong to an olefinic $-HC=$ group and an aliphatic $-HC-$ group.

The ^{13}C NMR studies show that dimerization of propene on NiO/silica gel catalysts occurs at about room temperature, involving the coordinatively unsaturated Ni^{2+} ions. The primary products in the coordinative mechanism of dimerization are 4-methylpent-1,2-ene and, to a lower extent, hex-1-ene. Other isomers are formed according to the isomerization of these primary products rather than in a direct dimerization process.

VI. THEORETICAL TREATMENT OF ADSORPTION COMPLEXES

To get a detailed characterization of the nature of bonds and the type of interaction, the theoretical analysis should proceed in two steps: (a) the quantum mechanical calculation of the structure of molecules interacting with active sites on the surface, and (b) the calculation of the NMR parameters on the basis of the structures derived.

A. Quantum mechanical treatment of the structure

A detailed discussion of the different methods for the theoretical treatment of adsorption phenomena can be found in references 132 and 133.

In the case of specific interaction between molecules and active sites of oxidic solids, e.g. molecules adsorbed in zeolites or on silica, quantum chemical methods can be applied to molecular clusters with a finite number of atoms.[134–136]

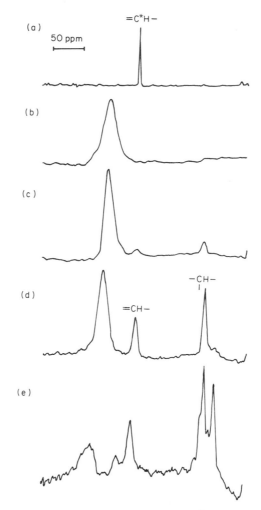

FIG. 23. ^{13}C NMR spectra of propene molecules $CH_2{=}{}^{*}CH{-}CH_3$ (the asterisk denotes an enrichment with ^{13}C of 90%) adsorbed on Aerosil (a) and on NiO/silica gel (1 mol %) in dependence on temperature: (b) 213 K; (c) 263 K; (d) 272 K; (e) 296 K (after a waiting time of 3 h).

In the practical realization, however, difficulties arise because the choice of a proper model and an adequate quantum chemical method are somewhat contradictory. Quantum chemical methods of reasonable accuracy (non-empirical methods which include an estimation of correlation energy) are confined to very small molecular clusters, so that for representative clusters at present only semi-empirical methods with their particular shortcomings are applicable.[137]

A reasonable way to overcome these difficulties is described in references 134 and 138, where the results of non-empirical SCF calculations for very simple molecular models are compared with the results of semi-empirical calculations in order to extend the semi-empirical methods to more representative clusters of sites in zeolites and silica. These calculations have been carried out on selected interaction sites in zeolites and silica, e.g. the aluminosiloxane anion and free and coordinated metal ions (see Fig. 24).

The optimal structures of complexes formed between the sites and the molecules are determined by calculating the interaction energy as the

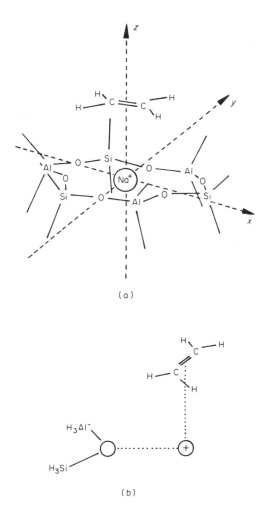

(a)

(b)

FIG. 24. Models of interaction between π hydrocarbons and a cation in zeolites: (a) six-membered ring (cation SII site, Fig. 1); (b) aluminosiloxane anion with a coordinated cation.

difference between the SCF energies of the complex and its components:

$$\Delta E = E^{\text{SCF}} \text{(complex)} - E^{\text{SCF}} \text{(molecule)} - E^{\text{SCF}} \text{(site)} \qquad (19)$$

At this level, correlation contributions to the interaction energy are neglected. The intersystem correlation energy is estimated by the London dispersion energy formula, using experimental values of the polarizibility. It is found that in polar systems the intersystem correlation energy accounts for up to 10% of the interaction energy, whereas for weak molecular complexes the dispersion energy might be comparable with the interaction energy ΔE.[134]

In Table 11 values for the calculated interaction energies of some complexes of unsaturated molecules with Na^+ and with Na^+ coordinated at a model cluster of a SII site in zeolites are shown. According to the discussion of the errors involved in the calculations, and comparison with related experimental values, the theoretical values calculated non-empirically are too low by about 20%.

The arrangement of the cation above the plane of the π-electron system is found to be the most stable one. This is the optimum structure for the electrostatic ion–quadrupole contribution which dominates the interaction in the cation–π-electron molecule complexes studied. The preference for this sandwich-like structure has been found to be in agreement with spectroscopic results[142-146] for π hydrocarbons adsorbed in zeolites. The charge transfer to the cation proves to be very small and amounts to a few hundredths of an electron.[138]

TABLE 11

Experimental and calculated interaction energies ΔE (in kilojoules per mole) for complexes of ethene, isobutene and benzene with Na^+ and a Na^+–SII site cluster.[138,139]

	ΔE (kJ mol^{-1})				
	Na^+			Na^+–SII site^e	
Molecule	Exp.^{a}	4–31^c	CNDO/2	Exp.^{b}	CNDO/2
Ethene	−55	−48.5	−140	−37	−94
Isobutene	−80	−60.3	−176	−48	−113
Benzene	−105	(−78)^d	−202	−70	−130

^{a} Derived from gas-phase enthalpies of the corresponding Li^+–molecule complexes taken from reference 140 according to the ratio of the ionic radii.[138]
^{b} Heat of adsorption on NaX, NaY zeolites.[141]
^{c} Non-empirical calculation with the split valence 4-31G basis set.[138]
^{d} Extrapolated value from non-empirical STO-3G calculation.[138]
^{e} Na^+–SII site as the $Na^+(Si_5AlO_{18}H_{12})^-$ cluster according to Fig. 24(a).[189]

On the basis of the non-empirical results, the capability of the semi-empirical CNDO/2-PSS[165] method for predicting the electronic structure of cation–molecule complexes can be checked.[77,147] Although both the calculated equilibrium distances and the absolute values of the interaction energies are too large, all trends are qualitatively correct (see Table 11). In contrast to CNDO/2-PSS, the semi-empirical PCILO method fails for the cation–hydrocarbon complexes.[49,138,148,149] This result for CNDO/2-PSS fits in with previous findings, according to which CNDO/2-PSS is claimed to be capable of reproducing small energy differences between σ and π complexes of protonated benzene,[150] and, moreover, to describe the different paths of approach towards benzene for protons on the one hand and for Li^+ and Na^+ on the other.[151]

The "tetrahedral" σ complex is calculated as the deepest minimum on the potential surface of protonated arenes in qualitative agreement with experimental results in solution.[150] In contrast to protonated benzene, it is found that the centrosymmetrical π structure is energetically favourable for benzene complexes with alkali cations (Li^+, Na^+).[151] This structure is supported by the experimental ^{13}C NMR spectra for arenes adsorbed in zeolites (see reference 152 and Section VI.B).

Therefore CNDO/2-PSS has been applied to the interactions of π hydrocarbons with Na^+ attached to different clusters cut off from faujasite-type zeolites.[139] As an example, the results for unsaturated hydrocarbons interacting with a cluster of Na^+-SII sites in zeolites are shown in Table 11. From these calculations, conclusions concerning the influence of the surroundings on interactions of π hydrocarbons with cations at different sites in faujasites can be drawn. In general it may be stated that the coordination of the cation by the aluminosilicate framework manifests itself in a pronounced reduction of the interaction energy.

For the stabilization energy of the benzene–Na^+ complex at the Na^+-SII site cluster, a value has been calculated, by means of the CNDO/2-PSS method, which is 35% smaller than for the isolated benzene–Na^+ complex, in qualitative agreement with the experimental values (see Table 11).

The electronic structure of SiO_2 clusters has been studied thoroughly by the CNDO method in the parametrization of Boyd and Whitehead (CNDO-BW).[153] Optimum parameters have been derived for the pseudo-atoms terminating the finite clusters as models of the infinite oxidic skeleton.[136] Applying this method, the valence angle of the SiOH bond on silica gel is calculated to be 145°, in good agreement with ^1H NMR data taken on the solid.[154] Furthermore, the structure of the adsorption complex formed by silanol has been studied using the cluster $(OA)_3SiOH$, where A represents pseudo-atoms.[136,155] For the $(OA)_3SiOH$–pyridine (see Fig. 25) and $(OA)_3SiOH$–acetone complexes, a linear structure is assumed for the OH bond interacting with the lone pairs of the nitrogen and oxygen atoms

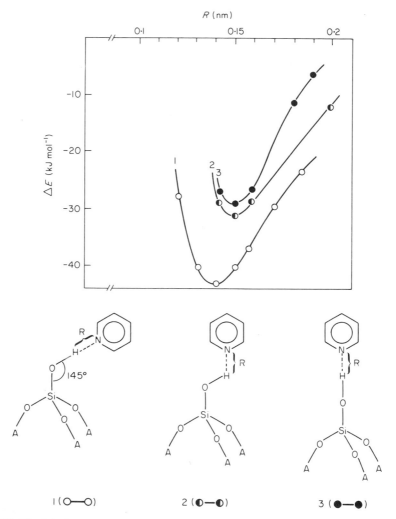

FIG. 25. Calculated potential curves for three different $(OA)_3SiOH$–pyridine structures by means of the CNDO-BW method[155] (interaction energy ΔE in kilojoules per mole).

of the complexed molecules.[155] It is found that the OH bond is weakened due to interactions with the molecules in accordance with experimental results obtained by IR and ^1H NMR.[71,156]

B. Calculations of nuclear shielding

To check the reliability of the structural models derived on the basis of quantum chemical calculations, one has to evaluate observable quantities and to compare them with experimental results. In particular, the observed shifts may be correlated with structural parameters such as electron density, spin density, bond length, hybridization, etc. The comparison between calculated and experimental data provides a sensitive test of the most probable models of interaction, especially for adsorbate–adsorbent systems where unusual structures can appear.

Numerous methods have been developed for the calculation of nuclear shielding.[157–159] For adsorbate–adsorbent systems only the simplest methods are used because of the large number of atoms involved. Shielding tensor calculations have been reported for ^{13}C,[160] ^{15}N[161] and ^{31}P[161] at the same level of approximation as that used for the energy calculations of adsorption complexes. The method employed applies a sum-over-states perturbation expression and uses CNDO wave functions, to give

$$\sigma_{\alpha\beta}(A) = \frac{\mu_0 e^2}{8\pi m} \sum_{\mu}^{A} p_{\mu\mu}^{A} \left\langle \chi_\mu \left| \frac{r^2 \delta_{\alpha\beta} - x_\alpha x_\beta}{r^3} \right| \chi_\mu \right\rangle$$

$$- \frac{\mu_0 e^2 \hbar^2}{4\pi m^2} \langle r^{-3} \rangle_{2p} \sum_{i}^{occ.} \sum_{j}^{unocc.} \frac{1}{\Delta E_{ij}} \left[(\vec{c}_i^A \times \vec{c}_j^A)_\alpha \sum_{c} (\vec{c}_i^c \times \vec{c}_j^c)_\beta \right.$$

$$\left. + (\vec{c}_i^A \times \vec{c}_j^A)_\beta \sum_{c} (\vec{c}_i^c \times \vec{c}_j^c)_\alpha \right] \tag{20}$$

where α, β refer to x, y, z components of the shielding tensor $\sigma_{\alpha\beta}(A)$ for nucleus A;

$$p_{\mu\mu}^A = 2 \sum_{i}^{occ.} c_{i\mu}^A c_{i\mu}^A,$$

which represents bond order matrix elements; χ_μ refers to a Slater-type orbital; $\langle r^{-3} \rangle_{2p}$ is the average value of r^{-3} over 2p atomic functions;

$$\Delta E_{ij} = \varepsilon_j - \varepsilon_i - J_{ij} + 2K_{ij}$$

is the transition energy from the occupied MO i to the unoccupied MO j (with energies of ε_i, ε_j respectively), J_{ij} is the Coulomb integral, K_{ij} is the exchange integral; and $\vec{c}_i$ represents $\{c_{i2p_x}; c_{i2p_y}; c_{i2p_z}\}$.

Equation (20) has been used to calculate ^{13}C shieldings of isolated complexes of Na^+ with unsaturated molecules such as olefins and arenes.[77,49,152] The ^{13}C shieldings computed for the energetically most favourable butene–Na^+ complexes show the same trend as the experimental data for adsorbed butene, but in some cases there are discrepancies in the sign and magnitude of the shifts.[77,152] Similar difficulties have been reported in a paper by Strong et al.[162] The reason for these differences is not yet known. In the case of adsorbed arenes in zeolites, the comparison between calculated and experimental ^{13}C shifts supports the centrosymmetrical arrangement of Na^+ above the π-electron system (see Table 12).

TABLE 12

Experimental and theoretical ^{13}C shifts of benzene and toluene adsorbed in NaX zeolite and in interaction with an isolated Na$^+$ ion, respectively, with respect to liquid (isolated) molecules.^{a}

	^{13}C shifts				
System	C-1	C-2,6	C-3,5	C-4	C-7
Benzene/NaX					
Experimental	0				
Calculated	0				
Toluene/NaX					
Experimental	2.9	⊢———1.8———⊣			−1.3
Calculated	2.6	⊢———2.5———⊣			0.2

^{a} Values are in ppm and are positive for high frequency shifts.

Measurement of the ^{13}C shielding tensor[171] offers new possibilities for studying the nature of chemical bonds, as has been demonstrated in the case of adsorbate–adsorbent systems for the benzene–hydroxyl interaction on silica gel.[41] The authors conclude from a comparison of the ^{13}C NMR spectra of benzene adsorbed on silica gel and on carbon black that no specific interaction occurs between benzene and OH groups on silica gel.[41] This is in contrast to the results obtained by other experimental methods.[2,71]

To estimate the influence of this (possibly weak) interaction on the ^{13}C shifts, the ^{13}C shielding tensor has been calculated for a benzene molecule which interacts with an OH group or a proton.[152]

The models chosen are shown in Fig. 26. They are the most stable complexes: a bridged structure in the case of [benzene–OH]$^{-}$ [163] and a σ complex in the case of the benzene–H$^+$ system.[150] Both complexes show favourable peripheral interactions, in contrast to the centrosymmetrical arrangement of the benzene–alkaline cation complexes.

The principal values of the ^{13}C shielding tensor, calculated for the [benzene–OH]$^{-}$ complex, are given in Table 13.

The calculated changes in comparison with the free benzene molecule are very small and are more significant in the tensor components than in the isotropic value (chemical shift). Comparing the calculated values with experimental data, one must bear in mind that the CNDO/2 method used yields excitation energies which are two to three times larger than the experimental ones. Therefore the calculated components of the shielding tensor are expected to be smaller by about the same factor, as indicated in equation (20). This relationship is reflected by the shielding anisotropy calculated for the benzene molecule in comparison with experimental values,

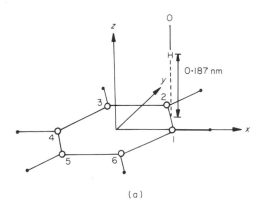

(a)

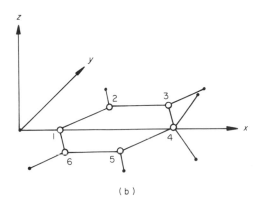

(b)

FIG. 26. Energetically most stable structure of the [benzene–OH]$^-$ complex and of the benzenium ion: (a) [benzene–OH]$^-$ complex, stabilization energy $E_{stab} = 20$ kJ mole^{-1};[163] (b) benzenium ion (σ complex), proton affinity $E^{PA} \approx 1000$ kJ mole^{-1}.[150]

TABLE 13

Calculated differences in the principal values of the shielding tensor ($\Delta\sigma_{ii}$) and isotropic value (δ) for the [benzene–OH]$^-$ complex (energetically most stable configuration according to CNDO/2-PSS; Fig. 26) with respect to an isolated benzene molecule.[a]

C-atom	$\Delta\sigma_{11}$	$\Delta\sigma_{22}$	$\Delta\sigma_{33}$	δ
C-1,2	1.2	−1.4	2.3	0.7
C-3,6	2.4	−2.4	0	0
C-4,5	−1.4	−5.5	0.2	−2.3

[a] Values are in ppm and are positive for high frequency shifts.

as shown in Table 14. For the same reason, some other parametrization, such as the CNDO-FK parametrization, which is also shown in Table 14, may be preferable.[152]

Bearing this in mind, the values given in Table 13 reveal that the anisotropy of the ^{13}C shielding tensor changes by only a few parts per million if benzene molecules interact with OH groups. This theoretical result is in agreement with experiment,[41] where no change of the ^{13}C spectrum of benzene adsorbed on silica gel surfaces has been observed within the limits of experimental error, i.e. 10 ppm. Hence, the calculations may explain the apparent contradiction between the interpretation of the measurements[41] and the experimentally found specific interaction between benzene molecules and OH groups on silica gel surfaces.

To extend this computation and to check its reliability, the ^{13}C shielding of the benzenium ion has been calculated.[180] Theoretical values are compared with experimentally determined shifts in Table 15. The total region of shifts calculated for the σ complex of the benzenium ion agrees fairly well with the experimental result. The values in detail, however, show remarkable deviations, especially with respect to the assignment of C-1 and C-3,5. The wrong sequence of ^{13}C lines for C-1 and C-3,5 corresponds to the incorrect prediction of electron densities, for the respective carbons, calculated by the CNDO-FK method. *Ab initio* calculations of electron density with an extended basis set[168] are in agreement with the expected correlation between the net charges at the different carbons and the measured sequence of ^{13}C lines (Table 15).

Pyridine, with a stronger proton affinity than benzene, is often used for the characterization of acidic properties of hydroxyl groups on solid surfaces (Section V.A).

To find the geometrical and electronic structures, the energetically most stable configuration of the [pyridine–H]$^+$ complex was calculated[173] by a geometry optimization procedure using the CNDO-FK method. This formalism has been applied successfully to the determination of the structure

TABLE 14

Experimental and calculated data on the anisotropy of the ^{13}C shielding tensor of the benzene molecule.[a]

Method	$\Delta\sigma_{11}$	$\Delta\sigma_{22}$	$\Delta\sigma_{33}$
SOS-CNDO/2-PSS[165]	35	3	−37
SOS-CNDO-FK[166]	90	2	−92
Experimental[167]	88.2	12	−127.2

[a] Values are given in ppm with respect to the isotropic value.

TABLE 15

Experimental and calculated ^{13}C shifts (δ) and calculated changes of charge density (ΔQ_C) in units of electronic charge, e, of the benzenium ion with respect to benzene.[a]

C-atom	δ		ΔQ_c (e)	
	Calculated	Experimental[164]	CNDO-FK	Ab initio[168]
C-1	16	48.4	0.220	0.230
C-2,6	−27	7.2	−0.107	−0.068
C-3,5	1	56.9	0.172	0.236
C-4	−74	−77.5	−0.087	−0.309

[a] σ complex according to CNDO-FK in connection with a complete geometry optimization;[150] Fig. 26(b). Values for δ are in ppm and are positive for high frequency shifts.

and charge distribution of protonated molecules.[150] The most stable configuration is found to correspond to the proton interacting with the lone pair electrons of the nitrogen atom. The results are shown in Table 16. The geometry of the pyridinium ion deviates much less from a regular hexagon than the pyridine molecule (see Table 16). The bond between nitrogen and the adjacent carbons is lengthened by about 0.005 nm and the CNC angle is reduced by 6° by comparison with the pyridine molecule. Most of the positive charge in the pyridinium ion (about 85%) is distributed over all of the H atoms, where the largest amount ($q = 0.22|e|$) belongs to the H atom at the nitrogen. A residual positive charge (15%) is found on the ring C atoms with a strongly alternating charge distribution (Table 16). The most negatively charged position is at the nitrogen atom ($q = -0.16|e|$). On the basis of these results the ^{13}C shieldings of the pyridinium ion have been calculated.[152] As shown in Table 17, the theoretical values are in good agreement with the experimental ^{13}C shifts.

Hence we may conclude that the quantum chemical calculation gives a realistic structure and charge distribution of the pyridinium ion. Because of discrepancies in the sign and magnitude of the shifts in some cases, e.g. for olefin–cation (Na$^+$, Ag$^+$) complexes, an attempt has been made to interpret the ^{13}C spectra by the well known empirical relationship between the calculated changes in the electron density and the ^{13}C shifts:[169]

$$\Delta\sigma = C_\sigma\Delta\rho_\sigma + C_\pi\Delta\rho_\pi \qquad (21)$$

where $\Delta\sigma$ is the change in the ^{13}C shift, $\Delta\rho_\sigma$ and $\Delta\rho_\pi$ are the calculated changes of σ and π electron densities in the adsorption complex relative to the free molecule, and C_σ and C_π are constants derived from the fit of the experimental ^{13}C shifts to the electron densities calculated for free ethene and but-1-ene ($C_\sigma = 150$ ppm per electron; $C_\pi = 300$ ppm per electron).

TABLE 16

Structure and charge density q of the pyridine molecule and the pyridinium ion calculated by means of the CNDO-FK method.[173] [a]

Quantity	Pyridine	Pyridinium ion
r(N-C2)	0.132	0.137
r(2-C3)	0.142	0.141
r(C3-C4)	0.142	0.143
∢C2NC2	127.5	121.9
∢NC2C3	117.6	120.9
∢C2C3C4	118.9	117.7
∢C3C4C3	119.4	120.8
q(N)	−0.148	−0.161
q(C2)	0.047	0.195
q(C3)	−0.048	−0.122
q(C4)	−0.017	0.155
q(H2)	0.034	0.126
q(H3)	0.034	0.132
q(H4)	0.030	0.132
q(H)	—	0.220

[a] The charge densities are in units of electronic charge (e); the bond lengths, r, are in nanometres; the bond angles are in degrees. The numbers 2, 3 and 4 denote the atoms in the *ortho*, *meta* and *para* positions, respectively, with respect to the nitrogen atom.

TABLE 17

[13]C shifts δ (in ppm) and charge density q (in units of electronic charge) of the pyridinium ion.[a]

C atom	δ		q
	Calculated	Experimental	(calculated)
C-2	0	0	0.195
C-3	−12	−13.4	−0.122
C-4	+6	+6	0.155

[a] δ is given with respect to the position of the resonance line of the C-2 atom and is positive for a high frequency shift. For details see Table 16.

The change of electron densities at the carbon atoms is evaluated by taking into account only the polarizing action of a cation and the surrounding framework of atoms at particular cation sites on the bond polarizabilities of butene molecules.[170] The calculated ^{13}C shifts are in good agreement with the experimental values for butene adsorbed in NaX and AgNaX zeolites. In particular, the typical shielding differences found in the experimental ^{13}C spectra can be interpreted by assuming a charge transfer to the cation in the case of Na^+–butene complexes, and a charge polarization within the butene molecule only in the case of Ag^+–butene complexes. The resulting structures for the cation–butene complexes on the basis of this model, however, do not agree with the optimum structures predicted by quantum chemical calculations (Section VI.A). The reason for this probably lies in the electrostatic model, wihich includes only ion-induced dipole interactions (bond polarizabilities), whereas the quantum chemical analysis shows that the ion–quadrupole term dominates in the cation–olefin interaction.

Summarizing the theoretical treatment of adsorption complexes, it can be stated that at present extensive models of surface structure containing, for example, some dozens of atoms are tractable only at the semi-empirical level. Unfortunately, semi-empirical methods are known to fail in the treatment of interaction problems and, in particular, van der Waals interactions are not correctly described.[137] In special cases, however, e.g. for the cation–molecule complexes or protonated species on the surface, the scmi-empirical CNDO/2 method may be applied successfully.

It is therefore necessary to check the applicability of semi-empirical methods by comparison with reliable experimental and/or theoretical data. A reasonable possibility is to study *simple* structural units by a non-empirical method supplemented by estimates of the dispersion energy. These theoretical results may be used for selecting suitable semi-empirical methods (Section VI.A). The semi-empirical method found in such a way may then be applied to more representative surface structures. The analysis of the extended model of a particular surface site can yield a simplified but improved model, taking into account the characteristic interaction mechanisms. In the case of cation sites in zeolites, e.g. the aluminosilicates, the framework should be simulated by point charges on the atoms in the Hamiltonian used for non-empirical calculations.[139] In the case of adsorbate–adsorbent interactions, rather simple methods for the calculation of nuclear shielding may be applied successfully (see Section VI.B), because only the *change* in the chemical shift between the free and adsorbed molecules is required. It appears that the sum-over-states method can be used together with various semi-empirical methods, at least to discriminate against unrealistic models of surface complexes.

ACKNOWLEDGMENTS

The authors are indebted to their colleagues Dr T. Bernstein, Dr D. Freude, Dr J. Kärger, Dr D. Michel and Professor Dr H. Winkler for reading the manuscript and helpful discussions.

REFERENCES

1. H. Pfeifer, *NMR – Basic Principles and Progress*, Vol. 7, Springer Verlag, Berlin, 1972, p. 53.
2. H. Pfeifer, *Phys. Reports*, 1976, **26**, 293.
3. H. Lechert, *Catalysis Rev.*, 1976, **14**, 1.
4. J. J. Fripiat, *J. Phys. (Paris)*, 1977, **38**, 44.
5. W. Derbyshire, *Specialist Periodical Report on NMR*, Vol. 9 (G. A. Webb, ed.), The Chemical Society, London, 1981, p. 256.
6. D. Freude, H. Pfeifer, W. Ploß and B. Staudte, *J. Mol. Catal.*, 1981, **12**, 1.
7. R. W. Vaughan, L. B. Schreiber and J. A. Schwarz, *ACS Symp. Ser.*, 1976, **34**, 275.
8. J. Kärger and J. Caro, *J. Chem. Soc. Faraday Trans. I*, 1977, **73**, 1363.
9. J. Kärger, H. Pfeifer, M. Rauscher and A. Walter, *J. Chem. Soc. Faraday Trans. I*, 1980, **76**, 717.
10. D. Geschke, *Z. Phys. Chem. (Leipzig)*, 1972, **249**, 125.
11. D. Michel, *Z. Phys. Chem. (Leipzig)*, 1973, **252**, 263.
12. D. Michel, A. Germanus, D. Scheller and B. Thomas, *Z. Phys. Chem. (Leipzig)*, 1981, **262**, 113.
13. D. Michel, A. Germanus and H. Pfeifer, *J. Chem. Soc. Faraday Trans. I*, 1982, **78**, 237.
14. G. E. Maciel and D. W. Sindorf, *J. Amer. Chem. Soc.*, 1980, **102**, 7606.
15. D. W. Breck, *Zeolite Molecular Sieves*, Wiley-Interscience, New York, 1974.
16. O. Grubner, P. Jiru and M. Ralek, *Molekularsiebe*, VEB Deutscher Verlag der Wissenschaften, Berlin, 1968.
17. R. M. Barrer, *Zeolite and Clay Minerals as Sorbents and Molecular Sieves*, Academic Press, London, 1978.
18. R. M. Barrer, *Endeavour*, 1964, **23**, 122.
19. J. A. Rabo, *Zeolite Chemistry and Catalysis*, ACS Monograph no. 171, American Chemical Society, Washington, D.C., 1976.
20. H. Knoll, *Chem. Technol.* 1974, **26**, 391.
21. H. Pfeifer, *Kristall und Technik*, 1976, **11**, 577.
22. R. Schöllner, *Mitteil. Chem. Ges. DDR*, 1980, **27**, 97.
23. W. J. Mortier, *Compilation of Extra Framework Sites in Zeolites*, Butterworth, London, 1982.
24. K. K. Unger, *Porous Silica*, Journal of Chromatography Library Vol. 16, Elsevier, Amsterdam, 1979.
25. H. Ferch, *Chem.-Ing. Tech.*, 1976, **48**, 922.
26. F. Janowski and W. Heyer, *Z. Chem.*, 1979, **19**, 1.
27. G. D. Parfitt and K. S. W. Sing, *Characterization of Powder Surfaces*, Academic Press, London, 1976.
28. V. Ja. Davydov, A. V. Kiselev, H. Pfeifer and I. Jünger, in preparation.
29. T. A. Egerton and R. D. Green, *Trans. Faraday Soc.*, 1971, **67**, 2699.
30. D. Graham and W. D. Phillips, *Shifts in the Proton Magnetic Resonance Frequencies of Organic Compounds Adsorbed on Carbon, Solid–Gas Interface*, Butterworth, London, 1957, p. 22.

31. A. V. Volkov, A. V. Kiselev, V. I. Lygin and V. B. Chlebnikov, *Zhur. Fis. Chim.*, 1972, **46**, 502.
32. A. Jutand, D. Vivien and J. Conard, *Surface Sci.*, 1972, **32**, 258.
33. T. Miyamoto and H.-J. Cantow, *Makromol. Chem.*, 1972, **162**, 43.
34. V. V. Mank and N. G. Vailev, *Ukrainskij Chim. Zhur.*, 1975, **41**, 92.
35. Z. Varga and M. Ràkos, *Czech. J. Phys. b*, 1974, **24**, 1028.
36. D. Geschke and H. Pfeifer, *Z. Phys. Chem. (Leipzig)*, 1966, **232**, 127.
37. D. Geschke, *Z. Phys. Chem. (Leipzig)*, 1969, **242**, 74.
38. M. Mehring, *NMR – Basic Principles and Progress*, Vol. 11, Springer Verlag, Berlin, 1976.
39. D. Doskočilová and B. Schneider, *Chem. Phys. Lett.*, 1970, **6**, 381.
40. H. Ernst, D. Fenzke and H. Pfeifer, *Ann. Phys.*, 1981, **38**, 257.
41. S. Kaplan, H. A. Resing and J. S. Waugh, *J. Chem. Phys.*, 1973, **59**, 5681; cf. D. Deininger, *Z. Phys. Chem. (Leipzig)*, 1981, **262**, 369.
42. E. Lippmaa, M. Alla and T. Tuherm, *Proc. XIXth congr. Ampére, Heidelberg*, 1976, p. 113.
43. M. M. Maricq and J. S. Waugh, *J. Chem. Phys.*, 1979, **70**, 3300.
44. R. E. Taylor, R. G. Pembleton, L. M. Ryan and B. C. Gerstein, *J. Chem. Phys.*, 1979, **71**, 4541.
45. E. O. Stejskal, J. Schaefer, J. M. S. Henis and M. K. Tripodi, *J. Chem. Phys.*, 1974, **61**, 2351.
46. D. Michel, W. Meiler, A. Gutsze and A. Wronkowski, *Z. Phys. Chem. (Leipzig)*, 1980, **261**, 953.
47. J. L. Bonardet and J. P. Fraissard, *J. Magn. Reson.*, 1976, **22**, 543.
48. F. H. A. Rummens, *NMR – Basic Principles and Progress*, Vol. 10, Springer Verlag, Berlin, 1975.
49. D. Deininger, Thesis, Leipzig, 1981.
50. K. Jackowski and W. T. Raynes, *Mol. Phys.*, 1977, **34**, 465.
51. I. D. Gay and J. F. Kriz, *J. Phys. Chem.*, 1978, **82**, 319.
52. B. Tiffon and J. E. Dubois, *Org. Magn. Reson.*, 1978, **11**, 295.
53. R. O. Duthaler and J. D. Roberts, *J. Amer. Chem. Soc.*, 1978, **100**, 4969.
54. M. Witanowski, L. Stefaniak, S. Szymanski and H. Januszewski, *J. Magn. Reson.*, 1977, **28**, 217.
55. M. Alei, A. E. Florin, W. M. Litchman and J. F. O'Brien, *J. Phys. Chem.*, 1971, **75**, 932.
56. C. R. Johnson, in *Advances in Magnetic Resonances*, (J. S. Waugh, ed.), Academic Press, London, New York, 1965.
57. M. Cohn and B. D. Nageswara Rao, *Bull. Magn. Reson.*, 1979, **1**, 38.
58. H. S. Gutowsky, D. W. McCall and C. P. Slichter, *J. Chem. Phys.*, 1953, **21**, 279.
59. M. M. McConnel, *J. Chem. Phys.*, 1958, **28**, 430.
60. L. M. Piette and W. A. Anderson, *J. Chem. Phys.*, 1959, **30**, 899.
61. D. Deininger, K. Hopf and H. Winkler, *Ann. Phys.*, 1981, **38**, 353.
62. V. Ju. Borovkov, G. M. Zhidomirov and V. B. Kazansky, *Zhur. Strukt. Chim.*, 1975, **16**, 308.
63. Th. Bernstein, P. Fink, D. Michel and H. Pfeifer, *J. Colloid Interface Sci.*, 1981, **84**, 310.
64. D. Michel, *Surface Sci.*, 1974, **42**, 453.
65. D. Michel, W. Meiler and E. Angelé, *Z. Phys. Chem. (Leipzig)*, 1974, **259**, 389.
66. D. Michel, W. Meiler and D. Hoppach, *Z. Phys. Chem. (Leipzig)*, 1974, **255**, 509.
67. D. Michel, E. Angelé and W. Meiler, *Z. Phys. Chem. (Leipzig)*, 1978, **259**, 92.
68. H. W. Quinn, in *Progress in Separation and Purification*, John Wiley & Sons, New York, 1971, p. 133.
69. H. W. Quinn, J. S. McIntyre and D. J. Peterson, *Canad. J. Chem.*, 1965, **43**, 2896.
70. D. W. Breck, *Zeolite Molecular Sieves*, Wiley–Interscience, New York, 1974, pp. 593–724.

71. A. V. Kiselev and V. I. Lygin, *IR Spectra of Surface Compounds and Adsorbed Species* [in Russian], Nauka, Moscow, 1972.
72. J. L. Carter, D. J. C. Yates, P. J. Lucchesi, J. J. Elliot and V. Kevorkian, *J. Phys. Chem.*, 1966, **70**, 1126.
73. H. Pfeifer, *ACS Symp. Ser.*, 1976, **34**, 36.
74. J. Kärger and D. Michel, *Z. Phys. Chem.* (*Leipzig*), 1976, **257**, 983.
75. W. Meiler, D. Michel and H. Pfeifer, in preparation.
76. D. Michel, W. Meiler and H. Pfeifer, *J. Mol. Catal.*, 1975, **1**, 85.
77. W. Meiler, Thesis, Leipzig, 1977.
78. H. Pfeifer, *Proc. Conf. "Ramis 77", Poznań*, 1977, p. 107.
79. O. M. Dzhigit, A. V. Kiselev, K. N. Mikos, G. G. Muttik and T. A. Rahmanova, *Trans. Faraday Soc.*, 1971, **67**, 458.
80. H. J. Herden, private communication.
81. M. D. Sefcik, *J. Amer. Chem. Soc.*, 1979, **101**, 2164.
82. M. D. Sefcik, J. Schaefer and E. O. Stejskal, *ACS Symp. Ser.*, 1977, **40**, 344.
83. J. B. Stothers, *Carbon-13 NMR Spectroscopy*, Academic Press, New York, London, 1972.
84. Y. BenTaarit, *Chem. Phys. Lett.*, 1979, **62**, 211.
85. A. Michael, W. Meiler, D. Michel and H. Pfeifer, *Chem. Phys. Lett.*, 1981, **84**, 30.
86. V. Bosáček, D. Freude, Th. Fröhlich, H. Pfeifer and H. Schmiedel, *J. Colloid Interface Sci.*, 1982, **85**, 502.
87. I. D. Gay and S. Liang, *J. Catal.*, 1976, **44**, 306.
88. I. D. Gay and J. F. Kriz, *J. Phys. Chem.*, 1975, **79**, 2145.
89. I. D. Gay, *J. Phys. Chem.*, 1974, **78**, 38.
90. Th. Bernstein, D. Michel, H. Pfeifer and P. Fink, *J. Chem. Soc. Faraday Trans. I*, 1981, **77**, 797.
91. Th. Bernstein, L. Kitaev, D. Michel, H. Pfeifer and P. Fink, *J. Chem. Soc. Faraday Trans. I*, 1982, **78**, 761.
92. Th. Bernstein, W. Meiler, D. Michel, H. Pfeifer, and H.-J. Rauscher, *Proc. 4th Specialized Colloque Ampere, Leipzig*, 1979, p. 149.
93. W. Meiler, in preparation.
94. R. M. Barrer and R. M. Gibbons, *Trans. Faraday Soc.*, 1963, **59**, 2569.
95. A. V. Kiselev, *Proc. 5th Internat. Conf. Zeolites, Naples*, 1980, p. 400.
96. R. E. Wasylishen and T. Schaefer, *Canad. J. Chem.*, 1973, **51**, 3087.
97. A. Germanus, Diplomwork, Leipzig, 1980.
98. G. A. Webb and M. Witanowski, *Nitrogen NMR*, Plenum Press, London, 1973, p. 204.
99. V. Bosáček, V. Patzelova, Z. Tvaruzkova, U. Lohse, W. Schirmer and H. Stach, *Proc. Workshop: Adsorption of Hydrocarbons in Zeolites, Berlin* 1979, Vol. 1, p. 157.
100. H. J. Keller, *NMR – Basic Principles and Progress*, Vol. 2, Springer Verlag, Berlin, 1970.
101. G. N. La Mar, W. De W. Horrock and R. H. Holm, *NMR of Paramagnetic Molecules*, Academic Press, New York, 1973.
102. V. Ju. Borovkov and V. B. Kazansky, *Kinet. Katal.*, 1972, **13**, 1356.
103. V. B. Kazansky, V. Ju. Borovkov and G. M. Zhidomirov, *J. Catal.*, 1975, **39**, 205.
104. D. Deininger, V. Ju. Borovkov and V. B. Kazansky, *J. Catal.*, 1977, **48**, 35.
105. C. J. Johnson, *Advance. Magn. Reson.*, 1965, **1**, 33.
106. V. N. Parmon and G. M. Zhidomirov, *Theoret. Eksp. Chim.*, 1975, **11**, 323.
107. I. J. Swift and R. E. Connick, *J. Chem. Phys.*, 1962, **37**, 307.
108. W.-D. Hoffmann, *Z. Phys. Chem.* (*Leipzig*), 1976, **257**, 817.
109. D. Deininger, D. Geschke, W.-D. Hoffmann and G. Wendt, *Z. Phys. Chem.* (*Leipzig*), 1977, **258**, 1169.
110. D. Deininger and G. Wendt, *React. Kinet. Catal. Lett.*, 1981, **17**, 277.
111. K. Hopf, Thesis, Leipzig, 1981.
112. O. I. Brotikovsky, V. A. Svec and V. B. Kazansky, *Kinet. Katal.*, 1972, **13**, 1342.

113. M. A. Enriquez and J. P. Fraissard, *J. Catal.*, 1982, **74**, 77.
114. M. A. Enriquez and J. P. Fraissard, *J. Catal.*, 1982, **74**, 89.
115. B. Bigot, M. A. Enriquez and J. P. Fraissard, *J. Catal.*, 1982, **74**, 84.
116. I. T. Ali and I. D. Gay, *J. Catal.*, 1980, **62**, 341.
117. G. A. Olah and D. J. Donavan, *J. Amer. Chem. Soc.*, 1977, **99**, 5026.
118. R. N. Young, *NMR Spectr.*, 1979, **12**, 000.
119. G. A. Olah and R. J. Spear, *J. Amer. Chem. Soc.*, 1975, **97**, 1539.
120. H.–J. Rauscher, D. Michel, D. Deininger and D. Geschke, *J. Mol. Catal.*, 1980, **9**, 369.
121. I. D. Gay, *J. Catal.*, 1977, **48**, 430.
122. H.-J. Rauscher, D. Michel and H. Pfeifer, *J. Mol. Catal.*, 1981, **12**, 159.
123. E. Breitmaier, G. Haas and W. Voelter, *Atlas of Carbon-13 NMR Data*, Heyden & Son, London, 1975.
124. G. Dombi, D. Michel and W. Meiler, *Z. Phys. Chem. (Leipzig)*, 1979, **260**, 587.
125. D. Michel, W. Meiler, H. Pfeifer and H.-J. Rauscher, *Proc. Symp. Zeolites, Szeged*, 1978, p. 221.
126. D. Michel, W. Meiler, H. Pfeifer, H.-J. Rauscher and H. Siegel, *J. Mol. Catal.*, 1979, **5**, 263.
127. J. B. Nagy, M. Guelton and E. G. Derouane, *J. Catal.*, 1978, **55**, 43.
128. H.-J. Rauscher, Thesis, Leipzig, 1980.
129. H.-J. Rauscher, Diplomwork, Leipzig, 1976.
130. J. F. Kriz and I. D. Gay, *J. Phys. Chem.*, 1976, **80**, 2951.
131. D. Deininger, in preparation.
132. H. H. Dunken and V. J. Lygin, *Quantenchemie der Adsorption an Festkörperoberflächen*, Deutscher Verlag für Grundstoffindustrie, Leipzig, 1978.
133. U. Landmann and G. G. Kleimann, in *Specialist Periodical Reports on Surface and Defect Properties of Solids*, Vol. 6, Royal Society of Chemistry, London, 1977.
134. J. Sauer, P. Hobza and R. Zahradnik, *J. Phys. Chem.*, 1980, **84**, 3318.
135. D. Geschke, W.-D. Hoffmann and D. Deininger, *Surface Sci.*, 1976, **57**, 559.
136. I. D. Micheikin, I. A. Abronin, A. I. Lumpov and G. M. Zhidomirov, *Kinet. Katal.*, 1978, **19**, 1050.
137. P. Hobza and R. Zahradnik, *Weak Intermolecular Interactions in Chemistry and Biology*, Elsevier, Amsterdam, 1980.
138. J. Sauer and D. Deininger, *J. Phys. Chem.*, 1982, **86**, 1327.
139. J. Sauer and D. Deininger, *Zeolites*, 1982, **2**, 113.
140. R. K. Staley and J. L. Beauchamp, *J. Amer. Chem. Soc.*, **97**, 5920.
141. A. G. Bezus, A. V. Kiselev, Z. Sedlacek and Pham Quang Du, *Trans. Faraday Soc.*, 1971, **67**, 468.
142. H. Förster and R. J. Seelman, *J. Chem. Soc. Faraday Trans. I*, 1978, **74**, 1435.
143. J. J. Freeman and M. L. Unland, *J. Catal.*, 1978, **54**, 183.
144. W.-D. Hoffmann, *Z. Phys. Chem. (Leipzig)*, 1976, **257**, 315.
145. H. Lechert and K.-P. Wittern, *Ber. Bunsenges. Phys. Chem.*, 1978, **82**, 1054.
146. C. J. Wright and C. Riekel, *Mol. Phys.*, 1978, **36**, 695.
147. W. Meiler, D. Deininger and D. Michel, *Z. Phys. Chem.*, 1977, **258**, 139.
148. R. Lochmann and W. Meiler, *Z. Phys. Chem. (Leipzig)*, 1977, **258**, 1059.
149. R. Lochmann, W. Meiler and K. Müller, *Z. Phys. Chem. (Leipzig)*, 1980, **261**, 165.
150. D. Heidrich and M. Grimmer, *Internat. J. Quantum Chem.*, 1975, **9**, 923.
151. D. Heinrich and D. Deininger, *Tetrahedron Lett.*, 1977, **42**, 3751.
152. D. Deininger, D. Michel and D. Heidrich, *Surface Sci.*, 1980, **100**, 541.
153. R. I. Boyd and M. A. Whitehead, *J. Chem. Soc. Dalton*, 1972, **73** and **78**.
154. Th. Bernstein, H. Ernst, D. Freude, I. Jünger, J. Sauer and B. Staudte, *Z. Phys. Chem. (Leipzig)*, 1981, **262**, 1123.
155. Th. Bernstein, Thesis, Leipzig, 1981.

156. M. L. Hair and W. Hertl, J. Phys. Chem., 1970, 74, 91.
157. W. T. Raynes, Nucl. Magn. Reson., 1972, 1, 1.
158. R. Ditchfield and P. D. Ellis, Topics ^{13}C NMR, 1974, 1, 1; R. Ditchfield, in Specialist Periodical Report on NMR, Vol. 5, Royal Society of Chemistry, London, 1979, p. 1.
159. B. R. Appleman and B. P. Dailey, Advanc. Magn. Reson., 1974, 7, 1.
160. H. Schwind, D. Deininger and D. Geschke, Z. Phys. Chem. (Leipzig), 1974, 255, 149.
161. T. Weller, D. Deininger and R. Lochmann, Z. Chem., 1981, 21, 105.
162. A. B. Strong, D. Ikenberry and D. M. Grant, J. Magn. Reson., 1973, 9, 145.
163. C. Doberenz, D. Geschke and W.-D. Hoffmann, Z. Phys. Chem. (Leipzig), 1974, 255, 666.
164. G. A. Olah, J. S. Staral, G. Asencio, G. Liang, D. A. Forsyth and G. D. Mateescu, J. Amer. Chem. Soc., 1978, 100, 6299.
165. J. A. Pople and D. L. Beveridge, Approximate MO Theory, McGraw Hill, New York, 1970.
166. H. Fischer and H. Kollmar, Theoret. Chim. Acta, 1969, 13, 213.
167. M. Linder, A. Höhener and R. R. Ernst, J. Magn. Reson., 1979, 35, 379.
168. W. C. Ermler, R. S. Mulliken and E. Clementi, J. Amer. Chem. Soc., 1976, 98, 388.
169. K. Salzer, Z. Phys. Chem. (Leipzig), 1978, 259, 795.
170. K. Salzer, Z. Phys. Chem. (Leipzig), 1981, 262, 1.
171. A. Pines, M. G. Gibby and J. S. Waugh, Chem. Phys. Lett., 1972, 15, 373.
172. I. Jünger, W. Meiler and H. Pfeifer, Zeolites, 1982, 2, 309.
173. H.-J. Rauscher, D. Heidrich, H. J. Köhler and D. Michel, Theoret. Chim. Acta, 1980, 57, 255.
174. V. Ju. Borovkov, W. Keith Hall and V. B. Kazansky, J. Catal., 1978, 51, 437.
175. A. D. H. Clague, I. E. Maxwell, J. P. C. M. Van Dongen and J. Binsma, Appl. Surface Sci., 1978, 288.
176. E. G. Derouane, P. Dejaifve and J. B. Nagy, J. Mol. Catal., 1978, 3, 453.
177. E. G. Derouane, P. Dejaifve and J. B. Nagy, Compt. Rend. Acad. Sci. (Paris), 1977, 284, 23.
178. E. G. Derouane, P. Dejaifve and J. B. Nagy, J. Mol. Catal., 1980, 10, 331.
179. G. M. Muha and J. C. Yates, J. Chem. Phys., 1968, 49, 5073.
180. D. Deininger, Z. Phys. Chem. (Leipzig), 1981, 262, 369.
181. V. Yu. Borovkov, A. V. Zaiko, V. B. Kazansky and W. K. Hall, J. Catal., 1982, 75, 219.
182. J. A. Ripmeester, J. Am. Chem. Soc., 1982, 105, 2925.